D0936400

STOCHASTIC STRUCTURAL DYNAMICS

Progress in Theory and Applications

Contributions in honor of the sixty-fifth birthday of Yu-Kweng Michael Lin on October 30, 1988

Professor Y. K. Lin

STOCHASTIC STRUCTURAL DYNAMICS
Progress in Theory and Applications

Edited by

S. T. ARIARATNAM
*Professor, Solid Mechanics Division, University of Waterloo,
Waterloo, Ontario, Canada*

G. I. SCHUËLLER
*Professor, Institute of Engineering Mechanics, University of Innsbruck,
Austria*

and

I. ELISHAKOFF
*Professor, Department of Aeronautical Engineering, Technion–Israel
Institute of Technology, Haifa, Israel*

ELSEVIER APPLIED SCIENCE
LONDON and NEW YORK

ELSEVIER APPLIED SCIENCE PUBLISHERS LTD
Crown House, Linton Road, Barking, Essex IG11 8JU, England

Sole distributor in the USA and Canada
ELSEVIER SCIENCE PUBLISHING CO., INC.
52 Vanderbilt Avenue, New York, NY 10017, USA

WITH 12 TABLES AND 148 ILLUSTRATIONS

© 1988 ELSEVIER APPLIED SCIENCE PUBLISHERS LTD

British Library Cataloguing in Publication Data

Stochastic structural dynamics.
1. Dynamics. Stochastic processes
I. Ariaratnam, S. T.   II. Schuëller, G. I.
III. Elishakoff, Isaac
531'.11'015192

ISBN 1-85166-211-1

Library of Congress Cataloging-in-Publication Data

Stochastic structural dynamics: progress in theory and applications/
edited by S. T. Ariaratnam, G. I. Schuëller, I. Elishakoff.
p. cm.
Bibliography: p.
Includes index.
ISBN 1-85166-211-1
1. Structural dynamics.   2. Stochastic processes.   3. Lin, Y. K.
(Yu-Kweng), 1923–   I. Ariaratnam, S. T.   II. Schuëller, Gerhart I.
III. Elishakoff, Isaac.
TA654.S755 1988   88-7268   CIP
624.1'71—dc19

No responsibility is assumed by the Publisher for any injury and/or damage to persons or property as a matter of products liability, negligence or otherwise, or from any use or operation of any methods, products, instructions or ideas contained in the material herein.

Special regulations for readers in the USA
This publication has been registered with the Copyright Clearance Center Inc. (CCC), Salem, Massachusetts. Information can be obtained from the CCC about conditions under which photocopies of parts of this publication may be made in the USA. All other copyright questions, including photocopying outside the USA, should be referred to the publisher.

Printed by The Universities Press (Belfast) Ltd.

Dedication to Professor Y. K. Lin

This book, which contains a series of original contributions in the area of Stochastic Dynamics, is dedicated to Professor Y. K. Lin by some of his friends, colleagues and former students on the occasion of his sixty-fifth birthday. It is a token of appreciation for the valuable things he has taught all of us by his lectures, discussions and writings.

Mike Lin was born on 30 October 1923 in Fukien Province, which now belongs to The People's Republic of China. After receiving his basic University education in China, he studied at Stanford University, California, USA where, in 1957, he earned a Ph.D. in structural engineering. During this period, he also worked as stress engineer for the Vertol Aircraft Corporation. From 1957 to 1958 he joined the Faculty of the Imperial College of Engineering in Ethiopia. Returning to the US he went to work with the Transport Division of the Boeing Company as research engineer. In 1960 he accepted an appointment as professor of aeronautical and astronautical engineering and of civil engineering at the University of Illinois at Urbana-Champaign where he stayed until 1983. He then accepted the Charles E. Schmidt Eminent Scholar Chair at the Florida Atlantic University in Boca Raton, Florida. Here he also serves as Director of the Center for Applied Stochastics Research (CASR).

During the period 1967–8 he was a National Science Foundation (NSF) Senior Post-Doctoral Fellow and Visiting Professor at the Massachusetts Institute of Technology. He also taught at two NSF sponsored institutes for college teachers, one at the University of New Mexico in 1966 and the other at the University of Illinois in 1970. In

1976 he was a Senior Visiting Fellow at the Institute of Sound and Vibration Research at the University of Southampton. In 1984 he was awarded the Alfred M. Freudenthal Medal for outstanding contributions to random vibration by ASCE.

Mike Lin's academic career was accompanied by consulting to major companies and US government laboratories including General Motors Corp., General Dynamics Corp., Boeing Co., TRW Corp., Air Force Material Laboratory, NASA and the General Thomas J. Rodman Laboratory. He is a registered PE in the states of Florida and Illinois and a fellow of many professional societies, including the Acoustical Society of America, the American Academy of Mechanics and ASCE.

The papers contributed to this volume demonstrate the impact of Mike Lin's research and teaching in the area of random vibration and structural dynamics. In particular, his book on *Probabilistic Theory of Structural Dynamics* has developed into an indispensable reference book for students, researchers and practising engineers working in this field. Most important, his valuable contributions in the application of Markov process theory to structural dynamics in stochastic stability and in nonlinear stochastic dynamics are considered to be landmarks in the development of the field.

The great number of publications which are still in press attest to his admirable continued activities. We consider ourselves fortunate to be able to share with him his findings in the years to come and wish Professor Lin health and strength in the continuation of his work.

THE CONTRIBUTORS

Contents

List of Contributors

Professor A. H-S. ANG
Department of Civil Engineering, 3118 Newmark Civil Engineering Laboratory, University of Illinois, Urbana, Illinois 61801, USA

Professor S. T. ARIARATNAM
Solid Mechanics Division, Faculty of Engineering, University of Waterloo, Waterloo, Ontario N2L 3G1, Canada

Dr C. G. BUCHER
Institut für Mechanik, Universität Innsbruck, Technikerstrasse 13, A-6020 Innsbruck, Austria

Professor S. H. CRANDALL
Department of Mechanical Engineering, Massachusetts Institute of Technology, Cambridge, Massachusetts 02139, USA

Dr M. F. DIMENTBERG
Institute for Problems of Mechanics, USSR Academy of Sciences, Prospect Vernadskogo 101, 117526 Moscow, USSR

Professor I. ELISHAKOFF
Department of Aeronautical Engineering, Technion–Israel Institute of Technology, Haifa 32000, Israel

ix

Dr Y. FUJIMORI
*National Aerospace Laboratory, 1133–32 AIHARA, Machida-shi,
Tokyo 194-02, Japan*

Mr H. GLUVER
*Department of Structural Engineering, Technical University of
Denmark, Lyngby, Denmark*

Professor M. GRIGORIU
*Department of Structural Engineering, Cornell University, Hollister
Hall, Ithaca, New York 14853-3501, USA*

Professor H.-K. HONG
*Department of Civil Engineering, National University of Taiwan,
Taipei, Taiwan*

Professor M. HOSHIYA
*Department of Civil Engineering, Musashi Institute of Technology,
1–28 Tamazutsumi, Setagaya-ku, Tokyo, Japan*

Professor R. A. IBRAHIM
*Department of Mechanical Engineering, Wayne State University,
Detroit, Michigan 48202, USA*

Professor R. N. IYENGAR
*Department of Civil Engineering, Indian Institute of Science, Ban-
galore 560 012, India*

Professor F. KOZIN
*Department of Electrical Engineering and Computer Science, Poly-
technic University, Route 110, Farmingdale, New York 11735, USA*

Professor S. KRENK
*Department of Structural Engineering, Technical University of
Denmark, Lyngby, Denmark*

Dr W. LI
*Department of Mechanical Engineering, Texas Tech University,
Lubbock, Texas 79409, USA*

Mr F. X. LONG
Department of Civil, Mechanical and Environmental Engineering, The George Washington University, Washington, DC 20052, USA

Mr E. LUBLINER
Department of Aeronautical Engineering, Technion–Israel Institute of Technology, Haifa 32000, Israel

Dr A. I. MENYAILOV
Institute for Problems of Mechanics, USSR Academy of Sciences, Prospect Vernadskogo 101, 117526 Moscow, USSR

Professor J. B. ROBERTS
School of Engineering and Applied Sciences, University of Sussex, Falmer, Brighton, Sussex BN1 9QT, UK

Dr A. H. SADEGHI
School of Engineering and Applied Sciences, University of Sussex, Falmer, Brighton, Sussex BN1 9QT, UK

Professor S. SARKANI
Department of Civil, Mechanical and Environmental Engineering, The George Washington University, Washington, D.C. 20052, USA

Professor G. I. SCHUËLLER
Institut für Mechanik, Universität Innsbruck, Technikerstrasse 13, A-6020 Innsbruck, Austria

Professor M. SHINOZUKA
Department of Civil Engineering and Operations Research, Princeton University, Engineering Quadrangle, Princeton, New Jersey 08544, USA

Dr A. A. SOKOLOV
Institute for Problems of Mechanics, USSR Academy of Sciences, Prospect Vernadskogo 101, 117526 Moscow, USSR

Professor R. VAICAITIS
Department of Civil Engineering, Columbia University, New York, New York 10027, USA

Professor W. V. WEDIG
Institut für Technische Mechanik, Universität Karlsruhe, Kaiserstrasse 12, D-7500 Karlsruhe, FRG

Professor Y. K. WEN
Department of Civil Engineering, 3118 Newmark Civil Engineering Laboratory, University of Illinois, Urbana, Illinois 61801, USA

Mr W.-C. XIE
Solid Mechanics Division, Faculty of Engineering, University of Waterloo, Waterloo, Ontario N2L 3G1, Canada

Dr F. YAMAZAKI
Ohsaki Research Institute, Shimizu Corporation, 2-2-2, Uchisaiwaicho, Chiyoda-ku, Tokyo 100, Japan

Professor Y. N. YANG
Department of Civil, Mechanical and Environmental Engineering, The George Washington University, Washington, DC 20052, USA

List of Publications of Y. K. Lin

Books or Monographs

Probabilistic Theory of Structural Dynamics, McGraw-Hill Book Company, New York, 1967. Reprint with revisions, R. E. Krieger Publishing Company, 1976.

Papers Published in Books

Random vibration of periodic and almost periodic structures. In *Mechanics Today*, ed. S. Nemat-Nasser. Pergamon Press, Oxford, 1976, pp. 93–124.

Dynamic characteristics of continuous skin-stringer panels. In *Acoustical Fatigue in Aerospace Structures*, ed. W. J. Trapp & D. M. Fornay, Jr. Syracuse University Press, 1965, pp. 163–84.

Response of linear and nonlinear continuous structures subject to random excitation and the problems of high level excursions. In *Structural Safety and Reliability*, ed. A. M. Freudenthal. Pergamon Press, Oxford, 1972, pp. 117–30.

Stochastic aspects of dynamic systems. In *Stochastic Problems in Mechanics*, ed. S. T. Ariaratnam & H. H. E. Leipholz. Solid Mechanics Division, University of Waterloo, 1974, pp. 145–63.

Response of periodic beam to supersonic boundary-layer pressure fluctuation (with S. Maekawa, H. Nijim, L. Maestrello). In *Stochastic Problems in Dynamics*, ed. B. L. Clarkson. Pitman Publishing Ltd, London, 1977, pp. 468–85.

Structural response under turbulent flow excitations. In *Random Excitations of Structures by Earthquakes and Atmospheric Turbulence*, ed. H. Parkus, Springer-Verlag, Wien, CISM 225, 1977, pp. 238–307.

Stochastic analysis of bridge motion in large scale turbulent winds. In *Wind Engineering*, ed. J. E. Cermak. Pergamon Press, Oxford, New York, 1980, pp. 887–97.

Random vibration of a cooling tower under earthquake excitation. In

Stochastic Methods in Structural Mechanics, ed. F. Casciati & L. Faravelli. Servizio Arti Grafiche, December 1983, pp. 63–78.

A physical interpretation of Markov approximation. In Probabilistic Mechanics and Structural Reliability, ed. Y. K. Wen. ASCE, January, 1984, pp. 410–13.

Applications of cumulant closure to random vibration problems (with W. F. Wu). In Random Vibration, ed. T. C. Huang & P. D. Spanos. AMD-65, ASME, 1984, pp. 113–25.

On modeling earthquake as nonstationary random process (with Y. Yong). In: Structural Safety and Reliability, ed. I. Konishi, A. H-S. Ang & M. Shinozuka. IASSAR, 1985, pp. II-207–16.

Orthogonal expansion of nonstationary random processes (with Y. Yong). In Methods of Stochastic Structural Mechanics, ed. F. Casciati & L. Faravelli. University of Pavia Press, 1986, pp. 49–67.

Some observations on spectral analysis (with Y. Yong). In Random Vibration, ed. I. Elishakoff & R. Lyon. Elsevier Science Publishers, Amsterdam, 1986, pp. 193–200.

Application of Markov process theory to nonlinear random vibration problems. In Stochastic Methods in Structural Dynamics, ed. G. I. Schuëller & M. Shinozuka. Martinus Nijhoff Publishers, Dordrecht/Boston/Lancaster, 1987, pp. 134–53.

Exact and approximate solutions for response of nonlinear systems under parametric and external white noise excitations (with Y. Yong, G. Q. Cai & A. Brückner). Proceedings of IUTAM Symposium on Nonlinear Stochastic Engineering Systems, ed. F. Ziegler & G. I. Schuëller. Springer Verlag, Berlin, 1988, pp. 323–34.

Energy dissipation balancing—An approximate solution for non-linearly damped system under random excitations (with G. Q. Cai). In Developments in Theoretical and Applied Mechanics, Vol. 14. School of Engineering, The University of Mississippi, Mississippi, 1988, pp. 381–8.

Papers Published in Technical Journals

Coupled vibrations of open thin-walled beams (with J. M. Gere). Journal of Applied Mechanics, 25 (3) (September 1958) 373–8.

Approximate buckling loads of open columns. Journal of Engineering Mechanics, ASCE Proceedings, EM 4, Paper 1793, October 1958, pp. 1793-1–1793-22.

Free vibrations of continuous skin-stringer panels. Journal of Applied Mechanics, 27 (4) (December 1960) 669–76.

Coupled bending and torsional vibrations of restrained thin-walled beams. Journal of Applied Mechanics, 27 (4) (December 1960) 739–40.

Vibrations of thin paraboloidal shells of revolution (with F. A. Lee). Journal of Applied Mechanics, 27 (4) (December 1960) 743–4.

Stresses in continuous skin-stiffner panels under random loading. Journal of Aero/Space Sciences, 29 (1) (January 1962) 67–86.

On torsion analysis of multicell tubes. Journal of Aero/Space Sciences, 29 (April 1962) 475–6.

Expansion of Jacobian elliptic functions about the modulus (with F. A. Lee). *Mathematics of Computation*, **XVI**(79) (July 1962) 372–5.

Response of a non-linear panel under periodic and randomly varying loading. *Journal of Aero/Space Sciences*, **29** (9) (September 1962) 1929–34.

Free vibrations of continuous beams on elastic supports. *International Journal of Mechanical Sciences*, **4** (September–October 1962) 409–23.

Nonstationary response of continuous structures to random loading. *Journal of Acoustical Society of America*, **35** (February 1963) 222–7.

Free vibrations of a finite row of continuous skin-stringer panels (with I. D. Brown & P. C. Duetschle). *Journal of Sound and Vibration*, **1** (January 1964) 14–27.

Probability distributions of stress peaks in linear and non-linear structures. *AIAA Journal*, **1** (5) (May 1963) 1133–8.

Application of nonstationary shot noise in the study of system response to a class of nonstationary excitations. *Journal of Applied Mechanics*, **30** (4) (December 1963) 555–8.

On nonstationary shot noise. *Journal of Acoustical Society of America*, **36** (1) (January 1964) 82–4.

Random vibration of a Myklestad beam. *AIAA Journal*, **2** (8) (August 1964) 1448–51.

Transfer matrix representation of flexible airplanes in gust response study. *Journal of Aircraft*, **2** (1965) 116–21.

Nonstationary excitation and response in linear systems treated as sequences of random pulses. *Journal of Acoustical Society of America*, **38** (1965) 453–60.

Dynamics of beam-type periodic structures (with T. J. McDaniel). *Journal of Engineering for Industry, ASME Transactions*, **91**(13–4) (November, 1969) 1133–41.

A brief survey of transfer matrix techniques with special reference to the analysis of aircraft panels (with B. K. Donaldson). *Journal of Sound and Vibration*, **10** (1) (1969) 103–43.

On first-excursion failure of randomly excited structures. *AIAA Journal*, **8** (4) (February 1970) 720–5.

On first-excursion failure of randomly excited structures, II. *AIAA Journal*, **8** (10) (October 1970) 1888–90.

Response of flight vehicles to nonstationary atmospheric turbulence (with L. J. Howell). *AIAA Journal*, **9** (11) (November 1971) 2201–7.

Spatial decay in the response of damped periodic beam (with R. Vaicaitis & K. Doi). *Journal of Aircraft*, **9** (1) (January 1972) 91–3.

Response of finite periodic beam to turbulent boundary layer pressure fluctuation (with R. Vaicaitis). *AIAA Journal*, **10** (8) (August 1972) 1020–4.

Analysis of airplane response to nonstationary turbulence including wing bending flexibility (with Y. Fujimori). *AIAA Journal*, **11** (3) (March 1973) 334–9.

Analysis of airplane response to nonstationary turbulence including wing bending flexibility, Part II (with Y. Fujimori). *AIAA Journal*, **11** (9) (September 1973) 1343–5.

Free vibration of disordered periodic beams (with J. N. Yang). *Journal of Applied Mechanics*, **41** (2) (June 1974) 383–91.

Frequency response functions of a disordered periodic beam (with J. N. Yang). *Journal of Sound and Vibration*, **38** (3) (1975) 317–40.

Statistical analysis of longevity of prosthetic aortic valves (with D. N. Ghista). *Journal of Applied Mechanics*, **43** (1) (1976) 2–7.

Decomposition of turbulence forcing field and structural response (with S. Maekawa). *AIAA Journal*, **15** (5) (May 1977) 609–10.

Deterministic stability analysis of a wind loaded structure (with P. J. Holmes). *Journal of Applied Mechanics*, **45** (1) (1978) 165–9.

Stochastic analysis of wind-loaded structures (with P. J. Holmes). *Journal of Engineering Mechanics Division, ASCE*, Em 2, **104** (April 1978) 421–40.

Rotor blade stability in turbulence flow, Part I (with Y. Fujimori & S. T. Ariaratnam). *AIAA Journal*, **17** (6) (June 1979) 545–52.

Rotor blade stability in turbulence flow, Part II (with Y. Fujimori & S. T. Ariaratnam). *AIAA Journal*, **17** (7) (July 1979) 673–8.

Motion of suspension bridges in turbulent winds. *Journal of Engineering Mechanics Division, ASCE*, **105** EM 6 (December 1979) 921–3.

Stability of bridge motion in turbulent winds (with S. T. Ariaratnam). *Journal of Structural Mechanics*, **8** (1) (1980) 1–15.

Tall building response to earthquake excitations (with J. N. Yang & S. Sae-Ung). *Journal of Engineering Mechanics Division, ASCE*, **106** (EM 4) (August 1980) 801–17.

Column response to horizontal vertical earthquakes (with T. Y. Shih). *Journal of Engineering Mechanics Division, ASCE* (December 1980) 1099–109.

Along-wind motion of multi-storey buildings (with J. N. Yang). *Journal of Engineering Mechanics Division, ASCE*, **107** (EM 2) (April 1981) 295–307.

Coupled motion of wind-loaded multi-storey building (with J. N. Yang & B. Samali). *Journal of Engineering Mechanics Division, ASCE*, **107** (EM 6) (December 1981) 1209–26.

Vertical seismic load effect on hysteretic columns (with T. Y. Shih). *Journal of Engineering Mechanics Division, ASCE*, **108** (EM 2) (April 1982) 242–54.

Vertical seismic load effect on building response (with T. Y. Shih). *Journal of Engineering Mechanics Division, ASCE*, **108** (EM 2) (April 1982) 331–43.

Variability of tall building response to earthquake with changing epicenter direction (with J. N. Yang & S. Sae-Ung). *Journal of Earthquake Engineering and Structural Dynamics*, **10** (2) (March-April 1982) 211–23.

Rotor blade lead-lag stability in turbulent flows (with J. E. Prussing). *Journal of American Helicopter Society*, **27** (2) (April 1982) 51–7.

Concepts of stochastic stability in rotor dynamics (with J. E. Prussing). *Journal of American Helicopter Society*, **27** (2) (April 1982) 73–4.

Multimode bridge response to wind excitations (with J. N. Yang). *Journal of Engineering Mechanics, ASCE*, **109** (2) (April 1983) 586–603.

On statistical moments of fatigue crack propagation (with J. N. Yang). *Journal of Engineering Fracture Mechanics*, **18** (1983) 243–56.

Coupled flap-torsional response of a rotor blade in forward flight due to atmospheric turbulence excitations (with J. S. Fuh, C. Y. R. Hong & J. E. Prussing). *Journal of American Helicopter Society*, **28** (3) (July 1983) 3–12.

A closed-form analysis of rotor blade flap-lag stability in hover and low-speed forward flight in turbulent flow (with J. E. Prussing). *Journal of American Helicopter Society*, **28** (3) (July 1983) 42–6.

A closed-form earthquake response analysis of multi-storey building on compliant soil (with W. F. Wu). *Journal of Structural Mechanics*, **12** (1) (January 1984) 87–110.

Along-wind motion of building on complaint soil (with W. F. Wu). *Journal of Engineering Mechanics, ASCE*, **110** (1) (1984) 1–19.

Cumulant-neglect closure for nonlinear oscillators under random parametric and external excitations (with W. F. Wu). *International Journal of Nonlinear Mechanics*, **19** (4) (1984) 349–62.

Rotor blade flap-lag stability and response in forward flight in turbulence flows (with J. E. Prussing & T. N. Shiau). *Journal of American Helicopter Society*, **29** (4) (October 1984) 81–7.

A stochastic theory of fatigue crack propagation (with J. N. Yang). *AIAA Journal*, **23** (1) (Jan. 1985) 117–24.

Free and random vibrations of column supported cooling towers (with Y. Yong). *Journal of Sound and Vibration*, **98** (4) (1985) 539–63.

Some observations on the stochastic averaging methods. *Probabilistic Engineering Mechanics*, **1** (1) (1986) 23–7.

Methods of stochastic structural dynamics (with F. Kozin, Y. K. Wen, F. Casciati, G. I. Schuëller, A. Der Kiureghian, O. Ditlevsen & E. H. Vanmarcke). *Structural Safety*, **3** (1986) 167–94.

Stochastic response of flexible rotor-bearing system to seismic excitations (with Ki Bong Kim & Jann N. Yang). *Probabilistic Engineering Mechanics*, **1** (3) (1986) 122–30.

On random pulse train and its evolutionary spectral representation. *Probabilistic Engineering Mechanics*, **1** (4) (1986) 219–23.

Evolutionary Kanai-Tajimi earthquake models (with Y. Yong). *Journal of Engineering Mechanics, ASCE*, **113** (EM 8) (August 1987) 1119–37.

Exact stationary-response solution for second order nonlinear systems under parametric & external white-noise excitations (with Y. Yong). *Journal of Applied Mechanics*, **54** (1987) 414–18.

Generalization of the equivalent linearization method for non-linear random vibration problems (with A. Brückner). *International Journal of Nonlinear Mechanics*, **22** (3) (1987) 227–35.

Application of complex stochastic averaging to nonlinear random vibration problems (with A. Brückner). *International Journal of Nonlinear Mechanics*, **22** (3) (1987) 237–50.

A note on spectral moments in nonstationary random vibration (with C. G. Bucher). *Journal of Probabilistic Engineering Mechanics*, **3** (1) (1988) 53–5.

Invited Papers Published in Proceedings of Congresses, Symposia or Handbooks, etc.

Random processes. *Applied Mechanics Reviews*, **22** (8) (August 1969) 825–31.

Stochastic stability of wind loaded structures. *Proceedings of Second Annual Engineering Mechanics Division Specialty Conference*, Raleigh, North Carolina, May 23–27, 1977, pp. 59–62.

Stochastic theory of rotor blade dynamics. *Transactions of the 24th Conference of Army Mathematicians*, ARO Report 79–1, 1979, pp. 377–94.

On stochastic cross-wind structural stability (with S. T. Ariaratnam). *Proceedings of Third US National Conference on Wind Engineering*, Gainesville, FL, February 26–March 1, 1978, pp. V5.1–4.

Self excited bridge motion in turbulent wind (with S. T. Ariaratnam). *Proceedings of Third US National Conference on Wind Engineering*, Gainesville, FL, February 26–March 1, 1978, pp. V6.1–4.

Dynamics of structures under random loadings. *Proceedings of International Conference on Probabilistic Safety of Structures*, Paris, France, September 8–9, 1980, pp. 29–54.

Coupled translational and rotational motions of a multi-storey building under wind excitations (with J. N. Yang). *Proceedings of Fourth US National Conference on Wind Engineering Research*, Seattle, WA, July 27–29, 1981, pp. 224–33.

Random vibration of some civil structures. *Proceedings of ASCE EMD/SID Symposium on Probabilistic Methods in Structural Engineering*, St. Louis, MO, October 26–30, 1981, pp. 210–28.

Turbulence-excited flapping motion of a rotor blade in hovering flight (with C.Y.R. Hong). In *Advances in Aerospace Structures and Materials*, ed. S. S. Wang & W. J. Renton. AD-01, ASME, 1981, pp. 149–53.

A stochastic theory of fatigue crack propagation (with J. N. Yang). *Proceedings of AIAA/ASME/ASCE/ASH 24th Structures, Structural Dynamics and Materials Conference*, Lake Tahoe, NV, May 1983, Part I, pp. 555–62.

Examples of physical systems stabilized and destabilized by random parametric excitations. *Proceedings of International Workshop on Stochastic Structural Mechanics*, University of Innsbruck, Inst. Engr. Mech., Report 1–83, June 1983, pp. 62–6.

Along-wind response of tall buildings on compliant soil (with W. F. Wu). *Proceedings of 4th International Conference on Applications of Statistics and Probability in Soil and Structural Engineering*, Vol. 1, June 13–17, 1983, pp. 527–38.

Random vibrations of structural systems. *Proceedings of 44th International Statistical Institute Meeting*, Madrid, Spain, September 12–22, 1983, I.P. 30.1–19.

Stochastic modeling of fatigue crack propagation (with W. F. Wu & J. N. Yang). *Proceedings of IUTAM Symposium on Probabilistic Methods in Mechanics of Solids and Structures*, Stockholm, June 19–21, 1984, Springer Verlag, Berlin-Heidelberg, pp. 103–10.

Response of cooling tower or containment shell structures under seismic and wind excitations. *Transactions of 8th International Conference on Structural Mechanics in Reactor Technology*, Brussels, Belgium, Aug. 19–23, 1985, Vol. M1-M2, pp. 129–34.

Turbulence effects on vortex-induced vibrations (with K. She). *Proceedings of International Conference on Vibration Problems in Engineering*, ed. Du-Qing Hua, 1986, pp. 481–6.

Structural stability in turbulent wind. In *Recent Trends in Aeroelasticity,*

Structures and Structural dynamics, ed. P. Hajela. Papers from the Professor R. L. Bisplinghoff Memorial Symposium, University of Florida, Feb. 6–7, 1986, University of Florida Press, 1987, pp. 259–70.

Exact solutions for nonlinear systems under parametric and external white-noise excitations (with Y. Yong & G. Cai). *Lecture Notes in Engineering*, No. 31, Springer-Verlag, Berlin 1987, pp. 264–76.

Application of stochastic averaging for nonlinear dynamical systems with high damping (with N. Sri Namachchivaya). *Lecture Notes in Engineering*, No. 31, Springer-Verlag, Berlin 1987, pp. 277–306.

Evolutionary Kanai-Tajimi type earthquake models (with Y. Yong). *Lecture Notes in Engineering*, No. 32, Springer-Verlag, Berlin, 1987, pp. 174–203.

Effect of turbulence on flow induced oscillations (with N. Sri Namachchivaya). *Proceedings of the 20th Midwestern Conference*, Purdue University, W. Lafayette, IN, Aug. 31–Sept. 2, 1987.

Recent advances in nonlinear random vibration, *Proceedings of the Third International Conference on Recent Advances in Structure Dynamics*, Southampton, England, July 18–22, 1988.

Technical Reports with Wide Distribution

Free vibration of continuous skin-stringer panels with non-uniform stringer spacing and panel thickness (with T. J. McDaniel, B. K. Donaldson, C. F. Vail & W. J. Dwyer). Report AFML-TR064-457, Air Force Systems Command, Wright-Patterson Air Force Base, Ohio, 48 pp. February 1965.

Response of multi-spanned beam and panel systems under noise excitation, Part I. Report AFML-TR-064-458, Air Force Systems Command, Wright-Patterson Air Force Base, Ohio, 41 pp., February 1965.

Approximate correlation function and spectral density of the random vibration of an oscillator with nonlinear damping. AFML-TR-66-62, 11 pp, Wright-Patterson Air Force Base, Ohio, March 1966.

A method for the determination of the matrix of impulse response functions with special reference to applications in random vibration problems. *Proceedings of Conference on Matrix Methods in Structural Mechanics*, Report AFFDL-TR-66-80, Wright-Patterson Air Force Base, Ohio, pp. 743–51, November 1966.

Response of nonlinear structures to random excitations (with T. C. Chen & C. Y. Yang). AFML-TR-66-416, Air Force Systems Command, Wright-Patterson Air Force Base, Ohio, March, 1967.

Response of multi-spanned beam and panel systems under noise excitation (with T. J. McDaniel), Part II, Report AFML-TR-64-348, Air Force Systems Command, Wright-Patterson Air Force Base, Ohio, September 1967.

Spatial decay in the response of damped periodic structures (with K. Doi). Report AFML-TR-69-308, November 1969, 52 pp. Air Force Materials Lab., Air Force Systems Command.

Spatial decay in the response of damped periodic structures, Part II (with R. Vaicaitis). Report AFML-TR-69-308, Air Force Systems Command, Wright Patterson Air Force Base, Ohio, March 1970, 23 pp.

Vibroacoustic response of structures and perturbation Reynolds stress near structure turbulence interface (with S. Maekawa). NASA CR-2876, September 1977, 110 pp.

Dynamic response of some tentative compliant wall structure to convected turbulence fields (with H. H. Nijim). NASA CR-2909, November 1977, 116 pp.

Parametric studies of frequency response of secondary systems under ground-acceleration excitations (with Y. Yong). NCEER-87-0012, National Center for Earthquake Engineering Research, June 1987, 33 pp.

Frequency response of secondary systems under seismic excitation (with J. A. HoLung & J. Cai). NCEER-0013, National Center for Earthquake Engineering Research, July 1987, 88 pp.

Publications in Press

Exact stationary-response solution for second order nonlinear systems under parametric and external white noise excitations—Part II (with G. Q. Cai). *Journal of Applied Mechanics.*

On exact stationary solutions of equivalent nonlinear stochastic systems (with G. Q. Cai). *International Journal of Nonlinear Mechanics.*

Equivalent stochastic systems (with G. Q. Cai). *Journal of Applied Mechanics.*

Stochastic stability of bridges considering coupled modes (with C. G. Bucher). *Journal of Engineering Mechanics Division, ASCE.*

A new approximate solution technique for randomly excited nonlinear oscillators (with G. Q. Cai). *International Journal of Nonlinear Mechanics.*

Stochastic stability of bridges considering coupled modes II (with C. G. Bucher). *Journal of Engineering Mechanics Division, ASCE.*

Experimental verification of the probability distribution of sampled wind spectra (with P. H. W. Prenninger & G. I. Schuëller). *Journal of Wind Engineering and Industrial Aerodynamics.*

Effect of spanwise correlation of turbulence field on the motion stability of long-span-bridges (with C. G. Bucher). *Journal of Fluids and Structures.*

1

Dynamic Snap-Buckling of Structures under Stochastic Loads

S. T. ARIARATNAM & W.-C. XIE

Solid Mechanics Division, Faculty of Engineering, University of Waterloo, Ontario, Canada

ABSTRACT

The paper deals with the dynamic stability of shallow structures such as arches and curved panels under stochastically fluctuating loads. Sufficient conditions guaranteeing the almost-sure stability in both symmetric and anti-symmetric modes of deformation are obtained. The results are illustrated by an example of a sinusoidal two-pinned arch subjected to a sinusoidally distributed stochastic loading.

1 INTRODUCTION

In shallow curved structures such as arches and shells under symmetrically distributed loads, usually only the symmetric mode is excited. However, under certain conditions, the unsymmetric mode which is at rest originally will also be excited and become unstable leading to snap-buckling, due to the inherent nonlinear coupling. To illustrate this phenomenon, we investigate in this paper the almost-sure (a.s.) asymptotic stability of a symmetric sinusoidal arch under symmetric sinusoidally distributed random load. The results are also applicable with suitable modification to shallow cylindrical panels under symmetrically distributed loads. The partial differential equations governing the motion are first obtained and then converted into a set of second-order ordinary differential equations by using modal expansion. These equations are nonlinear and possess both quadratic and cubic

1

nonlinearities. The equation for the amplitude in the principal symmetric mode has a random forcing term while all other modes are unforced. To investigate the a.s. asymptotic stability, we consider the linearized perturbed equations of motion, in which the forcing term appears in the coefficients of second-order differential equations as stochastic parameters. The probability density of the stochastic parameters can be obtained by solving the Fokker–Planck equation provided that the random load can be approximated by a Gaussian white noise process. With the knowledge of the probability density, it is possible to obtain sufficient a.s. asymptotic stability boundaries for both the excited mode and the rest mode, first by using the Schwarz inequality and then, for much sharper boundaries, by a numerical optimization method. In the numerical example, we investigate the stability boundaries by varying the initial central rise and the mean value of the load.

2 BASIC EQUATIONS

We consider the plane motion of a simply supported shallow arch of uniform cross-section. We assume that the axis of the unloaded arch has a form $w_0(x)$, on which the dynamic deflection $w(x, t)$ is superposed. There is a lateral loading $p(x, t)$ as shown in Fig. 1. The

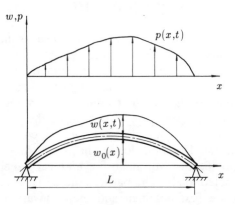

Fig. 1. Two-pinned arch and loading.

problem to be considered here is that of a low, half-sine, pinned arch loaded stochastically by a half-sine spatially distributed load. The initial shape is given by

$$w_0(x) = q_0 \sin \frac{\pi x}{L}$$

where q_0 is the initial rise parameter. The expression for the loading, directed upward, is given by

$$p(x, t) = -P(t) \sin \frac{\pi x}{L}$$

where $P(t)$ is a random process with mean value $E[P(t)] = P_0 \geqslant 0$. The deflection may be represented by an infinite sine series, each term of which satisfies the boundary conditions of simple support,

$$w(x, t) = \sum_{i=1}^{\infty} q_i \sin \frac{i\pi x}{L}$$

Then, the equations of motion are given by

$$\mu \ddot{q}_1 + \beta \dot{q}_1 + \left(\frac{\pi}{L}\right)^4 \left[EIq_1 + \frac{EA}{4}(q_0 + q_1)\left(\sum_{j=1}^{\infty} j^2 q_j^2 + 2q_0 q_1\right) \right] = -P(t)$$

$$\mu \ddot{q}_i + \beta \dot{q}_i + \left(\frac{\pi}{L}\right)^4 \left[i^4 EIq_i + \frac{EA}{4} i^2 q_i\left(\sum_{j=1}^{\infty} j^2 q_j^2 + 2q_0 q_1\right) \right] = 0, \tag{1}$$

$$i = 2, 3, \ldots$$

where μ is the mass per unit length, β is the viscous damping coefficient per unit length, EA and EI denote, respectively, the axial and flexural stiffness of the arch rib.

If $E[P(t)] = P_0 \neq 0$, we take $P(t) = P_0 + \xi(t)$, where $\xi(t)$ is a zero mean, wide-band process which can be approximated by a Gaussian white noise, i.e.

$$E[\xi(t)] = 0 \quad \text{and} \quad E[\xi(t)\xi(t + \tau)] = \kappa \delta(\tau) \tag{2}$$

where κ is the intensity of the load fluctuation, and $\delta(\tau)$ is the Dirac delta function.

3 STATIC CASE

Considering the static case first, which was studied by Fung & Kaplan[1] (see also Ashwell[2]), eqn (1) becomes

$$\left(\frac{\pi}{L}\right)^4\left[EIq_1 + \frac{EA}{4}(q_0 + q_1)\left(\sum_{j=1}^{\infty} j^2 q_j^2 + 2q_0 q_1\right)\right] = -P_0$$

$$i^4 EIq_i + \frac{EA}{4} i^2 q_i\left(\sum_{j=1}^{\infty} j^2 q_j^2 + 2q_0 q_1\right) = 0, \qquad i = 2, 3, \ldots \tag{3}$$

For notational convenience, we introduce the dimensionless parameters

$$\begin{aligned}
\lambda &= q_0/2\rho \\
a_i &= q_i/2\rho, \qquad i = 1, 2, \ldots \\
R &= P_0 L^4/2\pi^4 EI\rho \\
\rho^2 &= I/A
\end{aligned} \tag{4}$$

Then, eqns (3) become

$$a_1 + (\lambda + a_1)\left(\sum_{j=1}^{\infty} j^2 a_j^2 + 2\lambda a_1\right) = -R$$

$$a_i\left(i^2 + \sum_{j=1}^{\infty} j^2 a_j^2 + 2\lambda a_1\right) = 0, \qquad i = 2, 3, \ldots \tag{5}$$

If we take $\bar{w} = \sum_{i=1}^{\infty} b_i \sin\dfrac{i\pi x}{L} = w_0 + w$, so that

$$\begin{aligned}
b_1 &= q_0 + q_1 \\
b_i &= q_i, \qquad i = 2, 3, \ldots
\end{aligned} \tag{6}$$

or, in dimensionless form,

$$\begin{aligned}
B_1 &= \lambda + a_1 \\
B_i &= a_i, \qquad i = 2, 3, \ldots
\end{aligned} \tag{7}$$

where $B_i = b_i/2\rho$, then eqns (5) can be written as

$$B_1\left(\sum j^2 B_j^2 - \lambda^2 + 1\right) = -R + \lambda \tag{8}$$

$$B_i\left(\sum j^2 B_j^2 - \lambda^2 + i^2\right) = 0, \qquad i = 2, 3, \ldots$$

These equations have two possible solutions, the first being

$$B_2 = B_3 = \ldots = 0$$
$$B_1^3 - B_1(\lambda^2 - 1) = \lambda - R \tag{9}$$

In eqns (9), λ is proportional to the initial central rise, R to the load, and B_1 to the central rise when loaded. In Fig. 2, this relationship is shown graphically.

If the initial central rise corresponds to $\lambda < 1$, the deflection increases steadily with increasing load. If $\lambda = 1$, the B_1–R curve has a horizontal tangent. When $\lambda > 1$, the B_1–R curve has a maximum and a minimum, so that in some range there are more than one value of B_1 corresponding to a specified load R. At the maximum point, any further slight increment of load cannot be supported by the structure statically with any equilibrium configuration in the neighborhood of M; the structure is unstable and the deflection will increase dynamically until the point N is reached, the arch having *snapped through* into

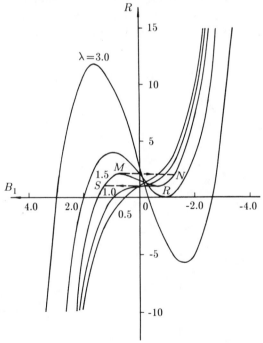

Fig. 2. Static load–deflection relationship $[B_1^3 - B_1(\lambda^2 - 1) = \lambda - R]$.

a nonadjacent stable equilibrium configuration. This phenomenon is the so-called *snap-through buckling* or oil-canning (Durchschlag). A similar snapping action will occur from R to S. Therefore, the equilibrium states of the structure and its load represented by MR are unstable.

The load corresponding to the point M, at which snap-through occurs for increasing load, is given by

$$R_{cr} = \lambda + \left[\frac{4}{27}(\lambda^2 - 1)^3\right]^{1/2} \tag{10}$$

which is the maximum point of the B_1–R curve.

The second possible solution to eqns (9) is

$$B_1 = (R - \lambda)/(i^2 - 1)$$
$$i^2 B_i^2 = \lambda^2 - i^2 - [(R - \lambda)/(i^2 - 1)]^2 \tag{11}$$

where i has only one value and all other B's are zero. From eqns (11), it is obvious that B_i can have a real value only within a certain range of values of R. The B_1–R and the B_i–R curves are shown in Fig. 3. Deflection in the ith mode first occurs at d, for which

$$B_1 = (\lambda^2 - i^2)^{1/2}, \qquad B_i = 0$$
$$R_{cr} = \lambda + (i^2 - 1)(\lambda^2 - i^2)^{1/2} \tag{12}$$

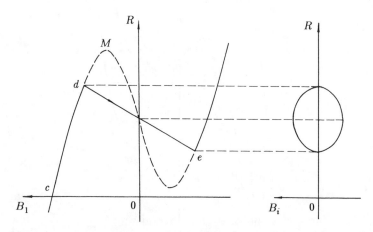

Fig. 3. Static load–deflection diagram (B_1 vs R and B_i vs R curves).

The dashed curve *dMe* is then replaced by the straight line *de* and the load R_{cr} at which the structure snaps corresponds to *d*, provided that

(1) R_{cr}, B_1 and B_i are real;
(2) R_{cr} given by eqn (12) is less than that given by eqn (10);
(3) *d* lies to the left of *M*;
(4) *i* is chosen for R_{cr} to be as small as possible.

These conditions are satisfied if

$$i = 2 \quad \text{and} \quad \lambda \geqslant \sqrt{5 \cdot 5} \tag{13}$$

Therefore, one can conclude that (for $i = 2$)

$$\begin{aligned}
&\text{for} \quad \lambda < 1, &&\text{there is no buckling}\\
&\text{for} \quad 1 < \lambda < \sqrt{5 \cdot 5}, &&R_{cr} = \lambda + (2/3\sqrt{3})(\lambda^2 - 1)^{3/2}\\
&\text{for} \quad \lambda > \sqrt{5 \cdot 5}, &&R_{cr} = \lambda + 3(\lambda^2 - 4)^{1/2}
\end{aligned} \tag{14}$$

4 DYNAMIC CASE

Now returning to the dynamic case, consider the effect of a stochastic load on the stability of the structure when $\lambda > \sqrt{5 \cdot 5}$ and $R < R_{cr}$ given by eqn (14), or

$$q_0 = 2\lambda\rho > 2\sqrt{5 \cdot 5}\rho$$
$$P_0 = 2EI\rho\left(\frac{\pi}{L}\right)^4 R < 2EI\rho\left(\frac{\pi}{L}\right)^4 [\lambda + 3(\lambda^2 - 4)^{1/2}] \tag{15}$$

The equilibrium states of the arch and its load are represented by *cd* in Fig. 3 and are stable. The static component of deflection is given by eqn (9), i.e.

$$B_{1s}^3 - B_{1s}(\lambda^2 - 1) = \lambda - R$$
$$B_{2s} = B_{3s} = \ldots = 0$$

or

$$q_{1s} = 2\rho a_{1s} = 2\rho(B_{1s} - \lambda)$$
$$q_{2s} = q_{3s} = \ldots = 0 \tag{16}$$

where the subscript s indicates static parameter.

We note that only the first mode is directly excited and the other modes are at rest. Then the first of eqns (1) becomes

$$\mu\ddot{q}_1 + \beta\dot{q}_1 + \left(\frac{\pi}{L}\right)^4 \left[EIq_1 + \frac{EA}{4}(q_0 + q_1)(q_1^2 + 2q_0q_1) \right] = -[P_0 + \xi(t)]$$

(17)

Let us assume

$$q_1 = q_{1s} + X_1(t)$$

(18)

where q_{1s} corresponds to the static load P_0 and $X_1(t)$ to the random load $\xi(t)$, the static deflection q_{1s} being given by

$$\frac{EA}{4}\left(\frac{\pi}{L}\right)^4 [\rho^2 q_{1s} + (q_0 + q_{1s})(q_{1s}^2 + 2q_0q_{1s})] = -P_0$$

(19)

Then, subtracting eqn (19) from eqn (17) leads to the equation for the deflection component corresponding to the fluctuational part of the load:

$$\mu\ddot{X}_1 + \beta\dot{X}_1 + \frac{EA}{4}\left(\frac{\pi}{L}\right)^4 [(4\rho^2 + 2q_0^2 + 6q_0q_{1s} + 3q_{1s}^2)X_1$$

$$+ (3q_0 + q_{1s})X_1^2 + X_1^3] = -\xi(t) \quad (20)$$

4.1 Two-Dimensional Fokker–Planck Equation
Suppose a noise process $X(t)$ satisfies a second-order fluctuation equation

$$\ddot{X} + b\dot{X} + f(X) = F(t)$$

(21)

where the excitation $F(t)$ is a stationary, mean zero, wide-band random process with correlation function

$$R_F(\tau) = E[F(t)F(t + \tau)]$$

If $F(t)$ can be approximated by a Gaussian white noise process with correlation function

$$R_F(\tau) = \kappa\delta(\tau)$$

where

$$\kappa = \int_{-\infty}^{+\infty} E[F(t)F(t + \tau)]\,d\tau$$

then the random displacement $X(t)$ and the random velocity $\dot{X}(t)$ are approximately the components of a Markov vector.

The first-order stationary probability density $p(x, \dot{x})$ of the Markov vector $(X, \dot{X})$ satisfies the Fokker–Planck equation

$$\dot{x} \frac{\partial p}{\partial x} - [b\dot{x} + f(x)] \frac{\partial p}{\partial \dot{x}} - bp - \frac{\kappa}{2} \frac{\partial^2 p}{\partial \dot{x}^2} = 0 \tag{22}$$

It is easy to show (see for example, Lin[3]) that for a second-order system with nonlinear stiffness and linear damping which is subjected to Gaussian white noise excitation, the stationary displacement and velocity are independent at the same instant and that the stationary solution of the Fokker–Planck eqn (22) is

$$p(x, \dot{x}) = C \exp\left\{ -\frac{b}{\kappa} \left[\dot{x}^2 + \int_0^x f(u) \, \mathrm{d}u \right] \right\} \tag{23}$$

where C is the normalization constant satisfying

$$\int_{-\infty}^{+\infty} \int_{-\infty}^{+\infty} p(x, \dot{x}) \, \mathrm{d}x \, \mathrm{d}\dot{x} = 1 \tag{24}$$

Therefore, for eqn (20), the stationary marginal probability density of the displacement process $X_1(t)$ is given by

$$p(x_1) = C_0 \exp\left\{ -\frac{\beta EA}{2\kappa} \left(\frac{\pi}{L} \right)^4 [\tfrac{1}{2}(4\rho^2 + 2q_0^2 + 6q_0 q_{1s} + 3q_{1s}^2)x_1^2 \right.$$

$$\left. + (q_0 + q_{1s})x_1^3 + \tfrac{1}{4}x_1^4] \right\} \tag{25}$$

where $E[\xi(t)\xi(t + \tau)] = \kappa\delta(\tau)$, and C_0 is the normalization constant satisfying the condition

$$\int_{-\infty}^{+\infty} p(x_1) \, \mathrm{d}x_1 = 1 \tag{26}$$

4.2 Perturbation Equation

To investigate the a.s. asymptotic stability of both the excited mode and the rest modes described by eqns (1), substituting the perturbed solution

$$q_1 = q_{1s} + X_1 + x_1$$
$$q_2 = 0 + 0 + x_2$$
$$\cdots\cdots$$
$$q_n = 0 + 0 + x_n \tag{27}$$
$$\cdots\cdots$$

into eqns (1), neglecting all nonlinear terms and making use of eqns (19) and (20) results in

$$\mu\ddot{x}_1 + \beta\dot{x}_1 + \frac{EA}{4}\left(\frac{\pi}{L}\right)^4 [(4\rho^2 + 2q_0^2 + 6q_0q_{1s} + 3q_{1s}^2)$$

$$+ 6(q_0 + q_{1s})X_1 + 3X_1^2]x_1 = 0$$

$$\mu\ddot{x}_i + \beta\dot{x}_i + \frac{EA}{4}\left(\frac{\pi}{L}\right)^4 i^2[(4i^2\rho^2 + q_{1s}^2 + 2q_0q_{1s}) \qquad (28)$$

$$+ 2(q_0 + q_{1s})X_1 + X_1^2]x_i = 0, \qquad i = 2, 3, \ldots$$

or

$$\ddot{x}_i + 2\zeta_i\omega_i\dot{x}_i + [\omega_i^2 + g_i(t)]x_i = 0, \qquad i = 1, 2, \ldots \qquad (29)$$

where

$$\zeta_i = \frac{\beta}{4\mu\omega_i}, \qquad i = 1, 2, \ldots$$

$$\omega_1^2 = \frac{EA}{4\mu}\left(\frac{\pi}{L}\right)^4 (4\rho^2 + 2q_0^2 + 6q_0q_{1s} + 3q_{1s}^2)$$

$$\omega_i^2 = \frac{EA}{4\mu}\left(\frac{\pi}{L}\right)^4 i^2(4i^2\rho^2 + q_{1s}^2 + 2q_0q_{1s}), \qquad i = 2, 3, \ldots \qquad (30)$$

$$g_1(t) = \frac{EA}{4\mu}\left(\frac{\pi}{L}\right)^4 3[2(q_0 + q_{1s})X_1 + X_1^2]$$

$$g_i(t) = \frac{EA}{4\mu}\left(\frac{\pi}{L}\right)^4 i^2[2(q_0 + q_{1s})X_1 + X_1^2], \qquad i = 2, 3, \ldots$$

Since the probability density of the random process $X_1, p(x_1)$, is known, it is possible to find the probability density of the random process g_i. According to the fundamental theorem (Papoulis[4]), the probability density of $g_i(t)$ is given by

$$p_{g_i}(g_i) = \frac{1}{\Delta_1}\left[p\left(\frac{-b_i + \Delta_i}{2a_i}\right) + p\left(\frac{-b_i - \Delta_i}{2a_i}\right)\right],$$

$$g_i \geq -\frac{b_i^2}{4a_i}, \qquad i = 1, 2, \ldots \qquad (31)$$

where $p(\cdot)$ is given by eqn (25), and

$$\Delta_i = (b_i^2 + 4a_i g_i)^{1/2}, \qquad i = 1, 2, \ldots$$

$$a_1 = \frac{3EA}{4\mu} \left(\frac{\pi}{L}\right)^4$$

$$a_i = \frac{i^2 EA}{4\mu} \left(\frac{\pi}{L}\right)^4, \qquad i = 2, 3, \ldots$$

$$b_1 = \frac{3EA}{2\mu} \left(\frac{\pi}{L}\right)^4 (q_0 + q_{1s})$$

$$b_i = \frac{i^2 EA}{2\mu} \left(\frac{\pi}{L}\right)^4 (q_0 + q_{1s}), \qquad i = 2, 3, \ldots$$

4.3 Sufficient Stability Boundary

Consider the second-order system

$$\ddot{x} + 2\zeta\omega\dot{x} + [\omega^2 + g(t)]x = 0 \qquad (32)$$

where $g(t)$ is an ergodic process (Infante[5]; Ariaratnam & Xie[6]). The transformation of the form $x = y \exp(-\zeta\omega t)$ converts eqn (32) to

$$\ddot{y} + [c + g(t)]y = 0$$

where $c = (1 - \xi^2)\omega^2$, or, in the state equation form,

$$\begin{aligned} \dot{y}_1 &= y_2 \\ \dot{y}_2 &= -[c + g(t)]y_1 \end{aligned} \qquad (33)$$

The square of the quadratic norm of y, $\|y\|^2$, may be taken in the form

$$\|y\|^2 = V = y^T A y$$

where A is a positive-definite matrix given by

$$A = \begin{bmatrix} \alpha_1^2 & \alpha_2 \\ \alpha_2 & 1 \end{bmatrix}, \qquad \alpha_1^2 - \alpha_2^2 > 0$$

where α_1, α_2 are parameters to be determined.

The time derivative of $V(y)$ along the trajectory of eqn (33) is

$$\dot{V}(y) = y^T B y$$

where

$$B = \begin{bmatrix} -2\alpha_2(c + g) & \alpha_1^2 - (c + g) \\ \alpha_1^2 - (c + g) & 2\alpha_2 \end{bmatrix}$$

Therefore, since A, B are real symmetric matrices, and A is positive definite,

$$\frac{\dot{V}}{V} = \frac{y^{\mathrm{T}}By}{y^{\mathrm{T}}Ay} \leqslant \lambda(BA^{-1}) \tag{34}$$

where $\lambda(t)$ is the maximum eigenvalue of BA^{-1}, i.e. λ is the maximum root of the determinantal equation $|B - \lambda A| = 0$ and is given by

$$\lambda = \left[\frac{(\alpha_1^2 - c)^2 + 4c\alpha_2^2 + (4\alpha_2^2 + 2c - 2\alpha_1^2)g + g^2}{\alpha_1^2 - \alpha_2^2} \right]^{1/2}$$

From (34),

$$V[y(t)] \leqslant V_0 \exp\left\{ \int_0^t \lambda(\tau)\, d\tau \right\}$$

where $V_0 = V[y(0)]$. Since $\| y \|^2 = V$,

$$\| y \|^2 \leqslant \| y_0 \|^2 \exp\left\{ \int_0^t \lambda(\tau)\, d\tau \right\} = \| y_0 \|^2 \exp\left\{ t \cdot \frac{1}{t} \int_0^t \lambda(\tau)\, d\tau \right\}$$

whose RHS, when $t \to \infty$, goes to

$$\| y_0 \|^2 \exp\{ tE[\lambda(t)] \}$$

with probability 1 (w.p.1) for ergodic process $g(t)$. Then, we have w.p.1 as $t \to \infty$.

$$\| x \| \leqslant \| y_0 \| \exp\left\{ \frac{t}{2}[-2\zeta\omega + E[\lambda(t)]] \right\}$$

Therefore, a sufficient condition for a.s. asymptotic stability is

$$-2\zeta\omega + E[\lambda(t)] < -\epsilon, \qquad \epsilon > 0 \tag{35}$$

By applying the Schwarz inequality to inequality (35), one has

$$E[\lambda^2(t)] < 4\zeta^2\omega^2$$

so that the stability boundary is

$$F = (\alpha_1^2 - c)^2 + 4c\alpha_2^2 - 4\zeta^2\omega^2(\alpha_1^2 - \alpha_2^2) + (4\alpha_2^2 + 2c - 2\alpha_1^2)\mu_g + \sigma_g^2 = 0 \tag{36}$$

where $\mu_g = E[g(t)]$, $\sigma_g^2 = E[g^2(t)]$.

Optimization of the stability boundary with respect to α_1, α_2 leads to

$$\frac{\partial F}{\partial(\alpha_1^2)} = 2(\alpha_1^2 - c) - 4\zeta^2\omega^2 - 2\mu_g = 0$$

$$\frac{\partial F}{\partial(\alpha_2^2)} = 8c\alpha_2 + 8\zeta^2\omega^2\alpha_2 + 8\alpha_2\mu_g = 0$$

$$(37)$$

or

$$\alpha_1^2 = \mu_g + (1 + \zeta^2)\omega^2$$
$$\alpha_2 = 0$$

$$(38)$$

Therefore a sufficient stability condition is

$$\sigma_g^2 < \mu_g^2 + 4\zeta^2\omega^2(\omega^2 + \mu_g) \tag{39}$$

Due to the complicated form of the probability density (25), it is not possible to obtain the sufficient stability boundary from (39) explicitly, however, we shall obtain it numerically in the following section.

Furthermore, since the distributional property of $g(t)$ is known by solving the Fokker–Planck equation, a numerical optimization method can be applied directly to obtain a much sharper sufficient stability boundary from (35), which is described in Section 5.2.

5 EXAMPLE

As an example, we consider a steel arch of rectangular cross-section having the following parameters:

density of material $= 7850$ kg m^{-3}
Young's modulus $E = 0{\cdot}2 \times 10^{12}$N m^{-2}
rectangular cross-section with height $= 0{\cdot}03$ m, width $= 0{\cdot}02$ m
length of the span $L = 1$ m

(i) Initial Central Rise $q_0 = 0{\cdot}2m$
This is a relatively steep arch, and, from eqns (4) and (14), $\lambda = 11{\cdot}547 > \sqrt{5{\cdot}5}$, $R_{cr} = 45{\cdot}664$, corresponding to $P_0 = 6{\cdot}934 \times 10^5$ N m^{-1}. Two values for the mean load P_0 are considered, namely $P_0 = 1 \times 10^5$ Nm^{-1}, 1×10^4 Nm^{-1}, corresponding to static deflections $q_{1s} = -0{\cdot}4275 \times 10^{-3}$ m, $-0{\cdot}426 \times 10^{-4}$ m, respectively.

(ii) Initial Central Rise $q_0 = 0.07$ m
This is a shallow arch, and, from eqns (4) and (14) $\lambda = 4.041 > \sqrt{5.5}$, $R_{cr} = 14.577$, corresponding to $P_0 = 2.213 \times 10^5$ N m^{-1}. The static load-deflection relationship is shown in Fig. 4. Again, two values for the mean load P_0 are considered, namely $P_0 = 1 \times 10^4$ Nm^{-1}, 5×10^4 Nm^{-1}, corresponding to static deflection $q_{1s} = -0.3421 \times 10^{-3}$ m, -0.17581×10^{-2} m, respectively.

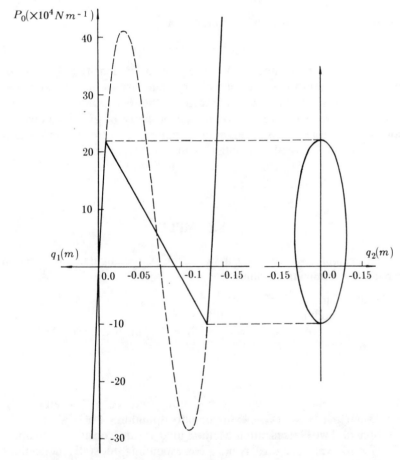

Fig. 4. q_1-P_0 and q_2-P_0 curves for arch with initial rise $q_0 = 0.07$ m.

5.1 Sufficient A.S. Asymptotic Stability Condition Using the Schwarz Inequality

Using the probability density of the random process $g(t)$, for any integrable function $f(g)$, one has

$$
\begin{aligned}
E[f(g)] &= \int_{-b^2/4a}^{+\infty} f(g) p_g(g) \, dg \\
&= \int_{-b^2/4a}^{+\infty} f(g) \frac{1}{(b^2 + 4ag)^{1/2}} \left[p\left(\frac{-b + (b^2 + 4ag)^{1/2}}{2a} \right) \right. \\
&\quad \left. + p\left(\frac{-b - (b^2 + 4ag)^{1/2}}{2a} \right) \right] dg
\end{aligned}
\tag{40}
$$

Changing to variable $ay^2 = g + \dfrac{b^2}{4a}$ results in

$$
E[f(g)] = \int_0^{+\infty} f\left(ay^2 - \frac{b^2}{4a} \right) \left[p\left(-\frac{b}{2a} + y \right) + p\left(-\frac{b}{2a} - y \right) \right] dy
\tag{41}
$$

where $p(\cdot)$ is given by eqn (25).

From eqn (41), one can obtain the expectation and variance of $g_i(t)$, $i = 1, 2, \ldots$,

$$
\mu_{g_i} = E[g_i(t)] = \int_0^{+\infty} a_i\left(y^2 - \frac{b_i^2}{4a_i^2} \right) \left[p\left(-\frac{b_i}{2a_i} + y \right) + p\left(-\frac{b_i}{2a_i} - y \right) \right] dy
$$

$$
\sigma_{g_i}^2 = E[g_i^2(t)] = \int_0^{+\infty} a_i^2\left(y^2 - \frac{b_i^2}{4a_i^2} \right)^2 \left[p\left(-\frac{b_i}{2a_i} + y \right) + p\left(-\frac{b_i}{2a_i} - y \right) \right] dy
\tag{42}
$$

It is now assumed that only the motions in the first and the second modes are significant, i.e. $x_3 = x_4 = \ldots = 0$. Substituting from eqns (42) into the sufficient a.s. asymptotic stability condition (39) for some structural damping β and solving the equation for κ leads to the sufficient a.s. asymptotic stability conditions expressed in terms of κ and ζ for the first and second modes, respectively. The numerical results are shown in Figs 5–8.

5.2 Sufficient A.S. Asymptotic Stability Condition Using a Numerical Optimization Method

Since the probability density function $p(x_1)$ is known by solving the Fokker–Planck equation, a sharper stability condition than (39) may

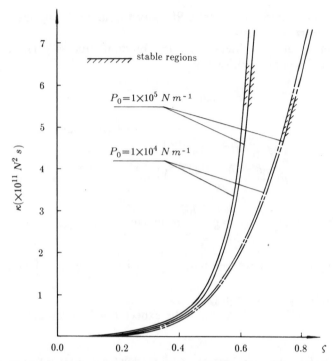

Fig. 5. Regions of almost-sure asymptotic stability ($q_0 = 0 \cdot 2$ m, first mode). – – –, stability boundary via Schwarz inequality; ——, stability boundary via optimization method.

be obtained using the 'complex' method of optimization due to Box,[7] see also Ariaratnam & Xie.[8]

In the optimization method, we use the stability condition (35) directly, which can be expressed as

$$\int_0^{+\infty} \lambda(y, a_i, b_i, \alpha_1, \alpha_2)\left[p\left(-\frac{b_i}{2a_i} + y\right) + p\left(-\frac{b_i}{2a_i} - y\right)\right] \mathrm{d}y - 2\zeta_i\omega_i < 0$$

(43)

and used as an implicit constraint in the optimization model.

Now the optimization model that may be solved by the complex method can be constructed as follows:

Maximize:

$$V = \kappa$$

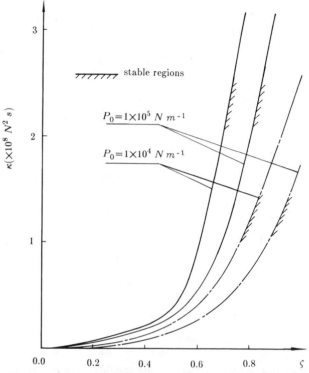

Fig. 6. Regions of almost-sure asymptotic stability ($q_0 = 0.2$ m, second mode). -----, stability boundary via Schwarz inequality; ——, stability boundary via optimization method.

Subject to constraints:

$$\underline{\kappa} \leq X(1) = \kappa \leq \bar{\kappa}$$

$$\underline{\alpha}_1 \leq X(2) = \alpha_1 \leq \bar{\alpha}_1$$

$$-|\alpha_1| + \delta \leq X(3) = \alpha_2 \leq |\alpha_1| - \delta$$

$$0 \leq X(4) = 2\zeta_i \omega_i - \int_0^{+\infty} \lambda_{\max}(y, a_i, b_i) \left[p\left(-\frac{b_i}{2a_i} + y \right) \right.$$

$$\left. + p\left(-\frac{b_i}{2a_i} - y \right) \right] dy \leq 2\zeta_i \omega_i$$

Since sufficient a.s. asymptotic stability boundaries have already been

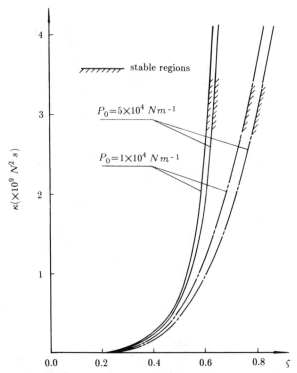

Fig. 7. Regions of almost-sure asymptotic stability ($q_0 = 0.07$ m, first mode). — – —, stability boundary via Schwarz inequality; ———, stability boundary via optimization method.

obtained by applying the Schwarz inequality, these results can be used not only as the initial points of the complex method, but also to decide the constraints of the optimization model; for example, $\underline{\kappa} = \kappa_0$, where κ_0 is the result found by using the Schwarz inequality. The range of α_1 is also decided by these results and δ is a positive small quantity.

Then the optimization model can be solved numerically; the results are plotted in Figs 5–8. The stability regions thus obtained are found to be considerably larger than those found by using the Schwarz inequality (39), which however is simpler to apply.

6 DISCUSSION AND CONCLUSION

For the sinusoidal arch studied in this paper, sufficient a.s. asymptotic stability regions for both the vibrating mode and the rest mode are

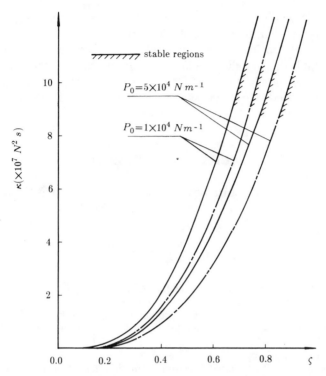

Fig. 8. Regions of almost-sure asymptotic stability ($q_0 = 0{\cdot}07$ m, second mode). $---$, stability boundary via Schwarz inequality; $---$, stability boundary via optimization method.

obtained by using the Schwarz inequality and an optimization method. The following conclusions may be drawn from this study:

(i) It is clear that for shallow symmetric structures under symmetrically distributed random load, usually only the symmetric mode is excited. However, under certain conditions, the unsymmetric mode that is at rest originally can be excited and become unstable due to autoparametric coupling. It also appears that the rest mode is less stable than the vibrating mode, especially for a relatively steep arch, so that the rest mode is of greater interest.

(ii) For a steep arch, the critical static load that the arch is able to support (eqn (14)) is larger than that of the shallow arch.

Moreover, the sufficient stability region of the steep arch is larger than that of the shallow arch.

(iii) Variation of the mean load has a distinct effect on the sufficient stability regions of the rest mode; the larger the mean load, the smaller the stability region, as may be expected. However, such an effect is not so important on the vibrating mode.

ACKNOWLEDGEMENT

This research was supported by the National Sciences and Engineering Research Council of Canada through Grant No. A-1815.

REFERENCES

1. Fung, Y. C. & Kaplan, A., Buckling of low arches or curved beams of small curvature, NACA Tech. Note 2840, 1952.
2. Ashwell, D. G., Nonlinear problems. *Handbook of Engineering Mechanics,* ed. W. Flügge, Chapter 45, McGraw-Hill, New York, 1962.
3. Lin, Y. K., *Probabilistic Theory of Structural Dynamics,* McGraw-Hill, New York, 1967, pp. 262–5.
4. Papoulis, A., *Probability, Random Variables, and Stochastic Processes,* McGraw-Hill Series in Electrical Engineering, Communications and Information Theory. McGraw-Hill, New York, 1984, pp. 95–6.
5. Infante, E. F., On the stability of some linear nonautonomous random systems. *J. Appl. Mech.,* **35** (1) (1968) 7–12.
6. Ariaratnam, S. T. & Xie, W.-C., Stochastic sample stability of oscillatory systems. To appear in *ASME J. Appl. Mech.*
7. Box, M. J., A new method of constrained optimization and a comparison with other methods. *Comp. J.,* **8** (1965) 42–52.
8. Ariaratnam, S. T. & Xie, W.-C., Effect of derivative process on the a.s. asymptotic stability of second-order linear stochastic systems. *Dynamics and Stability of Systems,* **3**(2) (1988).

2

Spatial Correlation in Structural Response to Wide-Band Excitation

STEPHEN H. CRANDALL

Department of Mechanical Engineering, Massachusetts Institute of Technology, Cambridge, Massachusetts, USA

ABSTRACT

When uniform lightly damped structural elements such as beams, cables, plates, shells, or membranes are driven by broad band random excitation the local response velocity and the local dynamic stress are stationary random fields with distinctive characteristics. In the temporal domain they are characterized by spectra with multiple narrow peaks. For plates, shells and membranes the number of resonances excited in practical applications can exceed a thousand. In the spatial domain they are characterized by a tendency toward uniformity of mean-square response as the bandwidth of the excitation increases. Here it is shown that in addition there is also an asymptotic tendency of the responses at different locations to be uncorrelated. The spatial correlations in typical structures are studied by obtaining exact solutions for rain-on-the-roof excitation and by using modal-sum and image-sum approximations for point-load excitation. The focussing phenomenon which produces zones or lanes of intensified mean-square response as exceptions to the general tendency toward uniform mean-square response also operates to produce zones of significant cross correlations as exceptions to the general tendency toward zero cross correlation.

1 INTRODUCTION

The subject of random fields is increasingly attracting interest from technical investigators.[1] A special class of random fields with interesting properties consists of the response fields of lightly damped uniform

21

structures to stationary wide-band random excitation.[2] The response of each structural mode is a narrow-band process but if the excitation is wide-band, a great many modes are excited so the response is a wide-band process whose spectrum consists of a large number of narrow peaks. In general, displacement responses are dominated by the lowest frequency modes and acceleration responses are dominated by the highest frequency modes, while velocity responses and most stress responses of interest tend to be broad-band processes with comparable contributions coming from every excited mode. It is known[2] that in the case of excitations with either no spatial correlation (rain on the roof) or with single point loading there is an asymptotic tendency for the mean-square velocity to become spatially uniform as the bandwidth of the excitation increases. It is additionally known[2] that in the case of a single point loading there may be exceptions to the tendency toward spatial uniformity in the form of increasingly narrow lanes or zones of intensified mean-square response which result from 'focussing effects' induced by appropriately symmetric boundary geometry.

In the present paper the spatial correlation of the velocity response in such cases is investigated. If $v(P_1)$ and $v(P_2)$ are the velocity responses at the same instant at two locations in the structure we examine the correlation coefficient

$$\rho_{12} = \frac{E[v(P_1)v(P_2)]}{\langle E[v^2(P)]\rangle} \qquad (1)$$

where the cross correlation is normalized by dividing by an appropriate spatial average of the mean-square velocity. If P_1 and P_2 coincide then ρ_{12} approaches unity[2] as the excitation bandwidth increases without limit. It will be shown below that when P_1 and P_2 are distinct then the asymptotic tendency is for ρ_{12} to vanish as the excitation bandwidth increases. Thus the general behavior of ρ_{12} for wide, but finite, excitation bandwidth is to be small for P_1 and P_2 distinct, and to be close to unity in the neighborhood of coincident P_1 and P_2. This general behavior is modified locally by focussing effects in the case of single point loading in structures with appropriate boundary symmetry. The analysis which follows is primarily for uniaxial structures such as beams or taut strings with a brief extension to rectangular biaxial structures such as plates or membranes. Spatially uncorrelated excitation is considered first, followed by excitation applied at a single point. The basic analysis technique is appropriate spatial averaging[2] applied to modal sum representations. For single point loading the image-sum procedure[2,3] provides an alternative technique which is especially convenient for determining the locations and magnitudes of intensifications due to focussing.

2 SPATIALLY UNCORRELATED EXCITATION OF A UNIAXIAL STRUCTURE

Consider a simply-supported uniform Bernoulli–Euler beam or taut string of length L and mass m subjected to a stationary random transverse load $f(x, t)$ per unit length with cross-spectral density function.

$$S_f(x_1, x_2, \omega) = S_0 \, \delta(x_2 - x_1)/L, \qquad |\omega| < \omega_c \qquad (2)$$

The excitation is spatially uncorrelated (rain on the roof) and temporally broad band with a band-limited white noise spectrum. If the structure is subjected to damping with uniform modal half-power bandwidth β the velocity response cross correlation is[2]

$$E[v(x_1)v(x_2)] = \frac{\pi S_0}{\beta m^2} \sum_{j=1}^{N} 2 \sin(j\pi x_1/L) \sin(j\pi x_2/L) \qquad (3)$$

where N is the number of modes excited by (2). For broad-band excitation the integer N is[3] essentially proportional to the cut-off frequency ω_c for taut strings, and proportional to $(\omega_c)^{1/2}$ for Bernoulli–Euler beams. When $x_1 = x_2 = x$, (3) reduces to the mean-square velocity

$$E[v^2(x)] = \frac{\pi S_0}{\beta m^2} \sum_{j=1}^{N} 2 \sin^2(j\pi x/L) = \frac{\pi S_0}{\beta m^2} s(x) \qquad (4)$$

For completeness (4) is discussed prior to (3). The nature of the distribution of $s(x)$ may be seen by summing the series to get

$$s(x) = N - \Delta(x) \qquad (5)$$

where the Δ-function, defined by

$$\Delta(x) = \sum_{j=1}^{N} \cos(2\pi x/L) = \frac{\sin\left[(2N+1)\pi x/L\right]}{2\sin(\pi x/L)} - \frac{1}{2} \qquad (6)$$

is periodic with period L, and has peaks of height N and width $(N + 1/2)^{-1}$ at $x = 0, \pm L, \pm 2L, \ldots$ and elsewhere fluctuates about the level $-1/2$ with an amplitude of order unity and wavelength of order $1/N$. The Δ-function is displayed in Fig. 11 of Ref. 2. Thus, except in the neighborhood of the ends, $x = 0$, L, where it vanishes, $s(x)$ oscillates about the level $N + 1/2$ with an amplitude of order unity. The shape of $s(x)$ for $N = 20$ is displayed by the dotted curve in Fig. 1. The asymptotic nature of $s(x)$ may alternatively be deduced by spatial averaging. Directly from (4) it follows that the spatial mean of $s(x)$ is

$$m_s = \langle s(x) \rangle_x = \frac{1}{L} \int_0^L s(x)\, dx = N \tag{7}$$

and that the spatial variance of $s(x)$ is

$$\sigma_s^2 = \langle s^2(x) - m_s^2 \rangle_x = N/2 \tag{8}$$

The relative fluctuation of $s(x)$ about its spatial mean

$$\sigma_s/m_s = 1/\sqrt{2N} \tag{9}$$

thus approaches zero as N increases without limit. This demonstrates the asymptotic tendency toward *uniform* mean-square velocity response.

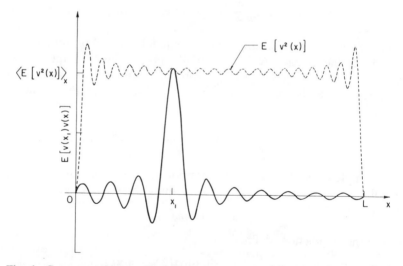

Fig. 1. Cross correlation of velocity response of a uniform beam or string to band-limited white noise rain-on-the-roof excitation which excites 20 modes.

Returning now to the cross correlation (3) we consider the corresponding correlation coefficient where for the denominator of (1) we use the value taken by (4) when $s(x)$ is replaced by its spatial mean (7)

$$\rho_{12} = \frac{1}{N} \sum_{j=1}^{N} 2 \sin(j\pi x_1/L) \sin(j\pi x_2/L) \tag{10}$$

The result of summing the series is

$$\rho_{12} = \frac{1}{N} (\Delta[(x_2 - x_1)/2] - \Delta[(x_1 + x_2)/2]) \tag{11}$$

from which it may be seen that ρ_{12} vanishes when either location x_1, x_2 approaches an end $x = 0$, L, that ρ_{12} is of order unity when x_2 and x_1 coincide, and that ρ_{12} is of order $1/N$ when x_2 and x_1 are separated. The shape of ρ_{12} for $N = 20$ is displayed in Fig. 1 where $E[v(x_1)v(x_2)]$ is plotted as a function of $x = x_2$ for a fixed value of x_1. When $x = x_1$, the cross correlation takes on the value of the mean square, but elsewhere it fluctuates about zero. It is interesting to note that the wavelength of the fluctuations in the cross correlation is twice that of the fluctuations in the mean square.

3 SINGLE POINT EXCITATION OF A UNIAXIAL STRUCTURE

Consider the same beam or string structure as above but now let the stationary random transverse load $f(x, t)$ per unit length have the cross-spectral density function

$$S_f(x_1, x_2, \omega) = S_0 \, \delta(x_1 - x_0) \, \delta(x_2 - x_0), \qquad |\omega| < \omega_c \tag{12}$$

The excitation is band-limited white noise applied at the single point $x = x_0$. For this excitation the velocity response cross-correlation for light damping is

$$E[v(x_1)v(x_2)] = \frac{\pi S_0}{\beta m^2} \sum_{j=1}^{N} 4 \sin^2(j\pi x_0/L) \sin(j\pi x_1/L) \sin(j\pi x_2/L) \tag{13}$$

When $x_1 = x_2 = x$, (13) reduces to the mean square velocity $E[v^2(x)]$ which has been discussed at length in Ref. 2. The spatial average with respect to independent variations of source location and receiver location x

$$\langle\langle E[v^2(x)]\rangle_x\rangle_{x_0} = \frac{\pi S_0}{\beta m^2} N \tag{14}$$

is the same as the spatial average of (4) with respect to receiver location. The special feature of the mean-square response to single point excitation is the appearance of localized zones of intensified response at the drive point $x = x_0$ and at its symmetric image point $x = L - x_0$. The shape of $E[v^2(x)]$ for $N = 20$ is displayed by the dotted curve in Fig. 2. There are several ways to estimate the degree of local intensification at $x = x_0$ or $x = L - x_0$.[2] One of the simplest is to set $x_1 = x_2$ equal to x_0 or $L - x_0$ in (13) and then to average with respect to x_0. The resulting mean values

$$\langle E[v^2(x_0)]\rangle_{x_0} = \langle E[v^2(L - x_0)]\rangle_{x_0} = \frac{3}{2}\frac{\pi S_0}{\beta m^2} N \tag{15}$$

are good asymptotic estimates of the mean-square at the intensified

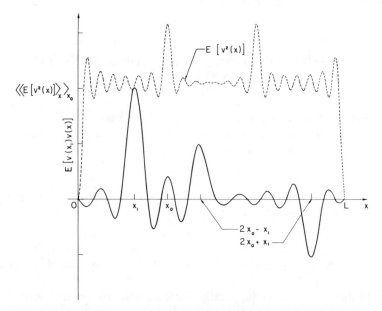

Fig. 2. Cross correlation of velocity response to band-limited white noise ($N = 20$) applied at single point $x = x_0$.

peaks for any particular driving point x_0 (that is not too close to the ends or center of the structure) because the relative fluctuation about the mean values (15) due to variation in driving point location approaches zero as $(13/8N)^{1/2}$.

A convenient correlation coefficient (1) corresponding to the cross-correlation (13) is obtained by using (14) for the denominator of (1). The resulting correlation coefficient so defined is

$$\rho_{12} = \frac{1}{N} \sum_{j=1}^{N} 4 \sin^2(j\pi x_0/L) \sin(j\pi x_1/L) \sin(j\pi x_2/L) \qquad (16)$$

The series can be summed to get

$$\rho_{12} = \frac{1}{N} (\Delta[(x_2 - x_1)/2] - \Delta[(x_2 + x_1)/2] - \Delta[(2x_0 + x_1 - x_2)/2]/2$$

$$- \Delta[(2x_0 - x_1 + x_2)/2]/2 + \Delta[(2x_0 + x_1 + x_2)/2]/2$$

$$+ \Delta[(2x_0 - x_1 - x_2)/2]/2) \qquad (17)$$

from which it may be seen that ρ_{12} vanishes when either location x_1, x_2 approaches an end $x = 0$, L, that ρ_{12} is of order unity when x_2 and x_1 coincide, that ρ_{12} is of the order $1/N$ for most values of x_1 and x_2 when they are separated, and that in addition ρ_{12} will have isolated peaks with amplitudes of order $1/2$ for any combination of values of x_1 and x_2 which makes an argument of one of the final four Δ-functions in (17) vanish. These latter peaks will be positive or negative according to the sign before the corresponding Δ-function in (17). The shape of ρ_{12} for $N = 20$ is displayed in Fig. 2 where $E[v(x_1)v(x_2)]$ is plotted as a function of $x = x_2$ for fixed values of x_1 and x_0. Note that when $x = x_1$ the cross correlation takes on the value of the mean square, and that away from this main peak the cross correlation generally fluctuates about zero with the exception of two half-size peaks. At $x = 2x_0 - x_1$ the argument of final Δ-function in (17) vanishes and there is a positive peak, while at $x = 2x_0 + x_1$ the argument of the third Δ-function in (17) vanishes and there is a negative peak. These additional half-size peaks in the cross correlation are a consequence of the same focussing mechanism which produces the local peaks at x_0 and $L - x_0$ in the mean-square response. As a consequence of the fact that the wavelength of the fluctuations in the cross correlation is twice that in the mean square the widths of the peaks in the cross correlation are twice those of the local peaks in the mean square.

To clarify the location and sign pattern of the half-size peaks in the

Stephen H. Crandall

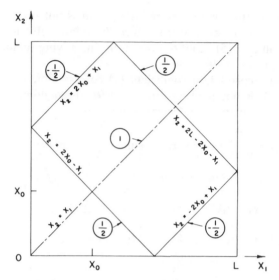

Fig. 3. Asymptotic peak values of $\rho_{12}(x_1, x_2)$.

correlation coefficient (17) the diagram in Fig. 3 has been constructed. The locations x_1 and x_2 for which ρ_{12} is to be evaluated are shown as Cartesian coordinates in the range $0 < x_1$, $x_2 < L$. Along the diagonal where $x_2 = x_1$, the value unity indicates that the asymptotic value of (17) is unity for these pairs of x_1 and x_2. The position and size of the inscribed rectangle, tipped at $45°$, is fixed by the location of the driving point $x = x_0$. The equations of the lines forming the sides of the rectangle are indicated. These are the loci where the arguments of the final four Δ-functions in (17) vanish (or, taking advantage of the periodicity of the Δ-function, have values $\pm L$, etc.). The values $1/2$ or $-1/2$ attached to these lines indicate the asymptotic value of (17) for these pairs of x_1 and x_2. Once x_0 is fixed and Fig. 3 has been constructed the asymptotic nature of a cross-correlation such as that displayed in Fig. 2 is given immediately for any choice of x_1. For example, if x_1 is selected to the left of x_0 in Fig. 3, as x_2 is increased from 0 to L with fixed x_1, it is seen that a unit peak occurs at $x_2 = x_1$, a positive half-unit peak occurs at $x_2 = 2x_0 - x_1$, and finally a negative half-unit peak occurs at $x_2 = 2x_0 + x_1$. This is an asymptotic description corresponding to Fig. 2. For finite N, the lines in Fig. 3 represent the centers of peaks whose width is of order $2/(N + 1/2)$ and the amplitude of the fluctuations of ρ_{12} about zero in the regions between the peaks is of order $1/N$. For other choices of x_1 and x_0 it may be seen from Fig. 3 that the distribution of the cross correlation as a function of x_2 might have positive and negative half-peaks, as in Fig. 2, or might have a pair of positive half-peaks, or might have a pair of negative half-peaks. The two half-peaks may be on either side of the unit peak or they may straddle the unit peak. The diagram of Fig. 3 may also be used to predict the asymptotic distribution of the mean-square response by considering only the diagonal along which $x_1 = x_2$. The unit value there corresponds to the asymptotic value (14) for the mean square at most locations. At the driving point x_0 and at its symmetric image $L - x_0$ the superposition of the unit and positive half-unit corresponds to the asymptotic value (15) for the intensified peaks. The diagram can also be used to predict special cases of coincidence. For example, if in Fig. 3 x_1 coincides with the drive point x_0, the form of ρ_{12} as a function of x_2 would include a peak of height $3/2$ at $x_2 = x_0$ plus a single negative half-peak at $x_2 = 2x_0 + x_1$. It is interesting to note that in this case there is an asymptotically *zero* correlation between the intensified response at the drive point x_0 and the equally intensified response at the symmetric image point $L - x_0$.

If the locations of the auxiliary peaks in ρ_{12} are known but their amplitudes are not, it is possible to predict their asymptotic amplitudes directly from (16) by using spatial averaging. For example, if it were suspected that ρ_{12} had a significant magnitude when $x_2 = 2x_0 - x_1$, one would insert this value of x_2 in (16) and average with respect to independent variations of x_0 and x_1 in the range from 0 to L. An elementary calculation yields

$$\langle\langle \rho_{12}(x_1, 2x_0 - x_1)\rangle_{x_0}\rangle_{x_1} = 1/2 \qquad (18)$$

in agreement with the asymptotic limit of the exact distribution (17) evaluated at $x_2 = 2x_0 - x_1$.

4 IMAGE-SUM PROCEDURE

An alternative procedure for predicting the asymptotic form of the correlation coefficient ρ_{12} for the velocity response of a finite structure excited at a single point is based on viewing the response as the response of an infinite structure to an infinite set of image sources.

This procedure has been used[2,3] to predict the distribution of mean-square response in strings, beams, membranes and plates. The procedure is well founded for uniaxial structures[2] but doubts about its rationale for application to two-dimensional structures have been expressed.[4] For the taut string or simply-supported beam treated above, the structure of length L is considered as a segment of an infinite structure, $-\infty < x < \infty$, and the effects of the boundary conditions at $x = 0$ and $x = L$ for the finite structure are simulated by introducing an infinite set of image sources. If the finite structure is driven by the wide-band random source S at $x = x_0$, the same excitation is simultaneously applied to the infinite structure at image sources $I+$ located at $x = x_0 + 2nL$ and the same excitation with reversed sign is simultaneously applied at the image sources $I-$ located at $x = -x_0 + 2nL$, $n = \ldots, -2, -1, 0, 1, 2, \ldots$. The net effect of the family of sources acting on the infinite structure is to make the stationary response in the segment $0 < x < L$ of the infinite structure the same as that for the finite structure with its boundary conditions when it responds to the single source S. Each source generates waves which propagate backwards and forwards. The waves are weakly attenuated due to the damping but otherwise propagate without alteration of spectral density. The velocity response at each point of the segment is thus a superposition of incoming waves with gradually decreasing contributions from increasingly distant sources. The situation is sketched in Fig. 4 where, by the device of drawing rays at 45°, it is possible to represent the distance from a source to any chosen point in the structure by the length of a vertical line down from the structure. For example, at the location $x = x_1$, the vertical line down to the point P_1 gives the distance to the negative source at $x = -x_0 + 2L$, the line down to P_2 gives the distance to the positive source at $x = x_0 - 2L$, etc. The asymptotic patterns deduced from this model are based on a simple either-or argument. When the excitation bandwidth increases without limit, two waves are taken to have zero correlation unless they have traveled exactly the same distance. In the latter case the correlation coefficient is plus one if the two sources have the same sign, and minus one if they have opposite sign. The relative magnitudes of asymptotic mean squares and cross correlations can then be predicted by accounting for the fractions of incoming waves that are mutually correlated. To obtain absolute magnitudes it is necessary to account for the attenuation[2] but here only relative magnitudes are discussed. For this purpose it is possible to take advantage of the $2L$

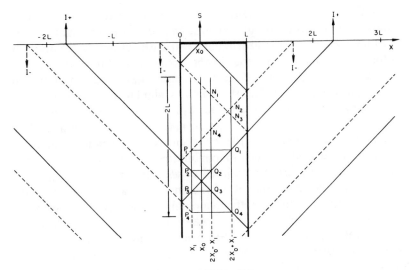

Fig. 4. Image-sum diagram.

periodicity of the source distribution and evaluate the mutual correlations of the four waves arriving from the sources located in an
arbitrary $2L$ interval of distance from the structure. The same pattern
of arriving waves repeats in subsequent $2L$ intervals but with
increasingly attenuated amplitudes. With light damping it is assumed
that differences in RMS amplitude within any one $2L$ interval can be
neglected. Thus in Fig. 4, for the $2L$ interval indicated, the four waves
arriving at $x = x_1$ travel different distances as indicated by the vertical
distances down from the structure to the points P_1, P_2, P_3 and P_4. It is
assumed that the travel distance differences are sufficient to cause the
waves to be mutually uncorrelated, but insufficient to cause significant
amplitude differences.

Before describing how this model is used to predict the asymptotic
distribution of cross correlation, we review the prediction of the
mean-square distribution. At the typical location $x = x_1$ four uncorrelated velocity waves of essentially equal strength arrive from the four
sources under consideration. Their contribution to the mean-square
velocity at x_1 is 4 units if the common mean-square velocity of each
wave is one unit. At most other locations the same four waves arrive
mutually uncorrelated and contribute the same 4 units to the mean-
square velocity. At the driving point x_0, however, the waves from the

positive sources at $x_0 - 2L$ and $x_0 + 2L$ have traveled exactly the same distance (as indicated by the crossing of the rays). The sum of these two identical signals is a double amplitude signal and its contribution to the mean square is 4 units. When the contribution from the other two uncorrelated waves is included the total contribution to the mean-square velocity at x_0 is 6 units. A similar argument applies at $x = L - x_0$, the point at which the rays from the negative sources at $x = -x_0$ and $x = -x_0 + 2L$ have their intersection. Thus from each $2L$ interval the contributions to the mean square are essentially uniform at all locations except at the drive point x_0 and its symmetric image $L - x_0$ where the contribution is greater in the ratio of 6 to 4. Summing the contributions of all such intervals with proper attention to the attenuation provides the same absolute asymptotic distribution of mean-square response as the modal-sum procedure.[2] The derivation of the relative distribution just given provides perhaps the most intuitively satisfying explanation of how the 'focussing effect' produces local zones of intensified mean-square response.

The cross correlation of the velocities at x_1 and x_2 when each velocity is viewed as the superposition of waves arriving from all the image sources involves the sum of cross correlations between individual waves arriving at x_1 and individual waves arriving at x_2 where the sum includes all possible combinations. Since the travel distances are generally different these individual cross correlations are generally zero. Only for certain special configurations are there systematic non-zero contributions. Consider again the location x_1 in Fig. 4. The distances to the four sources within the $2L$ interval selected are indicated by the vertical lines terminating in P_1, P_2, P_3 and P_4. For most locations of x_2 the corresponding distances to its four sources do not match any of the distances from x_1, and there is zero contribution to cross correlation from the interval. If however x_2 coincides with x_1 then all four distances match and the contribution to the cross correlation is the same as the contribution to the mean square: 4 units. In addition to this obvious case there are two other locations with non-zero cross correlations. These are the supplementary peaks due to the 'focussing effect'. In the present procedure these locations are determined in a simple manner by extending horizontal lines from the points P_i. For example, the horizontal line from P_1 intersects the ray from the positive source at $x_0 + 2L$ at point Q_1 located at $2x_0 + x_1$. This implies that if x_2 is taken at this location the wave arriving at x_2 from the source at $x_0 + 2L$ will always be equal and opposite to the wave arriving at x_1 from the source at $-x_0 + 2L$. The contribution to cross correlation from this individual cross correlation is one negative unit. It follows from Fig. 4 that the same argument applies to P_4 and Q_4 and the positive and negative sources at $x_0 - 2L$ and $-x_0 - 2L$, and that Q_4 is also located at $2x_0 + x_1$. The other two waves arriving at this location within the chosen interval are indicated by the rays intersecting N_2 and N_3. The distances to the sources involved do not match any of the source distances at x_1. Thus the total contribution to the cross correlation between velocity responses at x_1 and $x_2 = 2x_0 + x_1$ is two negative units. A similar argument applies to P_2 and P_3 and their matching points Q_2 and Q_3 located at $x_2 = 2x_0 - x_1$ except that here both sources are positive so that the contribution to the cross correlation is two positive units. Finally if these contributions from the $2L$ interval of Fig. 4 are normalized by dividing by the mean-square contribution of 4 units the result is the same asymptotic distribution as that for ρ_{12} predicted by Fig. 3 for $0 < x_1 < x_0$; i.e. a unit peak at $x_2 = x_1$, a positive half-peak at $2x_0 - x_1$, a negative half-peak at $x_2 = 2x_0 + x_1$, and zero cross correlation elsewhere.

An alternative technique for performing the image-sum analysis is based on using a typical pair of sources as the fundamental analysis element instead of the typical $2L$ interval in distance to sources. The contributions from the pair of sources located at $-x_0 + 2nL$ and $x_0 + 2nL$ for a particular n have the same relative distribution as those for any other n (except $n = 0$). The absolute magnitudes of the contributions diminish with increasing $|n|$ due to attenuation. The relative shape of the asymptotic distributions of mean square and cross correlation can thus be obtained by analyzing a single pair of sources. For example, consider in Fig. 4 the pair of sources located at $2L \pm x_0$ as the analysis element. The arrivals of waves from these sources at the location x_1 are indicated by the points P_1 and P_3. Having traveled different distances they are uncorrelated and their contribution to the mean-square velocity at x_1 is 2 units. The same situation prevails at most other locations. At $x = x_0$ however the wave from the source at $2L + x_0$ is reinforced by an identical wave from the source at $x_0 - 2L$, the amplitude is doubled and the contribution to the mean square is quadrupled. Since this is a joint contribution from one source in the analysis element and one outside source, the analysis element is credited with 2 units. To this is added the uncorrelated unit contribution from the source at $2L - x_0$ to obtain the total contribution to the mean square at x_0 from the analysis element of 3 units. The same intensification occurs at $L - x_0$ due to the reinforcement of the wave from the source at $2L - x_0$. The relative shape of the asymptotic mean-square distribution is thus once more produced.

The asymptotic distribution of the cross correlation is obtained in a similar manner. With x_1 fixed in Fig. 4 for most x_2 there is no match in the distances to the sources in the fundamental analysis element and therefore the contribution to the cross correlation is zero. When x_2 coincides with x_1, the distances match and the contribution is 2 units. To locate the supplementary peaks due to 'focussing', horizontal lines are extended from P_1 and P_3 to meet rays at Q_1 and Q_3 at the locations $2x_0 + x_1$ and $2x_0 - x_1$ respectively. When $x_2 = 2x_0 + x_1$ the arrivals at P_1 and Q_1 are equal with opposite signs producing a contribution of -1 unit while the arrivals at P_3 and N_2 are uncorrelated. The total contribution to the cross correlation for $x_2 = 2x_0 + x_1$ is thus -1 unit. When $x_2 = 2x_0 - x_1$ the arrival at P_3 from the source at $2L + x_0$ in the analysis element is identical to the arrival at Q_3 from the source at $-2L + x_0$ outside the analysis element producing a joint contribution of $+1$ unit. These same two sources produce an additional $+1$ unit contribution due to arrivals at Q_2 and P_2. Half of the total joint contribution of $+2$ units is credited to the analysis element. There is no contribution from the source at $2L - x_0$ since its arrivals at P_1 and N_4 are not matched by arrivals at x_2 and x_1 respectively. The total contribution to the cross correlation for $x_2 = 2x_0 - x_1$ is thus $+1$ unit. The relative shape of the asymptotic distribution of the cross correlation is again established. This technique based on a fundamental analysis element consisting of a source group readily extends to two-dimensional structures.[2]

5 TWO-DIMENSIONAL STRUCTURES

The preceding considerations are readily extended to uniform rectangular membranes and plates. In the case of spatially uncorrelated excitation where the cross-spectral density of the excitation is the two-dimensional extension of (2) the velocity response cross correlation is the two-dimensional extension of the modal sum (3). Because of the irregular sequence of modal patterns when the modes of two-dimensional structures are arranged in order of increasing natural frequency it does not appear to be possible to obtain a closed form expression comparable to (11) for the cross correlation. Approximate solutions may however be obtained by approximating the discrete sum by an integral or by making the separable sum approximation.[2] The result here is a direct extension of that for uniaxial structures. For fixed location P_1 the velocity cross correlation $E[v(P_1)v(P_2)]$ fluctuates about zero with an amplitude of order $\langle E[v^2]\rangle/N$ for all locations P_2 except in the neighborhood of P_1 where there is a single isolated peak equal to the local value of $E[v^2]$.

In the case of single-point excitation at the location P_0 where the cross-spectral density of the excitation is the two-dimensional extension of (12) the velocity response cross correlation is the two-dimensional extension of (13). Again it is not possible to obtain a closed form result comparable to (17) but approximate solutions are possible by approximating the discrete sum by an integral or by making the separable sum approximation. Let the spatial correlation coefficient $\rho_{12}(P_1, P_2)$ of (1) be defined by using the two-dimensional extension of (14) as the denominator. The asymptotic form of the distribution of ρ_{12} is most simply predicted by applying the image-sum procedure. Using a group of four sources as a fundamental analysis element,[2] it is found that for fixed P_1 with $0 < y_1 < y_0$, the spatial correlation coefficient is asymptotically zero for most locations P_2 except for the following nine locations where ρ_{12} takes the indicated values:

$$
\begin{array}{lll}
x_2 = x_1 & y_2 = y_1 & \rho_{12} = 1 \\
x_2 = 2x_0 - x_1 & y_2 = y_1 & \rho_{12} = 1/2 \\
x_2 = 2x_0 + x_1 & y_2 = y_1 & \rho_{12} = -1/2 \\
x_2 = x_1 & y_2 = 2y_0 - y_1 & \rho_{12} = 1/2 \\
x_2 = x_1 & y_2 = 2y_0 + y_1 & \rho_{12} = -1/2 \quad (19) \\
x_2 = 2x_0 - x_1 & y_2 = 2y_0 - y_1 & \rho_{12} = 1/4 \\
x_2 = 2x_0 - x_1 & y_2 = 2y_0 + y_1 & \rho_{12} = -1/4 \\
x_2 = 2x_0 + x_1 & y_2 = 2y_0 - y_1 & \rho_{12} = -1/4 \\
x_2 = 2x_0 + x_1 & y_2 = 2y_0 + y_1 & \rho_{12} = 1/4
\end{array}
$$

Once these locations are known the spatial averaging method[2] provides a simple means of estimating the associated asymptotic values of ρ_{12}. For example, with P_2 at the point with coordinates $x_2 = 2x_0 + x_1$ and $y_2 = 2y_0 + y_1$ one averages ρ_{12} with respect to independent variations of x_0, x_1, y_0, and y_1 to get the asymptotic estimate $\rho_{12} = 1/4$

in agreement with (19). The location and sign pattern for the half-peaks and quarter-peaks depends on the relative location of the points P_0 and P_1 and can be predicted by using a pair of diagrams similar to Fig. 3, one for the x-coordinates and one for the y-coordinates.

6 CONCLUSION

The cross correlation between velocity responses at the same time but at different locations in uniform one- and two-dimensional structures has been investigated for two types of stationary wide-band random excitation. When normalized by a convenient mean-square level, the general tendency of the correlation coefficient $\rho_{12}(P_1, P_2)$ is to be small, fluctuating about zero with an amplitude of order $1/N$ as the locations of the points P_1 and P_2 are varied. For rain-on-the-roof excitation the only exception occurs in the neighborhood of coincidence $P_1 \equiv P_2$ where ρ_{12} has a local peak with amplitude of order unity. In the case of single-point excitation there are additional local peaks due to focussing. In one dimension there are two additional peaks with amplitudes of order one-half. In two dimensions there are eight additional peaks, four with amplitudes of order one-half and four with amplitudes of order one-quarter. The locations and sign patterns of these supplementary peaks is most simply predicted by image-sum methods.

REFERENCES

1. Vanmarcke, E. H., *Random Fields: Analysis and Synthesis.* MIT Press, Cambridge, MA, 1983.
2. Crandall, S. H., Random vibration of one-dimensional and two-dimensional structures. In *Developments in Statistics, Vol. 2* ed. P. R. Krisnaiah. Academic Press, New York, 1979, pp. 1–82.
3. Crandall, S. H., Structured response patterns due to wide-band random excitation. In *Stochastic Problems in Dynamics,* ed. B. L. Clarkson. Pitman, London, 1977, pp. 366–89.
4. Crandall, S. H., Image sum procedure for estimating wide-band response of plates. *Engineering Science Preprint* 23-86053, August 25–27, 1986, Society of Engineering Science.

3

One-Dimensional Waves in Media with Slowly Varying Random Inhomogeneities

M. F. Dimentberg, A. I. Menyailov & A. A. Sokolov

Institute for Problems of Mechanics, USSR Academy of Sciences, Moscow, USSR

ABSTRACT

One-dimensional waves are considered, which propagate in a non-dispersive media with smooth random spatial variations of an apparent elastic modulus, such as a straight liquid-filled pipe with slow longitudinal variations of air bubble content within a liquid. The basic stochastic wave equation is solved by Krylov–Bogoliubov averaging with subsequent explicit analytical integration of the resulting shortened equations. This solution is used to analyse the effect of spatial cross-correlation decay due to random phase variations of propagating waves both for the cases of periodic and random-in-time waves. The conditions for applicability of the method of moments to waves in media with slow random variations of properties are derived. The implications of the solutions concerning identification of media properties from measured random responses are discussed.

In this paper we consider one-dimensional waves in a medium with slow spatial variations of an apparent elastic modulus. These waves are governed by the partial differential equation

$$\frac{\partial}{\partial x}\left\{[1 + \xi(x)]\frac{\partial u}{\partial x}\right\} = \frac{1}{c^2}\frac{\partial^2 u}{\partial t^2} \tag{1}$$

where c is the expected or mean propagation speed and $\xi(x)$ is a stationary zero-mean Gaussian random function with spectral density $\Phi_{\xi\xi}(p)$. The latter is assumed to be rapidly decaying so that $\Phi_{\xi\xi}(0) \gg \Phi_{\xi\xi}(k)$, where k is the wave number of any one of the travelling waves. This condition corresponds to the basic assumption of slow variations of the elastic modulus. Moreover, these variations are assumed also to be small; however, since the perturbation method will not be used here, this restriction is much less severe than in the case where it is applied. Actually, the following analysis is based on the Krylov–Bogoliubov averaging with a direct subsequent integration of the resulting shortened equation. This analysis reveals a principal effect in system (1), namely the decay of spatial cross-correlation caused by random phase variations of waves in system (1).

It should be noted, that previously[1] these authors had made a similar analysis for the system, governed by equation

$$\frac{\partial^2 u}{\partial x^2} = \frac{1 + \xi(x)}{c^2} \cdot \frac{\partial^2 u}{\partial t^2} \tag{2}$$

From the order-of-magnitude analysis may be anticipated that, since $\xi(x)$ is assumed to be both small and slowly varying, the above mentioned effects of spatial cross-correlation decay in systems (1) and (2) should be essentially the same.[1] This conclusion is rigorously proved in the following by the results of the direct actual analysis of system (1). These results, however, have revealed the additional restriction on $\xi(x)$, namely $|\xi(x)|$ should be less than unity, so that the apparent elastic modulus should be positive everywhere.

The present solution is used also for studying the possibility for applying the method of moments, which is based on the theory of diffusional processes, to waves in media with slow random variations of properties. The conditions for applicability of this method (which implies, in fact, white-noise approximation for slowly varying random functions) are derived. Furthermore, the influence of small frequency-independent damping in system (1) will be studied by addition of a viscous term to eqn (1). Finally, certain implications of the results are considered concerning possible approaches to identification of the medium's properties from the random response data both for the cases of periodic- and random-in-time waves. As for possible specific examples of system (1), two such examples may be mentioned here:[1] axial stress waves in an elastic rod with longitudinal variations of elastic modulus and waves in a straight pipe with incompressible liquid in presence of random longitudinal variations of air bubble content within liquid and/or pipe wall thickness. In the first of these examples the state variable $u(x, t)$ is the axial displacement of rod particles, in the second one, pressure perturbation in the liquid within the pipe.

System (1) will be analysed here within the half-axis $x > 0$. Propagating waves 'from left to right' are considered, whereas at the free end $x = 0$ the timewise variations of $u(0, t)$ are assumed to be prescribed (the case of given $\partial u(0, t)/\partial x$ may be treated similarly). Therefore, the solutions dependent on $t - x/c$ are retained here, whereas those dependent on $t + x/c$ are discarded. At first simple harmonic oscillations at $x = 0$ will be considered, that is

$$u(0, t) = A_0 \cos \omega t \qquad (3)$$

The solution to eqn (1) is sought in the form

$$u(x, t) = z(x) \exp(i\theta) + z_*(x) \exp(-i\theta)$$
$$\partial u/\partial x = ik[z(x) \exp(i\theta) - z_*(x) \exp(-i\theta)] \qquad (4)$$
$$\theta = kx - \omega t, \qquad k = \omega/c, \qquad i = \sqrt{-1}$$

where the asterisk denotes the complex conjugate variable. Solving (4) for $z(x)$, $z_*(x)$, differentiating and using the basic eqn (1) yields, up to terms with higher powers of $\xi(x)$

$$z' = [\exp(-i\theta)/ik]\{-ik\xi'(x)[z \exp(i\theta) - z_* \exp(-i\theta)]$$
$$+ k^2\xi(x)[z \exp(i\theta) + z_* \exp(-i\theta)]\} \qquad (5)$$

where primes denote differentiation with respect to x.

Since $\xi(x)$ and $\xi'(x)$ are small, the Krylov–Bogoliubov averaging[2] may be applied to eqn (5), resulting in a shortened equation

$$z' = -\xi'(x)z/2 - (ik/2)\xi(x)z \qquad (6)$$

(Throughout this averaging over 'fast' variable θ the slowly varying function $\xi(x)$ is held fixed. This procedure can be proved formally by assuming the argument of ξ to be vx, where v is a small parameter, and considering ξ as an additional state variable. The latter is governed then by equation $\xi' = v$, which is indeed a standard form equation.)

The solution to eqn (6), which satisfies boundary condition (3), may be written as

$$z(x) = \tfrac{1}{2}A_0 \exp[-(ik/2)\varepsilon(x) - \xi(x)/2]$$
$$\varepsilon(x) = \int_0^x \xi(x') \, dx' \qquad (7)$$

From solution (7) it is possible to obtain the spatial cross-correlation function $K_{zz}(l) = \langle z(x)z_*(x+l) \rangle$ of the random function $z(x)$. The derivation is based on the following well-known[3] formula for the characteristic function of the four-dimensional Gaussian distribution

$$\left\langle \exp\left(i \sum_{j=1}^{4} v_j \eta_j \right) \right\rangle = \exp\left(-\tfrac{1}{2} \sum_{j=1}^{4} \sum_{k=1}^{4} \sigma_j \sigma_k r_{jk} v_j v_k \right) \tag{8}$$

where σ_j and r_{jk} are r.m.s. values and normalized correlation coefficients of Gaussian random variables $\eta_j, j = 1, \ldots, 4$. To apply formula (8) one may put $\eta_1 = \varepsilon$, $v_1 = k/2$; $\eta_2 = \varepsilon_l$, $v_2 = -k/2$; $\eta_3 = \xi$, $v_3 = i/2$; $\eta_4 = \xi_l$, $v_4 = i/2$ (where $\varepsilon \equiv \varepsilon(x)$, $\varepsilon_l \equiv \varepsilon(x+l)$, $\xi(x) \equiv \xi$, $\xi_l \equiv \xi(x+l)$) and express the r.m.s. values and correlation coefficients of the values of random function $\varepsilon(x)$ and its derivative $\xi(x)$ at two different points x and $x+l$ in terms of the spectral density $\Phi_{\xi\xi}(p)$ of $\xi(x)$. This yields

$$K_{zz}(l) = (A_0^2/4)\kappa(l, \omega)\mu(l, \omega) \tag{9}$$

$$\kappa(l, \omega) = \exp\left[-\frac{k^2}{2} \int_0^\infty \Phi_{\xi\xi}(p)p^{-2}(1 - \cos pl)\,\mathrm{d}p \right] \tag{10}$$

$$\mu(l, \omega) = \exp(v_r + iv_i), \qquad v_r = \int_0^\infty \Phi_{\xi\xi}(p)(1 + \cos pl)\,\mathrm{d}p$$
$$v_i = k \int_0^\infty \Phi_{\xi\xi}(p)p^{-1} \sin pl\,\mathrm{d}p \tag{11}$$

Formula (10) for the principal cross-correlation decay factor $\kappa(l, \omega)$ is found to be the same as for system (2), for which the counterpart of the shortened eqn (6) is[1]

$$z' = (ik/2)\xi(x)z \tag{12}$$

The additional decay factor $\exp v_r$, as defined by (11), is indeed of minor importance compared with $\kappa(l, \omega)$ in view of the rapid decay of $\Phi_{\xi\xi}(p)$. This becomes especially clear if the case of exponentially correlated $\xi(x)$ is considered, where

$$\Phi_{\xi\xi}(p) = \frac{\sigma^2\gamma}{\pi} \cdot \frac{1}{\gamma^2 + p^2}, \qquad \gamma > 0 \tag{13}$$

with the condition for slow variations of $\xi(x)$ being $\gamma \ll k$. Then, from (10) and (11)

$$\kappa(l, \omega) = \exp[-(k^2\sigma^2/4\gamma^2)(\gamma l + \exp(-\gamma l) - 1)] \tag{14}$$

$$v_r = -(\sigma^2/2)[1 + \exp(-\gamma l)], \qquad v_i = -(ik\sigma^2/2\gamma)[1 - \exp(-\gamma l)] \quad (15)$$

Thus, $\exp v_r$ may indeed be regarded as being close to unity compared with $\kappa(l, \omega)$, so that systems (1) and (2) indeed may be regarded as essentially equivalent (as suggested in Ref. 1), as long as the effect of cross-correlation decay is considered; as for the term $\exp(iv_i)$ in (11), it leads to some shift of the maximum of the cross-correlation function $\langle u(x, t)u(x + l, t + \tau)\rangle$, which is found to be negligibly small for sufficiently small $l\sigma^2/2\gamma$ (this will be shown later).

On the other hand, the 'damping' term with $\xi'(x)$ in (1) introduces the additional restriction on $\xi(x)$, namely $|\xi(x)|$ should always stay within the convergence radius of the power-series expansion of $(1 + \xi)^{-1}$, or $|\xi(x)| < 1$. This can be clearly seen if the exact representation of this term is used in the derivation of eqn (5) ($\xi'/(1 + \xi)$ instead of ξ'). Then the shortened eqn (6) is found to be

$$z' = -\frac{\xi'(x)z}{2[1 + \xi(x)]} - \frac{ik}{2}\xi(x)z \quad (6')$$

and it has the solution

$$z(x) = \frac{A_0}{2\sqrt{1 + \xi(x)}} \exp\left[-\frac{ik}{2}\varepsilon(x)\right] \quad (7')$$

This solution is indeed equivalent to (7) for the case of small variances of $\xi(x)$, as far as second-order moments of these solutions are concerned, however, rigorously speaking, the condition $|\xi(x)| < 1$ is required here for every x. This is, in fact, quite a natural requirement of the positive apparent elastic modulus of the system (the medium should be 'passive').

We consider now eqn (12) and obtain the solution for $K_{zz}(l)$ by the method of moments[1] (the discussion of the applicability of this method for those cases, which cannot be directly reduced to a Cauchy problem, may be found in Refs 1 and 4). Since this method is based on a white-noise approximation for slowly varying $\xi(x)$, certain doubts may arise concerning its applicability. It will be shown, however, that it leads to correct results for separation lengths l which are much higher than the correlation length l_0 of $\xi(x)$.

Introducing a new variable $V = zz_*$ and using eqn (12) together with its counterpart for z_* (in these equations $\xi(x)$ is assumed to be a white noise with intensity factor $D = 2\pi\Phi_{\xi\xi}(0)$) yields the deterministic equation $m'_V = 0$ for $m_V = \langle V \rangle$. Therefore, $m_V = \text{const} = A_0^2/4$. Now, according to the general procedure[1] the new argument $l = 0$ and the new state variable $y(l)$ are introduced, such that

$$dy/dl = 0, \qquad y(0) = z \qquad (16)$$

Denoting $v = yz_*$ we derive from eqn (16) together with the equation for z_* the following stochastic differential equation (SDE) for $v(l)$

$$dv/dl = (ik/2)\xi(l)v \qquad (17)$$

Transforming this 'physical' SDE to the Ito SDE (the corresponding correction term being $-k^2 Dv/8$) and applying the operator of mathematical expectation yields the desired deterministic equation for $\langle yz_* \rangle = K_{zz}(l)$, namely

$$dK_{zz}/dl = -(k^2 v Dv/8)K_{zz}, \qquad K_{zz}(0) = A_0^2/4 \qquad (18)$$

so that

$$K_{zz}(l) = (A_0^2/4)\kappa(l, \omega), \qquad \kappa(l, \omega) = \exp - (k^2 \pi l/4)\Phi_{\xi\xi}(0) \qquad (19)$$

Formula (19) for $\kappa(l, \omega)$ exactly coincides with that obtained from (10) for the case $\Phi_{\xi\xi}(p) = \text{const} = \Phi_{\xi\xi}(0)$. Thus, the method of moments may indeed be applicable to media with slow random variations of parameters. To obtain explicitly the conditions for its applicability consider the case of exponentially correlated $\xi(x)$. Then, inserting (13) into (19) yields $\kappa(l, \omega) = \exp(-k^2 \sigma^2 l/4\gamma)$. This result coincides with (14) provided that $\gamma l \gg 1$. Since $\gamma = 1/l_0$ where l_0 is the correlation length of $\xi(x)$, the conditions for validity of the results of the method of moments may be written as

$$\lambda \ll l_0 \ll l, \qquad \lambda = 2\pi/k \qquad (20)$$

The first of these inequalities is the condition for slow variations of $\xi(x)$, whereas the second one is that for 'white-noise' approximation for $\xi(x)$.

From (4) and (9)–(11) it is possible, similar to Ref. 1, to obtain the cross-correlation function $K_{uu}(l, \tau) = \langle u(x, t)u(x + l, t + \tau)\rangle$ of $u(x, t)$ as

$$K_{uu}(l, \tau) = \tfrac{1}{2}A_0^2 \kappa(l, \omega) \exp(v_r) \cos\left[kl - \omega\tau - k\int_0^\infty \frac{\Phi_{\xi\xi}(p)}{p} \sin pl \, dp\right] \qquad (21)$$

It can be seen from (21) that cross-correlation between values of $\varepsilon(x)$ and its derivative $\xi(x)$ at two different points leads to an additional shift of the point of absolute maximum of $K_{uu}(l, \tau)$ compared with the case of system (2). This shift, however, is found to be small for sufficiently small variances of $\xi(x)$ and not very small separation distances l; for example, in case of exponentially correlated $\xi(x)$ this shift is negligible provided $l \gg \sigma^2/2\gamma$.

The physical interpretation of the solution (21) is in fact the same as that presented in Ref. 1 for the system (2): whereas the amplitude of the wave changes very little along its path (the medium is non-dissipative), the cross-correlation between responses at two different points may decay rather rapidly because of the phase variations of the wave due to spatial variations of the propagation speed (and the cross-correlation here implies spatial averaging).

The expression (21) with $A_0 = 1$ provides, in fact, the averaged transfer functions of the medium. They may be used to obtain the solution for the case where $u(0, t)$ is a zero-mean random process with spectral density $\Phi_0(\omega)$. Similarly to Ref. 1 we obtain for this case

$$K_{uu}(l, \tau) = \int_{-\infty}^{\infty} \Phi_0(\omega)\kappa(l, \omega) \exp(v_r) \cos(kl - \omega\tau - v_i)\, d\omega, \qquad k = \omega/c$$

(22)

this cross-correlation function being based on both temporal and spatial averaging. Neglecting, for simplicity, the small influence of v_r and v_i, it is possible to study the variations of the cross correlation with increasing separation distance l, which are governed by the decay factor $\kappa(l, \omega)$. This effect becomes especially clear for the case of a 'Gaussian' input spectrum, or

$$\Phi_0(\omega) = (\sigma_0^2 \tau_0 / \sqrt{2\pi}) \exp(-\omega^2 \tau_0^2 / 2),$$
$$K_0(\tau) = \langle u(0, t) \cdot u(0, t + \tau) \rangle$$
$$= 2 \int_0^{\infty} \Phi_0(\omega) \cos \omega\tau\, d\omega = \sigma_0^2 \exp(-\tau^2 / 2\tau_0^2)$$

(23)

Inserting (23) into (22) yields

$$K_{uu}(l, \tau) = \frac{\sigma_0^2}{\sqrt{1 + 2f(l)/\tau_0^2}} \exp\left\{ -\frac{(\tau - l/c)^2}{2[2f(l) + \tau_0^2]} \right\}$$
$$f(l) = -\frac{\ln \kappa(l, \omega)}{\omega^2} = \frac{1}{2c^2} \int_0^{\infty} \frac{\Phi_{\xi\xi}(p)}{p^2} (1 - \cos pl)\, dp$$

(24)

It can be seen from (24) that for sufficiently small l, where $\kappa(l, \omega) \approx 1$ or $f(l) \ll \tau_0^2$, the cross-correlation function $K_{uu}(l, \tau)$ almost exactly reproduces the input autocorrelation function $K_0(\tau)$. On the other hand, increasing l leads to distortions of the cross-correlation function. In particular, for identification purposes it is usually interesting to estimate the point of absolute maximum and the corresponding maximal value of $K_{uu}(l, \tau)$. From (24) it can be seen, that for fixed l $K_{uu}(l, \tau)$ always reaches its maximum at $\tau = \tau_m = l/c$, that is, when its time shift coincides with the expected transit time along the separation length, and

$$\frac{\max_\tau K_{uu}(l, \tau)}{\max_\tau K_{uu}(0, \tau)} = \frac{K_{uu}(l, l/c)}{K_{uu}(0, 0)} = \frac{1}{\sqrt{1 + 2f(l)/\tau_0^2}} \quad (25)$$

Therefore, τ_m provides an estimate of mean propagation speed from measured random responses as $c = l/\tau_m$, whereas the ratio (25) provides an estimate of $\kappa(l, \omega)$ and thus of the level of the random spatial variations of the propagation speed. The latter, of course, may in general be estimated directly from purely temporal cross-correlation measurements at various locations, such measurements providing a sample of random local values $c_i = l/\tau_{mi}$. This direct approach requires, however, sufficiently small separations l with virtually constant values of c_i between each pair of sensors, otherwise the interpretation of the results would be rather difficult. Therefore, formula (25), which is free from such a restriction, may indeed provide a useful alternative, especially in view of the fact that the accuracy of such estimates of c_i may decrease with decreasing l.

The spatial averaging implied in the above cross-correlational analysis is possible in general only provided that damping in the medium may be ignored. This restriction is quite severe since rather significant decay of waves in real media may be possible within a high averaging length. In this case the response is not spatially homogeneous, its cross-correlation function $K_{uu} = \langle u(x, t)u(x + l, t + \tau) \rangle$ being dependent on both x and l (as well as on τ). In this case K_{uu} may be estimated from measured responses only through the averaging over the ensemble of waveguides (elastic rods, pipes with bubbly liquid, etc.). The simplest case for identification is that of small frequency-independent damping, which may be introduced into system (1), for example, through the additional term $\eta\, \partial u/\partial t$ in the right-hand-side, where $0 < \eta \ll 1$. Then, repeating the derivation, we obtain the additional factor $\exp(-c\eta x/2)$ in the solution (7) for $z(x)$, and the modified formula (21) for K_{uu} becomes

$$K_{uu}(x, l, \tau) = \tfrac{1}{2}A_0^2\kappa(l, \omega)\exp(-c\eta x - \tfrac{1}{2}c\eta l)\cos(kl - \omega\tau) \quad (21')$$

(for simplicity the small corrections due to ν_r and ν_i have been discarded here). The same decay factor $\exp(-c\eta x - c\eta l/2)$ should be added to K_{uu} in the case of random-in-time waves (formula (22)). It can be seen therefore, that for identification purposes the influence of such damping can be excluded by normalizing $\max K_{uu}(x, l, \tau)$ to $K_{uu}(x, 0, 0)K_{uu}(x + l, 0, 0)^{1/2}$ (cf. with (25)). It should be stressed though, that in the case of frequency-dependent damping the identification procedure would become more complicated, with the exponential decay factor entering the integrand in (22).

REFERENCES

1. Dimentberg, M. F., *Statistical Dynamics of Nonlinear and Time-Varying Systems*. Research Studies Press, 1988.
2. Bogoliubov, N. N. & Mitropolsky, Iu. A., *Asymptotic Methods in the Theory of Nonlinear Oscillations* (in Russian). Moscow, Nauka, 1974.
3. Levin, B. R., *Theoretical Foundations of Statistical Radio Engineering, Part 1* (in Russian). Moscow, Sovietskoye Radio, 1966.
4. Sobczyk, K., *Stochastic Wave Propagation*. PWN-Elsevier, 1984.

4

On Random Vibration of Structures with Complicating Effects

ISAAC ELISHAKOFF*

Department of Mechanical Engineering, Naval Postgraduate School, Monterey, California, USA

&

ELIEZER LUBLINER

Department of Aeronautical Engineering, Technion–Israel Institute of Technology, Haifa, Israel

ABSTRACT

Random vibration studies incorporating effects of shear deformation and rotary inertia are reviewed. Emphasis is placed on the convergence behavior of the mean square stress response of structures under timewise ideal white-noise excitation. Use of approximate differential equation, taking into account effects of shear deformation and rotary inertia, but neglecting their interactive contribution, is discussed. A two-degree-of-freedom model structure is analysed in detail, elucidating use of the above approximation.

1 INTRODUCTION

Random vibration of structures has been dealt with in a number of papers. Apparently the first extensive study, by Samuels and Eringen,[1] appeared 30 years ago. It considered the random vibration of the simply supported beam under timewise white-noise excitation. The beam possessed combined transverse and rotary dampings. It has been shown that taking into account both dampings yields a convergent

---

* On leave from the Technion–Israel Institute of Technology, Haifa 32000, Israel.

series for the mean-square stress, whereas neglecting rotary damping predicts the divergent mean-square stress. For the mean-square displacements the differences between numerical values calculated with the aid of classical beam theory

$$EI\frac{\partial^4 w}{\partial x^4} + \rho A \beta_0 \frac{\partial w}{\partial t} + \rho A \frac{\partial^2 w}{\partial t^2} = q(x, t) \tag{1}$$

and that calculated via the so-called Timoshenko beam theory

$$EI\frac{\partial^4 w}{\partial x^4} + \rho A \frac{\partial^2 w}{\partial t^2} - \rho I\left(1 + \frac{E}{k'G}\right)\frac{\partial^4 w}{\partial x^2 \partial t^2} + \frac{\rho^2 I}{k'G}\frac{\partial^4 w}{\partial t^4}$$

$$- \rho I\left(\frac{E\beta_0}{k'G} + \beta_1\right)\frac{\partial^3 w}{\partial x^2 \partial t} + \frac{\rho^2 I}{k'G}(\beta_0 + \beta_1)\frac{\partial^3 w}{\partial t^3}$$

$$+ \frac{\rho^2 I \beta_0 \beta_1}{k'G}\frac{\partial^2 w}{\partial t^2} + \rho A \beta_0 \frac{\partial w}{\partial t}$$

$$= q(x, t) + \frac{1}{k'GA}\left(-EI\frac{\partial^2 q}{\partial x^2} + \rho I\frac{\partial^2 q}{\partial t^2} + \rho I \beta_1 \frac{\partial q}{\partial t}\right) \tag{2}$$

turned out to be less than 5%. In eqns (1) and (2) standard notation is employed, and, moreover,

$$\beta_0 = \frac{C_0}{\rho A}, \ \beta_1 = \frac{C_1}{\rho I} \tag{3}$$

are the transverse and rotary damping coefficients, respectively.

Subsequent analysis was performed by Crandall & Yildiz.[2] They demonstrated that the response of a uniform beam to stationary random excitation depended largely on the dynamical model postulated (either Bernoulli–Euler, Rayleigh, pure shear, or Timoshenko beams), or the damping model employed (transverse viscous, rotary viscous, Voigt damping or any of their combinations) and on the spectral content of the random excitation. In this careful study Crandall and Yildiz considered as an excitation, also a band-limited white noise with cut-off frequency ω_c. Because of the complexity of the integrals involved, their results are formulated in terms of order-of-magnitude values with respect to damping parameters and the serial number of the term in modal series.

Numerical results for the influence of axial force on the response of Timoshenko beams were reported by Banerjee & Kennedy,[3] whereas

Cederkirst[4] studied the influence of application of the refined theory on the optimization of randomly vibrating beams. The plate problem has been tackled by Witt & Sobczyk[5] with prediction of about 30% difference in prediction of the response by classical and refined theories.

In Ref. 6, Elishakoff & Livshits studied the random vibration of simply supported beams under time- and spacewise ideal white noise. Modal analysis was used, and due to spacewise white-noise loading the cross-correlation terms[7] vanished identically, and therefore the response was written down as the sum of responses due to excitation 'configurational' to the normal modes; in other words, the response is the weighted sum of those of an infinite number of simple oscillators. Now, the response of the simple oscillation due to timewise white-noise excitation is well known (see, for example, monographs by Crandall & Mark[8] and Lin[9]):

$$E(|\dot{X}|^2) = \frac{\pi S_0}{2\zeta \omega_0^3 m^2} \qquad (4)$$

that is, the higher the natural frequency, the lower its contribution to the total random vibration response. This implies that for timewise white-noise excitation a simpler theory can be used than that attributed to Timoshenko.[10]

The Timoshenko beam theory predicts two series of natural frequencies (see, for example, discussion of this subject in Refs 11–13, and experimental verification provided by Barr[16]): one a lower series and the other a higher series. As shown above, we are interested in the contribution of lower natural frequencies. Therefore, the simplified theory, obtainable through neglecting of the terms underlined in eqn (2), can be utilized:

$$EI \frac{\partial^4 w}{\partial x^4} + \rho A \frac{\partial^2 w}{\partial t^2} - \rho I \left(1 + \frac{E}{k'G}\right) \frac{\partial^4 w}{\partial x^2 \partial t^2} - \rho I \left(\frac{E\beta_0}{k'G} + \beta_1\right) \frac{\partial^3 w}{\partial x^2 \partial t}$$

$$+ \frac{\rho^2 I \beta_0 \beta_1}{k'G} \frac{\partial^2 w}{\partial^2 t} + \rho A \beta_0 \frac{\partial w}{\partial t}$$

$$= q(x, t) + \frac{1}{k'GA} \left(-EI \frac{\partial^2 q}{\partial x^2} + \rho I \frac{\partial^2 q}{\partial t^2} + \rho I \beta_1 \frac{\partial q}{\partial t}\right) \qquad (5)$$

Random vibrations of beams utilizing this equation were given by Elishakoff & Livshits[6] for the case of the timewise ideal white-noise excitation, and by Elishakoff & Lubliner[17] for the timewise band limited white-noise excitation with two lower and upper cut-off frequencies.

To elucidate this use of the approximate differential equation, with the attendant lower series of natural frequencies, we will use the two-degree-of-freedom model structure, in which the complicating effect studied is represented by the rotary inertia of the mass attached to the massless cantilever.

2 FREE VIBRATION

Let us consider the mechanical system shown in Fig. 1. It consists of a cantilever beam of length l, with the absolutely rigid mass m. The moment of inertia of the mass with respect to the axis perpendicular to the plane of the figure, passing through the centroid C, is denoted by I. The damping coefficient is c. The applied point force $F(t)$ is a weakly stationary random function in time, which is taken for simplicity to be an ideal white noise, with constant spectral density

$$S_F(\omega) = S_0 \tag{6}$$

The system possesses two degrees of freedom, corresponding to the vertical displacement of point C and the rotation of the mass. Note that an analogous system was considered in Lin's monograph (see Ref. 9, pp. 164–9) to elucidate random vibration of multi-degree-of-freedom systems. Let us denote the vertical displacement by $x(t)$ and the rotation angle by $\theta(t)$.

Consider first the free vibration problem. Equations of free oscillations read:

$$x(t) = F(t)\,\delta_{11} + M(t)\,\delta_{12} \tag{7}$$

$$\theta(t) = F(t)\,\delta_{21} + M(t)\,\delta_{22} \tag{8}$$

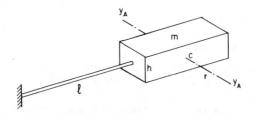

Fig. 1.

where

$$F(t) = -m\ddot{y}(t) \qquad (9)$$

$$M(t) = -J\ddot{\theta}(t) \qquad (10)$$

and $\delta_{\alpha\beta}$ are the flexibility (influence) coefficients; δ_{11} and δ_{21} are the vertical displacements, respectively, and the angle of rotation due to unit vertical force, whereas δ_{12} and δ_{22} are the vertical displacement, respectively, and the rotation angle, due to unit moment, so that the governing equations are in the matrix form:

$$[\delta][M]\{\ddot{x}\} + \{x\} = \{0\} \qquad (11)$$

where $[\delta]$ is a flexibility matrix, $[M]$ is a generalized mass matrix and $\{x\}$ is a generalized displacement column

$$\{x\} = \{y(t) \quad \theta(t)\}^{\mathrm{T}} \qquad (12)$$

$$[\delta] = \begin{bmatrix} \delta_{11} & \delta_{12} \\ \delta_{21} & \delta_{22} \end{bmatrix}, \qquad [M] = \begin{bmatrix} m & 0 \\ 0 & J_0 \end{bmatrix} \qquad (13)$$

Premultiplication from the left by $[\delta]^{-1} = [K]$, where $[K]$ is a stiffness matrix, leads to a standard form of the equation

$$[M]\{\ddot{x}\} + [K]\{x\} = \{0\} \qquad (14)$$

with

$$K = \frac{1}{\delta_{11}\,\delta_{22} - \delta_{12}^2} \begin{bmatrix} \delta_{22} & -\delta_{12} \\ -\delta_{12} & \delta_{11} \end{bmatrix} \qquad (15)$$

since $\delta_{12} = \delta_{21}$.

For free vibration we stipulate

$$y(t) = A \exp(i\omega t), \qquad \theta(t) = B \exp(i\omega t) \qquad (16)$$

which leads to the frequency equation

$$\det \begin{bmatrix} \dfrac{\delta_{22}}{\Delta} - \omega^2 & \dfrac{-\delta_{12}}{\Delta} \\[2mm] \dfrac{-\delta_{12}}{\Delta} & \dfrac{\delta_{11}}{\Delta} - J_0\omega^2 \end{bmatrix} = 0 \qquad (17)$$

where

$$\Delta = \delta_{11}\,\delta_{22} - \delta_{12}^2 \qquad (18)$$

The frequency equation becomes

$$mJ_0\omega^4 - \frac{1}{\Delta}(m\,\delta_{11} + J_0\,\delta_{22})\omega^2 + \frac{1}{\Delta} = 0 \tag{19}$$

yielding

$$\omega_{1,2}^2 = \frac{1}{2mJ_0\Delta}[m\,\delta_{11} + J_0\,\delta_{22} \mp \sqrt{(m\,\delta_{11} - J_0\,\delta_{22})^2 + 4mJ_0\,\delta_{12}^2}] \tag{20}$$

The influence coefficients are found by multiplying the moment diagrams, associated with the unit loading:

$$\delta_{11} = \int_{r/2}^{L+r/2} \frac{M_1^2\,dz}{EI} = \int_{r/2}^{L+r/2} \frac{z^2\,dz}{EI} \tag{21}$$

$$= \frac{L^3}{3EI}\left[\left(1 + \frac{1}{2L}\right)^3 - \left(\frac{2}{2L}\right)^3\right]$$

Denoting

$$\frac{r/2}{L} = \xi \tag{22}$$

we get

$$\delta_{11} = \frac{L^3}{3EI}a(\xi) \tag{23}$$

where

$$a(\xi) = (1 + \xi)^3 - \xi^3 = 1 - 3\xi - 3\xi^2 \tag{24}$$

Now

$$\delta_{22} = \int_{r/2}^{L+r/2} \frac{M_2^2\,dz}{EI} = \int_{r/2}^{L+r/2} \frac{z^2\,dz}{EI} = \frac{L}{EI} \tag{25}$$

$$\delta_{12} = \int_{r/2}^{L+r/2} \frac{M_1(z)M_2(z)\,dz}{EI} = \int_{r/2}^{L+r/2} \frac{z\,dz}{EI} = \frac{L^2}{2EI}\left[\left(1 + \frac{r}{2L}\right)^2 - \left(\frac{r}{2L}\right)^2\right] \tag{26}$$

or, with

$$b(\xi) = (1 + \xi)^2 - \xi^2 = 1 + 2\xi \tag{27}$$

we have

$$\delta_{12} = \delta_{21} = \frac{L^2}{2EI}b(\xi) \tag{28}$$

Natural frequencies become

$$\omega^2_{1,2} = \frac{EI}{mL^3} \frac{a/3 + \eta - \sqrt{(a/3 - \eta)^2 + \eta b^2}}{2\eta(a/3 - b^2/4)} \tag{29}$$

where the moment of inertia was represented as

$$J_0 = \eta mL^2 \tag{30}$$

Consider some particular cases. For the concentrated mass we have:

$$r \to 0, \quad \xi \to 0, \quad J_0 \to 0, \quad \eta \to 0 \tag{31}$$

which results in

$$\delta_{11} = \frac{L^3}{3EI}, \quad \delta_{22} = \frac{L}{EI}, \quad \delta_{12} = \frac{L^2}{2EI} \tag{32}$$

$$a(\xi) = 1, \quad b(\xi) = 1 \tag{33}$$

$$\omega^2_1 = \frac{3EI}{mL^3}, \quad \omega^2_2 \to \infty \tag{34}$$

For a particular case

$$\xi = 0 \cdot 5, \quad \eta = 0 \cdot 125 \tag{35}$$

Biderman[18] studied the free vibration of this system. We get

$$\delta_{11} = \frac{L^3}{3EI}[(1 + \tfrac{1}{2})^3 - (\tfrac{1}{2})^3] = \frac{13}{12}\frac{L^3}{EI} \tag{36}$$

$$\delta_{22} = \frac{L}{EI}, \quad \delta_{12} = \frac{L^2}{EI} \tag{37}$$

$$\Delta = \frac{L^4}{12E^2I^2}, \tag{38}$$

with

$$\omega^2_{1,2} = \frac{1 \cdot 208\,333\,3 \mp 1 \cdot 190\,967\,2}{0 \cdot 020\,833\,3}\frac{EI}{mL^3} \tag{39}$$

$$\omega^2_1 = 0 \cdot 193\frac{EI}{mL^3} \tag{40}$$

$$\omega^2_2 = 10 \cdot 193\frac{EI}{mL^3} \tag{41}$$

and coincide with Biderman's results. As we see, the system possesses widely separated natural frequencies, namely $\omega_2 \gg \omega_1$.

The modal matrix $[\psi]$ satisfying the condition of orthonormality

$$[\psi]^{T}[M][\psi] = [I] \tag{42}$$

(where $[I]$ is an identity matrix) is given by

$$\psi_{11} = \frac{1}{\mu_1}, \qquad \psi_{12} = \frac{1}{\mu_2} \tag{43}$$

$$\psi_{21} = \frac{1}{\mu_1 \eta LB}(x + \sqrt{x^2 + \eta B^2}), \qquad \psi_{22} = \frac{1}{\mu_2 \eta LB}(x - \sqrt{x^2 + \eta B^2}) \tag{44}$$

with

$$\mu_1 = \sqrt{m}\left[1 + \frac{(x + \sqrt{x^2 + \eta B^2})^2}{\eta B^2}\right]^{1/2} \tag{45}$$

$$\mu_2 = \sqrt{m}\left[1 + \frac{(x - \sqrt{x^2 + \eta B^2})^2}{\eta B^2}\right]^{1/2} \tag{46}$$

$$x = \eta - \frac{A}{3} \tag{47}$$

For the normalized modal matrix we have

$$[\psi]^{T}[K][\psi] = [\ulcorner \omega_r^2 \urcorner] \tag{48}$$

where $[\ulcorner \omega_r^2 \urcorner]$ is a diagonal matrix having at its diagonals the natural frequencies squared ω_j^2.

3 RANDOM VIBRATION

Consider now the random vibration of the system under stationary vertical force $F(t)$. The system possesses a transverse damping c_1 and the rotary damping c_2. Equations of motion read

$$y(t) = (-m\ddot{y} - c_1\dot{y} + F)\,\delta_{11} - (-J_0\ddot{\theta} - c_2\dot{\theta})\,\delta_{12} \tag{49}$$

$$\theta(t) = (-m\ddot{y} - c_1\dot{y} + F)\,\delta_{12} - (-J_0\ddot{\theta} - c_2\dot{\theta})\,\delta_{22} \tag{50}$$

or, in terms of the second time derivatives of the generalized displacements

$$m\ddot{y} + c_1\dot{y} + \frac{\delta_{22}}{\delta_{11}\delta_{22} - \delta_{12}^2}y - \frac{\delta_{12}}{\delta_{11}\delta_{22} - \delta_{12}^2}\theta = F(t) \qquad (51)$$

$$J_0\ddot{\theta} + c_2\dot{\theta} + \frac{\delta_{11}}{\delta_{11}\delta_{22} - \delta_{12}^2}\theta - \frac{\delta_{12}}{\delta_{11}\delta_{22} - \delta_{12}^2}y = 0 \qquad (52)$$

which can be put in the matrix form

$$\begin{bmatrix} m & 0 \\ 0 & J_0 \end{bmatrix}\begin{Bmatrix} \ddot{y} \\ \ddot{\theta} \end{Bmatrix} + \begin{bmatrix} c_1 & 0 \\ 0 & c_2 \end{bmatrix}\begin{Bmatrix} \dot{y} \\ \dot{\theta} \end{Bmatrix} + \begin{bmatrix} k_{11} & k_{12} \\ k_{21} & k_{22} \end{bmatrix}\begin{Bmatrix} y \\ \theta \end{Bmatrix} = \begin{Bmatrix} F \\ 0 \end{Bmatrix} \qquad (53)$$

with $k_{\alpha\beta}$ being the element of the stiffness matrix, given in eqn (51)

$$k_{11} = \frac{\delta_{22}}{\delta_{11}\delta_{22} - \delta_{12}^2}, \; k_{12} = k_{21} = -\frac{\delta_{12}}{\delta_{11}\delta_{22} - \delta_{12}^2}$$

$$k_{22} = \frac{\delta_{11}}{\delta_{11}\delta_{22} - \delta_{12}^2}$$

Equation (53) can be put in the matrix form

$$[M]\{\ddot{x}\} + [C]\{\dot{x}\} + [K]\{x\} = \{F\} \qquad (54)$$

As far as the damping coefficients are concerned, we assume that

$$\frac{c_1}{m} = \frac{c_2}{J_0} = \alpha \qquad (55)$$

so that we deal with the proportional damping $[c] = \alpha[M]$:

$$[M]\{\ddot{x}\} + \alpha[M]\{\dot{x}\} + [K]\{x\} = \{F\} \qquad (56)$$

We use the principal coordinates $\{q\}$:

$$[M]\{x\} = [\psi]\{q\} \qquad (57)$$

which upon being substituted into eqn (56) and results premultiplied by $[\psi]^T$ yields

$$[\psi]^T[M][\psi]\{\ddot{q}\} + \alpha[\psi]^T[M][\psi]\{\dot{q}\} + [\psi]^T[K][\psi]\{q\} = [\psi]^T\{q\} \qquad (58)$$

which by virtue of eqns (42) and (48) becomes

$$\{\ddot{q}\} + \alpha\{\dot{q}\} + [\diagdown\omega^2\diagdown]\{q\} = [\psi]^T\{F\} \qquad (59)$$

or

$$\ddot{q}_1 + 2\zeta_1\omega_1\dot{q}_1 + \omega_1^2 q_1 = \frac{1}{\mu_1} F(t) \tag{60}$$

$$\ddot{q}_2 + \zeta_2\omega_2\dot{q}_2 + \omega_2^2 q_2 = \frac{1}{\mu_2} F(t) \tag{61}$$

with

$$\zeta_1 = \frac{\alpha}{2\omega_1}, \qquad \zeta_2 = \frac{\alpha}{2\omega_2} \tag{62}$$

For the mean-square values $E(q_1^2)$, $E(q_2^2)$ and the correlation moment $E(q_1 q_2)$ we get

$$E(q_1^2) = \frac{\pi S_0}{2\mu_1^2 \zeta_1 \omega_1^3} \tag{63}$$

$$E(q_2^2) = \frac{\pi S_0}{2\mu_2^2 \zeta_2 \omega_2^3} \tag{64}$$

$$E(q_1 q_2) = \frac{4\pi(\omega_1 + \omega_2)}{\mu_1 \mu_2 \{ (\omega_1^2 - \omega_2^2)^2 + 4[\zeta_1\zeta_2\omega_1\omega_2(\omega_1^2 + \omega_2^2) + (\zeta_1^2 + \zeta_2^2)\omega_1^2\omega_2^2] \}} \tag{65}$$

Since, however, the natural frequencies are widely separated the cross-correlation term $E(q_1 q_2)$ can be neglected. For the direct correlations we get finally

$$E(q_1^2) = \frac{\pi S_0 m L^3}{EIc_1} \left[1 + \frac{(x + \sqrt{x^2 + \eta B^2})^2}{\eta B^2} \right]^{-1} \left[\frac{x - 2A/3 - \sqrt{x^2 + \eta B^2}}{2\eta(A/3 - B^2/4)} \right]^{-1} \tag{66}$$

$$E(q_2^2) = \frac{\pi S_0 J_0 L^3}{EIc_2} \left[1 + \frac{(x + \sqrt{x^2 + \eta B^2})^2}{\eta B^2} \right]^{-1} \left[\frac{x + 2A/3 + \sqrt{x^2 + \eta B^2}}{2\eta(A/3 - B^2/4)} \right]^{-1} \tag{67}$$

For the mean-square values in the original coordinates we get

$$E(y^2) = \psi_{11}^2 E(q_1^2) + \psi_{12}^2 E(q_2^2) + 2\psi_{11}\psi_{12} E(q_1 q_2)$$
$$\cong \psi_{11}^2 E(q_1^2) + \psi_{12}^2 E(q_2^2) \tag{68}$$

$$E(\theta^2) = \psi_{21}^2 E(q_1^2) + \psi_{22}^2 E(q_2^2) + 2\psi_{21}\psi_{22} E(q_1 q_2)$$
$$\cong \psi_{21}^2 E(q_1^2) + \psi_{22}^2 E(q_2^2) \tag{69}$$

or in the nondimensional form

$$E(\bar{y}^2) = 3\left[1 + \frac{(x + \sqrt{x^2 + \eta B^2})}{\eta B^2}\right]^{-2}\left[\frac{x + 2A/3 - \sqrt{x^2 + \eta B^2}}{2\eta(A/3 - B^2/4)}\right]^{-1}$$
$$+ \gamma\eta\left[1 + \frac{(x + \sqrt{x^2 + \eta B^2})}{\eta B^2}\right]^{-2}\left[\frac{x + 2A/3 + \sqrt{x^2 + \eta B^2}}{2\eta(A/3 - B^2/4)}\right]^{-1} \quad (70)$$

where

$$E(\bar{y}^2) = \frac{E(y^2)}{\pi S_0 L^3/3EIc_1} \quad (71)$$

$$\gamma = \frac{3c_1 L^2}{c_2} \quad (72)$$

γ signifies the nondimensional damping ratio, the denominator in the expression for $E(\bar{y}^2)$ represents the mean-square displacement of a single degree of freedom system obtainable from our problem when $\xi \to 0$ (corresponding to concentrated mass at the tip of the cantilever). In complete analogy, for $E(\theta^2)$ we get

$$E(\bar{\theta}_1^2) = 3\left[1 + \frac{(x + \sqrt{x^2 + \eta B^2})}{\eta B^2}\right]^{-2}\left[\frac{x + 2A/3 - \sqrt{x^2 + \eta B^2}}{\eta B^2}\right]^2$$
$$\times \left[\frac{x + 2A/3 - \sqrt{x^2 + \eta B^2}}{2\eta(A/3 - B^2/4)}\right]^{-1} + \gamma\eta\left[\frac{x - \sqrt{x^2 + \eta B^2}}{\eta B^2}\right]^2$$
$$\times \left[1 + \frac{(x + \sqrt{x^2 + \eta B^2})}{\eta B^2}\right]^{-2}\left[\frac{x + 2A/3 - \sqrt{x^2 + \eta B^2}}{2\eta(A/3 - B^2/4)}\right]^{-1} \quad (73)$$

where

$$E(\bar{\theta}_1^2) = \frac{E(\theta^2)}{\pi S_0 L/3EIc_1} \quad (74)$$

The results of calculations according to eqns (70) and (73) are listed in Tables 1 and 2. $E(y_1^2)$ and $E(\theta_1^2)$ denote the contribution of the first mode whereas $E(y_2^2)$ and $E(\theta_2^2)$ are the contributions of the second mode; the damping ratio was fixed at 0·1 and β at $\sqrt{2}/2$.

For J_0 we have (see Fig. 1 for dimensions)

$$J_0 = J_{y_1 y_1} = \tfrac{1}{12}m(h^2 + r^2) = \tfrac{1}{12}mr^2\left[1 + \left(\frac{h}{r}\right)^2\right] \quad (75)$$

Table 1

ξ	$E(\bar{y}_1^2)$	$E(\bar{y}_2^2)$	$E(\bar{y}^2)$
0	1	0	1
0·1	1·317 856	$3·339\,840 \times 10^{-9}$	1·317 856
0·2	1·669 491	$5·669\,127 \times 10^{-7}$	1·669 491
0·3	2·054 587	$9·479\,962 \times 10^{-6}$	2·054 588
0·5	2·474 389	$6·224\,281 \times 10^{-5}$	2·474 395
0·6	3·423 531	$7·313\,247 \times 10^{-4}$	2·930 345
0·7	3·954 851	$1·759\,342 \times 10^{-3}$	3·423 604
0·8	4·524 853	$3·670\,557 \times 10^{-3}$	4·525 220
0·9	5·133 927	$6·893\,595 \times 10^{-3}$	5·134 617
1	5·782 343	$1·194\,984 \times 10^{-2}$	5·783 538

Using the notation

$$\beta = \frac{h}{r}, \qquad \xi = \frac{r}{2L} \tag{76}$$

we obtain

$$J_0 = \tfrac{1}{12}mr^2(1 + \beta^2) \tag{77}$$

$$\eta = \frac{J_0}{mL^2} = \tfrac{1}{3}\xi^2(1 + \beta^2) \tag{78}$$

so that $\beta = \sqrt{2}/2$ corresponds, for $\xi = 1/2$, to $\eta = 1/8$, as was the case for the free vibration analysis in Ref. 18.

Table 2

ξ	$E(\bar{\theta}_1^2)$	$E(\bar{\theta}_2^2)$	$E(\bar{\theta}^2)$
0	2·250 083	0	2·250 083
0·1	2·424 019	$7·264\,484 \times 10^{-5}$	2·424 026
0·2	2·513 343	$9·416\,203 \times 10^{-4}$	2·513 437
0·3	2·547 684	$3·776\,140 \times 10^{-3}$	2·548 062
0·4	2·550 694	$9·436\,394 \times 10^{-3}$	2·551 638
0·5	2·537 348	$1·836\,203 \times 10^{-2}$	2·539 185
0·6	2·516 233	$3·071\,680 \times 10^{-2}$	2·519 305
0·7	2·492 036	$4·652\,445 \times 10^{-2}$	2·496 689
0·8	2·467 336	$6·574\,906 \times 10^{-2}$	2·473 811
0·9	2·443 093	$8·833\,500 \times 10^{-2}$	2·451 927
1	2·402 090	0·114 225	2·431 632

As seen from the tables the contribution of the first mode is at least 500 times that of the second mode in forming the mean-square displacement, and at least 20 times in forming the mean-square of the rotation angle. This suggests that in the final analysis the system may be treated as *single* degree of freedom with natural frequency ω_1, which approximately takes into account the rotational degree of freedom too, within *simplified refined theory* neglecting the fourth order time derivative. (For a more detailed discussion of this point, see Appendix.) As we see from Table 1, the nondimensional mean-square $E(\bar{y}^2)$ exceeds unity for $\xi > 0$ and equals unity for $\xi = 0$. For example, for $\xi = 0.5$ (the spread of the mass equal to the length of the cantilever) the response of the system is about three times its counterpart obtained by concentrating the mass at the tip. Omission of the rotational degree of freedom would yield an error of the order of 60%; moreover it would be on the unsafe side, as the response would be underestimated.

ACKNOWLEDGEMENT

One of us (I.E.) gratefully appreciates the support of this project by the Research Foundation of the Naval Postgraduate School and the fund of promotion of research by the Technion-I.I.T.

REFERENCES

1. Samuels, J. C. & Eringen, A. C., Response of a simply supported Timoshenko beam to a purely random Gaussian process. *Journal of Applied Mechanics*, **25** (1958) 496–500.
2. Crandall, S. H. & Yildiz, A., Random vibrations of beams. *Journal of Applied Mechanics*, **31** (1962) 267–75.
3. Banerjee, J. R. & Kennedy, D., Response of an axially loaded Timoshenko beam to random loads. *Journal of Sound and Vibration*, **101**(4) (1985) 481–7.
4. Cederkirst, J., Design of beams subjected to random loads. *Journal of Structural Mechanics*, **10** (1982) 49–65.
5. Witt, M. & Sobczyk, K., Dynamic response of laminated plates to random loading. *International Journal of Solids and Structures*, **16** (1980) 231–8.
6. Elishakoff, I. & Livshits, D., Some closed form solutions in random vibration of Timoshenko beams. *Proceedings of the 2nd International Conference on Recent Advances in Structural Dynamics*, eds M. Petyt & H. F. Wolfe. University of Southampton Press, 1984, pp. 639–48.

7. Elishakoff, I., *Probabilistic Methods on the Theory of Structures*, Wiley-Interscience, New York, 1983.
8. Crandall, S. H. & Mark, W. D., *Random Vibration in Mechanical Systems*, Academic Press, New York, 1963.
9. Lin, Y. K., *Probabilistic Theory of Structural Dynamics*, McGraw-Hill, New York, 1967 (2nd edition, R. Krieger, Malabar, FL, 1976).
10. Timoshenko, S. P., On the correction for shear of the differential equation for transverse vibrations of prismatic bar. *Philosophical Magazine*, Ser. 6, **41**/245, 1921, 744–6; Timoshenko, S. P., *The Collected Papers*. McGraw-Hill, New York, 1953, pp. 288–90; Timoshenko, S. P., *Vibration Problems in Engineering*, D. van Nostrand Co., 1928, pp. 227–8, 231–2.
11. Abbas, B. A. H. & Thomas, J., The second frequency spectrum of Timoshenko beams. *Journal of Sound and Vibration*, **51** (1977) 123–7.
12. Bhashyan, G. R. & Prathap, G., The second frequency spectrum of Timoshenko beams. *Journal of Sound and Vibration*, **76** (1981) 407–20.
13. Stephen, N. G., The second frequency spectrum of Timoshenko beams. *Journal of Sound and Vibration*, **80** (1982) 578–82.
14. Levinson, M. & Cooke, D. W., On the two frequency spectra of Timoshenko beams. *Journal of Sound and Vibration*, **84**(3) (1982) 319–26.
15. Hathout, I., Leipholz, H. & Singhal, K., Sensitivity of the frequencies of damped Timoshenko beams. *Journal of Engineering Science, University of Riyadh*, **6** (1980) 113–21.
16. Barr, A., Personal communications, 1984, 1986.
17. Elishakoff, I. & Lubliner, E., Random vibration of structures via classical and nonclassical theories. In *Probabilistic Methods in the Mechanics of Solids and Structures*, ed. S. Eggwertz & N. C. Lind. Springer, Berlin, 1984, pp. 455–64.
18. Biderman, V. L., *Applied Theory of Mechanical Vibrations*, 'Vyschaya Shkola' Publishing House, Moscow, 1972.

APPENDIX

The pair of differential equations (14) is equivalent to a resultant equation

$$mJ_0 \frac{\mathrm{d}^4\psi}{\mathrm{d}t^4} + \frac{1}{\Delta}(m\delta_{11} - J_0\,\delta_{22}) \frac{\mathrm{d}^2\psi}{\mathrm{d}t^2} + \frac{1}{\Delta}\,\psi = 0 \qquad (A.1)$$

where ψ is a potential function through which our original generalized displacements are found as

$$y = \frac{\delta_{12}}{\Delta}\,\psi, \qquad \theta = m\frac{\mathrm{d}^2\phi}{\mathrm{d}t^2} + \frac{\delta_{22}}{\Delta}\,\psi \qquad (A.2)$$

In particular, substitution

$$\psi = \mathrm{e}^{\mathrm{i}\omega t} \qquad (A.3)$$

leads to eqn (19). Neglecting the fourth order derivative, as it was described in the introduction, in perfect analogy with what we did with Timoshenko beam equations in Refs 6 and 17, we get

$$(m\delta_{11} + J_0\,\delta_{22})\frac{\mathrm{d}^2\psi}{\mathrm{d}t^2} + \psi = 0 \tag{A.4}$$

and with substitution (A.3) we are left with (for $\omega = \omega_1$):

$$- (m\delta_{11} + J_0\,\delta_{22})\omega^2 + 1 = 0 \tag{A.5}$$

or

$$\omega_1 = \frac{1}{m\delta_{11} + J_0\,\delta_{22}} \tag{A.6}$$

Equation (A.6) is also obtainable from eqn (20) by asymptotic evaluation of the square root in eqn (20) as

$$\sqrt{(m\delta_{11} + J_0\,\delta_{22})^2 - 4mJ_0\Delta} \approx (m\delta_{11} + J_0\,\delta_{22})\left[1 - \frac{2mJ_0\Delta}{(m\delta_{11} + J_0\,\delta_{22})^2}\right]$$
$$\tag{A.7}$$

For the first natural frequency we then obtain eqn (A.6).

In the case of the random vibration, we use eqns (51) and (52) to express $y(t)$ and $\theta(t)$ as follows

$$y(t) = J_0\frac{\mathrm{d}^2\psi}{\mathrm{d}t^2} + c_2\frac{\mathrm{d}\psi}{\mathrm{d}t} + k_{22}\psi, \qquad \theta(t) = -k_{21}\psi \tag{A.8}$$

to yield

$$mJ_0\overset{....}{\psi} + (mc_2 + J_0c_1)\overset{...}{\psi} + (mk_{22} + J_0k_{11})\ddot{\psi} + (c_1k_{22} + c_2k_{11})\dot{\psi}$$
$$+ (k_{11}k_{22} - k_{12}k_{21})\psi = F \quad (A.9)$$

or with eqn (55),

$$mJ_0\overset{....}{\psi} + 2\alpha mJ_0\overset{...}{\psi} + (mk_{22} + J_0k_{11})\ddot{\psi} + \alpha(mk_{22} + J_0k_{11})\dot{\psi}$$
$$+ (k_{11}k_{22} - k_{12}k_{21})\psi = F \quad (A.10)$$

Again omitting the term containing the product mJ_0, we have

$$\ddot{\psi} + \alpha\dot{\psi} + \omega_1^2\psi = \Delta\omega_1^2 F \tag{A.11}$$

which again leads to the results for a one-degree-of-freedom system, where the natural frequency is found within the simplified *refined* theory.

5

Stochastic Aspects of Structural Dynamics in Aerospace Vehicles

YOSHINORI FUJIMORI

National Aerospace Laboratory, Tokyo, Japan

ABSTRACT

The article intends to overview the stochastic aspects of aerospace vehicles' structural problems. Presented are stationary response of the panel under boundary layer pressure fluctuations, nonstationary response of the airplane in atmospheric turbulence, stability of a rotor blade in turbulent flow and derivation and dispersion analysis of the damping coefficient for the plate. Methodologies summarized have been tested by numerical examples, proving their usefulness to practical applications.

1 INTRODUCTION

The need for stochastic structural dynamics has arisen through the past few decades from the demand of engineering practitioners to understand the physical phenomena in more general terms. Randomness experienced in aerospace vehicles, ocean-going vessels, ground vehicles, ground structures and other numerous systems we build and use daily give the theoreticians the impetus to construct the proper mathematical model of the problem. In this context, it is noted that Professor Y. K. Lin summarized the topics of interest and laid the solid foundation in this field of science.[1]

Confining ourselves here to the problems of aerospace systems, those awaiting further study are the panel response under boundary layer pressure fluctuations, the airplane gust response and the rotor blade stability in the atmospheric turbulence, to name only a few. Although the variety of the problems necessitates virtually different approaches in the practical applications, theorization has been executed in a rather unified manner that the system we consider is to be represented by linear physical differential equation (DEs), and randomness appears either or both in homogeneous and inhomogeneous terms of DEs. The achievements so far acquired seem far short of the real demand from the engineers because of the immaturity of this field of science. Nevertheless, the proposed methodologies can supply valuable knowledge to help the practitioners make decisions in the process of the system design, serving a proper guideline for action.

This article overviews the panel response under boundary layer pressure fluctuations, airplane nonstationary gust response, rotor blade stability in turbulent flow and the damping problem in the successive sections independently. Throughout the sections, some notations are used for different purposes, i.e. the meaning differs from one section to another, albeit most of them are obvious and can be understood clearly without confusion.

2 PLATE RESPONSE UNDER BOUNDARY LAYER PRESSURE FLUCTUATIONS

The boundary layer pressure fluctuation is one of the aerodynamic noises which excites the exterior panel members of the aerospace vehicles causing acoustic fatigue of the structure or inducing the excessive interior noise levels. Its characteristics are clarified by many experiments to a certain extent that the fluctuation is homogeneous in space and can be regarded as ergodic.

This problem exemplifies the stationary response of the panel under stationary pressure fluctuations, which is reduced to the mathematical model of the DE with deterministic coefficients in homogeneous terms and a random coefficient, i.e. a stochastic input, in the inhomogeneous term.

The governing equation of plate motion used in this chapter is written as follows,

$$\frac{D}{M}\nabla^4 w(\gamma,\, t) + 2\beta\frac{\partial w(\gamma,\, t)}{\partial t} + \frac{\partial^2 w(\gamma,\, t)}{\partial t^2} = \frac{p(\gamma,\, t)}{M} \qquad (2.1)$$

where γ expresses plate coordinates x and y.

As the input to the structure $p(\gamma, t)$ is a time-varying random function, the solution of eqn (2.1) is a time-varying random function whose properties are more easily understood when they are described in the stochastic manner. In the present analysis, the weak ergodicity of $p(\gamma, t)$ and $w(\gamma, t)$ is tacitly assumed, and only the steady response due to homogeneous-in-space pressure load $p(\gamma, t)$ is handled, the transient response will not be treated.

Putting aside the mathematical rigor, consider the truncated function about the excitation $p(\gamma, t)$ and the induced response $w(\gamma, t)$, and denote $p_T(\gamma, t)$, $w_T(\gamma, t)$ respectively. $w_T(\gamma, t)$ is defined as the solution of eqn (2.1) when $p(\gamma, t)$ is replaced by the truncated function $p_T(\gamma, t)$. Then additionally it is assumed that random functions are absolutely integrable and their spectral decomposition is possible. Some definitions given in Refs 1 and 2 are recalled here for the sake of readers' convenience.

Assuming spatial homogeneity and weak ergodicity of a random process spanned by the time-varying pressure fluctuations, the relation between power spectral density $\theta_P(\gamma_1 - \gamma_2, \omega)$ and cross-correlation function $K_P(\gamma_1 - \gamma_2, \tau)$ is written as follows,

$$\theta_p(\gamma_1 - \gamma_2, w) = \int_{-\infty}^{\infty} K_p(\gamma_1 - \gamma_2, \tau) e^{-j\omega\tau} \, d\tau \qquad (2.2)$$

$$K_p(\gamma_1 - \gamma_2, \tau) = \frac{1}{2\pi} \int_{-\infty}^{\infty} \theta_p(\gamma_1 - \gamma_2, \omega) e^{j\omega\tau} \, d\omega \qquad (2.3)$$

where

$$K_p(\gamma_1 - \gamma_2, \tau) \equiv E[p_T(\gamma_1, t_1) p_T(\gamma_2, t_2)]$$

$$= \lim_{T \to \infty} \frac{1}{2T} \int_{-T}^{T} p_T(\gamma_1, t_1) p_T(\gamma_2, t_2) \, dt, \qquad \tau = t_1 - t_2 \quad (2.4)$$

$$\theta_p(\gamma_1 - \gamma_2, \omega) = \lim_{T \to \infty} \frac{1}{2T} [p_T(\gamma_1, \omega) p_T^*(\gamma_2, \omega)] \qquad (2.5)$$

$$P_T(\gamma, \omega) = \int_{-T}^{T} p_T(\gamma, t) e^{-j\omega t} \, dt \qquad (2.6)$$

Equation (2.6) is a definition of the well-known 'Truncated Fourier Transformation'. The same relations are valid for the cross-correlation function and spectrum of response displacement in spite of non-homogeneity of $w_T(\gamma, t)$ in space.

$$\theta_w(\gamma_1, \gamma_2, \omega) = \int_{-\infty}^{\infty} K_n(\gamma_1, \gamma_2, \tau)e^{-j\omega\tau}\,d\tau \qquad (2.7)$$

$$K_w(\gamma_1, \gamma_2, \tau) = \frac{1}{2\pi}\int_{-\infty}^{\infty} \theta_n(\gamma_1, \gamma_2, \omega)e^{j\omega t}\,d\omega \qquad (2.8)$$

$$K_w(\gamma_1, \gamma_2, \tau) = E[w_T(\gamma_1, t_1)w_T(\gamma_2, t_2)]$$

$$= \lim_{T\to\infty}\frac{1}{2T}\int_{-T}^{T} w_T(\gamma_1, t_1)w_T(\gamma_2, t_2)\,dt \qquad (2.9)$$

$$\tau = t_1 - t_2$$

$$\theta_w(\gamma_1, \gamma_2, \omega) = \lim_{T\to\infty}\frac{1}{2T}[w_T(\gamma_1, \omega)w_T^*(\gamma_2, \omega)] \qquad (2.10)$$

$$W_T(\gamma, \omega) = \int_{-T}^{T} w_T(\gamma, t)e^{-j\omega t}\,dt \qquad (2.11)$$

Now $P_T(\gamma, \omega)$, $W_T(\gamma, \omega)$ can be described by the eigen function of the plate $\Psi(\gamma)$ as follows.

$$P_T(\gamma, \omega) = \Sigma\, b_i(\omega)\Psi_i(\gamma)$$
$$W_T(\gamma, \omega) = \Sigma\, a_i(\omega)\Psi_i(\gamma) \qquad (2.12)$$

$$\int_R \Psi_i(\gamma)\Psi_l(\gamma)\,d\gamma = \delta_{il} \qquad (2.13)$$

R = Domain where $\Psi_i(\gamma)$ is defined.

Performing the Fourier Transform of eqn (2.1), substituting eqn (2.12) into the result and using the relation of eqn (2.13), yield the following;

$$a_i(\omega) = -b_i(\omega)H_i(\omega) \qquad (2.14)$$

where

$$H_i(\omega) = \frac{1}{M(\omega^2 - 2j\beta\omega - D/Mk_i^4)} \qquad (2.15)$$

$$k_i^4 = \int_R \nabla^4(\Psi_i(\gamma))\Psi_i(\gamma)\,d\gamma$$

Considering the definition of power spectral density of pressure fluctuations within the plate, the important quantity is derived as follows:

$$\lim_{T \to \infty} \frac{1}{2T} b_i(\omega) b_l^*(\omega)$$

$$= \int_R \int_R \lim_{T \to \infty} [P_T(\gamma_1, \omega) P_T(\gamma_2, \omega)] \Psi_i(\gamma_1) \Psi_l(\gamma_2) \, d\gamma_1 \, d\gamma_2$$

$$= \int_R \int_R \theta_p(\gamma_1 - \gamma_2, \omega) \Psi_i(\gamma_1) \Psi_l(\gamma_2) \, d\gamma_1 \, d\gamma_2$$

$$= I_{il}(\omega) \tag{2.16}$$

Equation (2.16) is the generalized spectrum that is the spectrum of the generalized forces and can be significant only over the domain of the vibrating structure.

The cross-correlation function of displacement is given by the next formula:

$$E[w(\gamma_1, t_1) w(\gamma_2, t_2)] = \frac{1}{2\pi} \int_{-\infty}^{\infty} \theta_w(\gamma_1, \gamma_2, \omega) e^{j\omega t} \, d\omega$$

$$= \frac{1}{2\pi} \int_{-\infty}^{\infty} \sum_i \sum_l I_{il}(\omega) H_i(\omega) H_l^*(\omega) \Psi_i(\gamma_1) \Psi_l(\gamma_2) e^{j\omega \tau} \, d\omega \tag{2.17}$$

Equation (2.17) stipulates the input–output relation of the plate here considered.

One can evaluate the response cross-correlation knowing the eigen function of the structure and the generalized spectrum of the pressure fluctuations. It is valid for the linear system that the statistical properties of response along the temporal axis depend on those of the input to the structure. Homogeneity along the space axis of the external forces does not necessarily lead to the same aspect of the displacement response because of the finiteness of panel dimensions.

The first step toward the goal is to obtain the generalized spectrum $I_{il}(\omega)$ in eqn (2.16). It is desirable that a space–time cross-correlation of pressure fluctuations $K_p(\gamma_1 - \gamma_2, \tau)$ is known with the analytically simple expression, but it is generally impossible to obtain the function $K_p(\gamma_1 - \gamma_2, \tau)$ through mathematical procedures. The possible formula containing the δ-function along the space axis, exponential damping along the time axis has been proposed as the most suitable one to describe important characteristics of pressure fluctuations.

$$K_p(\xi, \eta, \tau) = E[p^2]2\pi L_x L_y \delta(\xi - \upsilon\tau)\delta(\eta)e^{-|\tau|/\zeta} \qquad (2.18)$$

where

L_x, L_y are X- and Y-wise eddy size;
$E[p^2]$ is mean square pressure fluctuation;
ζ is eddy time constant;
$\xi = x_1 - x_2$, $\eta = y_1 - y_2$.

Equation (2.18) describes the X-wise flow of pressure fluctuations and parameters L_x, L_y, ζ are assumed to be mode-dependent, then pressure level $E[p^2]$ is assumed to be also a variable that depends on a vibration mode.

The final form of the response can be given for each vibration mode by taking out the summation from eqn (2.17). Obviously, summarizing all modal responses yields the total response.

The analytical procedures are described in detail in Ref. 3 including the expressions for the degeneration response of the simply-supported square plate and those due to oblique flow. Numerical examples of the simply-supported plate are given in Ref. 4.

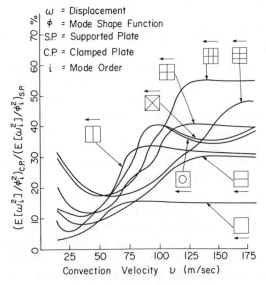

Fig. 1. Ratio of normalized response.

For the case of the clamped plate, the equations mentioned above can be also utilized in an approximate sense with somewhat lengthy manipulation of problem formulation and residue calculus as well. In practice, the eigen functions of the clamped plate are assumed by the double series expansion of those for the clamped beam which hold mathematical rigor. Numerical examples are given in Ref. 5. Comparison is given in Fig. 1 of the ratios of the normalized responses of the simply-supported and clamped plates. The results are shown as the percentage versus the eddy convection velocity. The arrows indicate the direction of the flow and the nodal lines of the modes are drawn inside the squares. The figure shows that the boundary constraint gets less influential as the mode order increases.

Comments are in order here on the expression of eqn (2.18). This analytical formula describes only part of the boundary layer pressure fluctuation characteristics. A more precise representation and associated response calculus are proposed by Lin & Maekawa.[6]

3 NONSTATIONARY RESPONSE OF AIRPLANE IN ATMOSPHERIC TURBULENCE

Although the spectrum analysis of the stationary time series seems very powerful in many engineering problems, it can not offer adequate means to describe some phenomena such as airplane gust response, building response due to ground turbulence or earthquake and so forth. Those are predominantly of a nonstationary nature.

In this section, will be proposed the spectrum analysis of the multi-modal linear system's response in terms of an evolutionary cross spectrum that includes the stationary analysis as a special case[7] and its application to airplane gust response will be illustrated.[10-12]

Considering the input $\vec{X}$ and output $\vec{Y}$ of the linear system specified by the impulse response function matrix $[h(t)]$, the response cross correlation of the output, the second order statistics has the form of

$$E[Y_j(t_1)Y_k(t_2)] = \sum_m \sum_l \int_0^{t_1} \int_0^{t_2} E[X_m(\tau_1)X_l(\tau_2)]h_{jm}(t_1 - \tau_1)$$
$$\times h_{kl}(t_2 - \tau_2)\,d\tau_1\,d\tau_2 \qquad (3.1)$$

Here a set of random processes $\vec{X}$ is supposed to be nonstationary. Being based on amplitude modulation, each random process X_m can be factored into the product of a deterministic envelope function and a stationary random process

$$X_m(t) = A_m(t, \omega)X_{S,m}(t) \tag{3.2}$$

The spectral decomposition theorem gives the following expression for the stationary process $X_{S,m}(t)$

$$X_{S,m}(t) = \int_{-\infty}^{+\infty} e^{i\omega t}\, d\chi_m(\omega) \tag{3.3}$$

Next it is assumed that the nonstationary process $X_m(t)$ has an expression similar to eqn (3.3),

$$X_m(t) = \int_{-\infty}^{\infty} A_m(t, \omega)e^{i\omega t}\, d\chi_m(\omega) \tag{3.4}$$

where $d\chi_m(\omega)$ is one of the stationary random processes.

After some manipulation, eqn (3.1) can be rewritten as

$$E[Y_j(t_1)Y_k(t_2)] = \sum_m \sum_l \int_{-\infty}^{+\infty} e^{i\omega(t_1-t_2)}M_{jm}(t_1, \omega)M_{kl}^*(t_2, \omega)\Phi_{X_{S,ml}}(\omega)\, d\omega \tag{3.5}$$

where

$$M_{jm}(t, \omega) = \int_0^t A_m(t - \tau, \omega)h_{jm}(\tau)e^{-i\omega\tau}\, d\tau \tag{3.6}$$

and $\Phi_{X_{S,ml}}(\omega)$ stands for the cross spectrum of stationary processes $X_{S,m}(t_1)$ and $X_{S,l}(t_2)$.

The integrand of eqn (3.5) is newly defined to be the evolutionary cross spectrum (evolutionary cross-spectral density function).

$$\bar{\psi}_{jk}(t_1, t_2, \omega) = \sum_m \sum_l M_{jm}(t_1, \omega)M_{kl}^*(t_2, \omega)\Phi_{X_{S,ml}}(\omega) \tag{3.7}$$

It can be shown in the sequel that eqn (3.7) includes the conventional definition as a special case.

Taking the Taylor series of $A_m(t - \tau, \omega)$ in the neighborhood of $\tau = 0$

$$A_m(t - \tau, \omega) = \sum_{r=0}^{\infty} \frac{(-\tau)^r}{r!} A_m^{(r)}(t, \omega) \tag{3.8}$$

where $A_m^{(r)} = \partial^r A_m / \partial t^r$, and introducing the notation

$$\hat{H}(t, \omega) = \int_0^t h(\tau) e^{-i\omega\tau} \, d\tau \qquad (3.9)$$

eqn (3.5) yields

$$E[Y_j(t_1)Y_k(t_2)] = \sum_m \sum_l \int_{-\infty}^{+\infty} e^{i\omega}(t_1 - t_2) \sum_{r=0}^{\infty} \sum_{s=0}^{\infty} \frac{A_m^{(r)}(t_1, \omega)A_l^{(s)*}(t_2, \omega)}{r! \, s!}$$
$$\times \hat{H}_{jm}^{(r)}(t_1, \omega)\hat{H}_{kl}^{(s)*}(t_2, \omega)\Phi_{X_{s,ml}}(\omega) \, d\omega \qquad (3.10)$$

where

$$(-1)^r \, \partial^r \hat{H} / \partial \omega^r = (-1)^r \hat{H}^{(r)}(t, \omega) = \int_0^t \tau^r h(\tau) e^{-i\omega\tau} \, d\tau$$

For the single-degree-of-freedom problem, i.e. $j = k = 1$, $m = l = 1$ and $t_1 = t_2 = t$, this results in

$$E[Y^2(t)] = \int_{-\infty}^{+\infty} \sum_{r=0}^{\infty} \sum_{s=0}^{\infty} \frac{A^{(r)}(t, \omega)A^{(s)*}(t, \omega)}{r! \, s!}$$
$$\times \hat{H}^{(r)}(t, \omega)\hat{H}^{(s)*}(t, \omega)\Phi_{X_s}(\omega) \, d\omega \qquad (3.11)$$

Thus the evolutionary spectrum of $Y(t)$ is as follows,

$$\Psi_Y(t, \omega) = [A(t, \omega)A^*(t, \omega)\hat{H}(t, \omega)\hat{H}^*(t, \omega)$$
$$+ \dot{A}(t, \omega)A^*(t, \omega)\hat{H}'(t, \omega)\hat{H}^*(t, \omega)$$
$$+ A(t, \omega)\dot{A}^*(t, \omega)\hat{H}(t, \omega)\hat{H}'^*(t, \omega) + 0\{|(A(t, \omega)|^2\}] \times \Phi_{X_s}(\omega)$$
$$(3.12)$$

In eqn (3.12), if taking the first term only, i.e. to assume $\dot{A}(t, \omega) \simeq 0$, then

$$\Psi_Y(t, \omega) = |A(t, \omega)|^2 \, |\hat{H}(t, \omega)|^2 \, \Phi_{X_s}(\omega) \qquad (3.13)$$

This corresponds to the widely-used popular definition.[8] The evolutionary spectra of the input is simply given by,

$$\Psi_X(t, \omega) = |A(t, \omega)|^2 \, \Phi_{X_s}(\omega) \qquad (3.14)$$

For the expressions of eqns. (3.13) and (3.14) to be physically meaningful it is required that the process should be slowly varying with respect to time. However, for the expression of eqn (3.7) to be applicable such a condition is not necessary. The application of eqns

(3.13) and (3.14) to the airplane gust response was formulated by Howell and Lin.[9]

In the same manner, the Kth order statistics are derived as,

$$E[Y_{n_1}(t_1)Y_{n_2}(t_2) \ldots Y_{n_K}(t_K)] = \underbrace{\int_{-\infty}^{+\infty} \int_{-\infty}^{+\infty} \ldots \int_{-\infty}^{+\infty}}_{K\text{-1 fold}}$$

$$\times e^{i\{\omega_1 t_1 + \omega_2 t_2 + \ldots + \omega_{K-1} t_{K-1} - (\omega_1 + \omega_2 + \ldots + \omega_{K-1}) t_K\}}$$

$$\times \Psi_{n_1 n_2 \ldots n_K}(t_1, \omega_1; t_2, \omega_2; \ldots; t_{K-1}, \omega_{K-1}; t_K)$$

$$\times d\omega_1 d\omega_2 \ldots d\omega_{K-1} \tag{3.15}$$

The evolutionary Kth spectrum has the form of

$$\Psi_{n_1 n_2 \ldots n_K}(t_1, \omega_1; t_2, \omega_2; \ldots; t_{K-1}, \omega_{K-1}; t_K)$$

$$= \sum_{m_1} \sum_{m_2} \ldots \sum_{m_k} M_{n_1 m_1}(t_1, \omega_1) M_{n_2 m_2}(t_2, \omega_2) \ldots M_{n_{K-1} m_{K-1}}(t_{K-1}, \omega_{K-1})$$

$$\times M_{n_K m_K}(t_K, -\omega_1 - \omega_2 - \ldots - \omega_{K-1})$$

$$\times \Phi_{X_{s,m_1 m_2 \ldots m_K}}(\omega_1, \omega_2, \ldots, \omega_{K-1}) \tag{3.16}$$

where

$$1 \leq n_1, n_2, \ldots, n_K \leq n$$

$$1 \leq m_1, m_2, \ldots, m_K \leq n$$

n gives the dimensions of the system.

The application of eqn (3.7) to the airplane gust response will be illustrated. Airplane dynamics are modeled by an elastic free-free beam with an attached mass at its center. The response can be the displacement, the velocity, the acceleration, the force per unit area over the wing, the shear force, the moment and so forth.

An example of the acceleration response spectrum is shown in Fig. 2 where two modes, i.e. the rigid vertical motion and the first bending, are taken into account. The time-wise spectrum change can be clearly understood by this figure. Immediately after the airplane enters the gust, it responds as a rigid mass. As time passes, the flexible mode response eventually builds up.

Figure 3 shows the corresponding mean square response where the transient peak takes the highest value, 20% higher than the stationary response at infinite time. This failed to appear in the conventional stationary analysis.

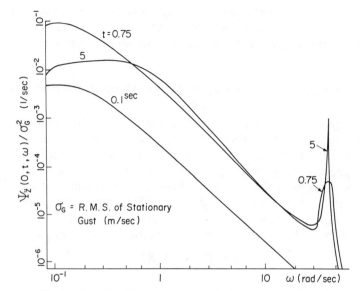

Fig. 2. Acceleration spectrum at airplane center of gravity.

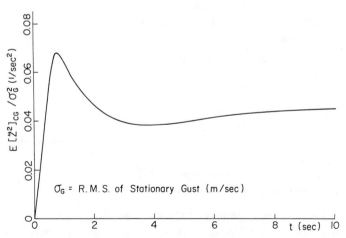

Fig. 3. Mean square acceleration at airplane center of gravity.

The examples of the swept-back wing modeled by three degrees of freedom[10-12] show almost the same results as Figs 2 and 3. In Refs 10–12, the pitching motion is included in an approximate sense.

4 ROTOR BLADE STABILITY IN TURBULENT FLOW

Rotor dynamics in the turbulent flow features the DE with random coefficients in homogeneous terms, which exhibits the distinct difference from the materials in the previous sections where randomness appeared only in the inhomogeneous term. Consequently our primary concern is directed toward the system stability prior to the discussion of the absolute response values. The new methodology will be proposed to cope with this type of problem in general.

Consider the linear DE with the parametric excitation in the form of

$$\{Z\} = [D]\{Z\} + [R]\{Z\} \qquad (4.1)$$

where

$\{Z\}$ = response vector
$[D]$ = determinate matrix
$[R]$ = stochastic matrix

The general procedure will be derived which can convert eqn (4.1) into the DEs of $\{E[Z]\}$ and $\{E[Z^2]\}$ by making use of the stochastic averaging and Ito differential rule.

Component-wise eqn, (4.1) can be rewritten as

$$\frac{d}{dt} Z_i = \sum_{j=1}^{N} D_{ij} Z_j + \sum_{j=1}^{N} R_{ij} Z_j \qquad i, j = 1, 2, \ldots, N \qquad (4.2)$$

where

D_{ij} = the (i, j) element of $[D]$
$R_{ij} = \sum_{l=1}^{M} \tau_{ij,l} \varepsilon_l$, the (i, j) element of matrix $[R]$
$\tau_{ij,l}$ = periodic coefficient of the lth random excitation in the (i, j) element of $[R]$
ε_l = a set of random excitations $l = 1, 2, \ldots, M$
M = number of random excitations

If ε_l is a broad band random excitation, eqn (4.2) can be regarded as a stochastic DE in the Stratonovich sense. The assumption that ε_l is a white noise reduces the stochastic averaging to be mere addition of the

Wong and Zakai correction term to eqn (4.2), resulting in a stochastic DE in Ito sense.

$$d\{Z\} = [\bar{B}]\{Z\}\,dt + [\sigma]\{dW\} \tag{4.3}$$

where

$$[\bar{B}] = [D] + [C] \tag{4.4}$$

$\{dW\}$ is a set of unit Wiener processes and the element of $[C]$ has the form of

$$C_{ij} = \pi \sum_{k}^{M} \sum_{m}^{M} \sum_{n}^{M} \tau_{ik,m} \tau_{kj,n} \Phi_{mn} \tag{4.5}$$

In eqn (4.5), Φ_{mn} expresses the cross spectrum and Φ_{mm} $(m = n)$ the autospectrum. When ε_l is not a white noise but has a broad band spectrum, the elements of $[C]$ can be derived through basically the same procedure. The characteristics of $[\sigma]$ will be needed in the form of $[\sigma\sigma^{\mathrm{T}}]$ whose element is described by

$$(\sigma\sigma^{\mathrm{T}})_{ij} = 2\pi \sum_{k}^{N} \sum_{l}^{N} \sum_{m}^{M} \sum_{n}^{M} \tau_{ik,m} \tau_{jl,n} \Phi_{mn} Z_k Z_l \tag{4.6}$$

The Ito differential rule is applied to the product of Z_i and Z_j, resulting in

$$d(Z_i Z_j) = \left[\sum_{k}^{N} (\bar{B}_{ik} Z_k Z_j + \bar{B}_{jk} Z_k Z_i) + (\sigma\sigma^{\mathrm{T}})_{ij} \right] dt$$
$$+ \sum_{m}^{M} (Z_j \sigma_{im} + Z_i \sigma_{jm})\,dW_m \tag{4.7}$$

Denoting $Z_i Z_j$ and $(\sigma\sigma^{\mathrm{T}})_{ij}$ by Y_l and S_l as follows

$$Y_l = Z_i Z_j \tag{4.8}$$

$$\bar{S}_l = (\sigma\sigma^{\mathrm{T}})_{ij} \tag{4.9}$$

and expressing the relationship between l and (i, j) in the form of

$$l = l(i, j) = N(i - 1) - \tfrac{1}{2}(j - 1)(j - 2) + j - i + 1 \tag{4.10}$$

when

$$1 \leq l \leq N(N + 1)/2 \quad \text{for} \quad j \geq i$$

it can be shown that

$$\tilde{S}_l = \sum_{m=1}^{N(N+1)/2} S_{lm} Y_m \tag{4.11}$$

Symmetry of $Z_i Z_j$ and $(\sigma\sigma^{\mathrm{T}})_{ij}$ with respect to i and j eliminates the need to consider $l(i, j)$ for $j \leqslant i$.

Rearranging the first and second terms of the right hand side of eqn (4.7) yields the lth row of the matrix $[A]$ as

$$\sum_{k=1}^{N} (\bar{B}_{ik} Z_k Z_j + \bar{B}_{jk} Z_k Z_i) = \sum_{k=1}^{N} (\bar{B}_{ik} Y_{l(k,j)} + \bar{B}_{jk} Y_{l(k,i)}) = \sum_{m=1}^{N(N+1)/2} A_{l(i,j),m} Y_m \tag{4.12}$$

Then eqn (4.7) can be rewritten as

$$\mathrm{d}\{Y\} = [[A] + [S]]\{Y\}\,\mathrm{d}t + [\tilde{\sigma}]\{\mathrm{d}W\} \tag{4.13}$$

Thus the DEs which govern the first and second moments are given by taking the expectations of eqns (4.3) and (4.7) respectively.

$$\frac{\mathrm{d}}{\mathrm{d}t}\{E[Z]\} = [[D] + [C]]\{E[Z]\} = [\bar{B}]\{E[Z]\} \tag{4.14}$$

$$\frac{\mathrm{d}}{\mathrm{d}t}\{E[Y]\} = [[A] + [S]]\{E[Y]\} = [\bar{\bar{B}}]\{E[Y]\} \tag{4.15}$$

Clearly the two eqns (4.14) and (4.15) are deterministic DEs, hence eigen values of matrices $[\bar{B}]$ and $[\bar{\bar{B}}]$ give the stability boundaries of the first and second moments. Their computation looks rather straightforward. When the periodic functions are involved, use may be made of the Floquet transition matrix method to find stability boundaries.

Figure 4 summarizes the procedure explained above and the conventional approach. The flow on the right column newly proposed uses only one assumption, that the input is a wide band random process. The left column, the conventional approach, requires the additional assumption that a very small parameter should exist in the physical DE so that the time averaging is possible.

The rotor blade dynamics problem is to be handled next as an example of application. Two linear cases of the flapping motion and the coupled flap–torsion motion are treated.[13,14] It is noticed that the atmospheric turbulence is specified by a random velocity. Usually three perpendicular velocity components are considered.

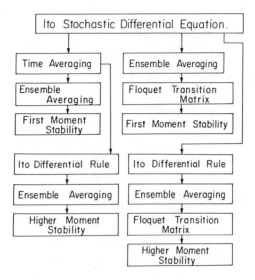

Fig. 4. Determination of moment stability conditions.

The numerical example of flapping motion stability boundaries is illustrated in Fig. 5 where the isotropic turbulences, one in the flight path direction, another in the lateral direction, are considered and no correlation exists between the two velocity components. Notations p, γ, μ and Φ stand for the flapping frequency normalized by blade angular velocity, Lock number, the advance ratio and the spectrum level respectively. The suffixes η and ξ indicate the flight direction and one perpendicular to it.

An example of the coupled flap–torsion motion is shown in Fig. 6 where unidirectional turbulence with various azimuth angles is considered. Here the turbulence is factored into two perpendicular velocity components by

$$\eta(t) = v(t) \cos \theta, \qquad \xi(t) = v(t) \sin \theta \qquad (4.16)$$

and spectra can be written as

$$\Phi_{\eta\eta} = \Phi_{vv} \cos^2 \theta, \qquad \Phi_{\xi\xi} = \Phi_{vv} \sin^2 \theta, \qquad \Phi_{\eta\xi} = \Phi_{vv} \sin \theta \cos \theta \qquad (4.17)$$

Figure 7 illustrates the meaning of the above two equations.

Figures 5 and 6 exhibit clearly that the instability region will expand

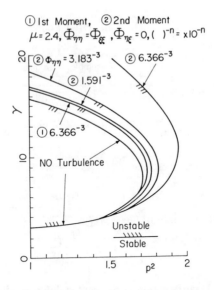

Fig. 5. First and second moment stability boundaries of flap motion.

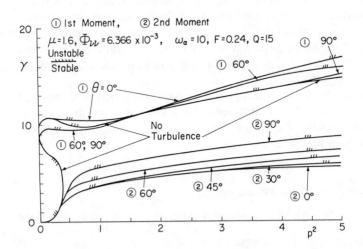

Fig. 6. First and second moment stability boundaries of coupled flap–torsion motion.

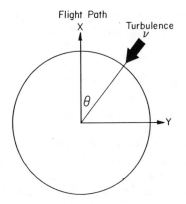

Fig. 7. Horizontal turbulence from arbitrary direction.

in the presence of the turbulence. The deviation from the deterministic problem (zero turbulence) can be well understood by the first and second moment stability boundaries. It is noted that the vertical turbulence velocity appears only in the inhomogeneous term of both the flapping and the coupled flap–torsion motions, and has nothing to do with the system stability.

Next an attempt[15] will be made to clarify the effect of three-dimensional atmospheric turbulence on the stability of the coupled flap–leadlag motion which features nonlinear differential equations with stochastic coefficients. The stability analysis will be carried out through the following steps: first the equilibrium point for the deterministic problem is determined based on the assumption that the random parametric excitation is infinitesimally small. Second the linear differential equation governing perturbations from the equilibrium solutions with random parametric excitations is derived. Finally the 1st and 2nd moment stabilities are examined by the general procedure mentioned previously.

To make clear the approximate method, the fundamentals will be presented. Consider the coupled flap–leadlag dynamics in the general form of a nonlinear DE as follows

$$\ddot{x} + g(x, \dot{x}) = f_1(t) + \{f_2^i(t) + F^i(x, \dot{x})\} R_i(t) \qquad (4.18)$$

where R_i are random functions and f_1 and f_2 the deterministic functions. i stands for the summation convention over i. The terms $F^i R_i$ express the parametric excitations, the sum of $f_1 + f_2^i R_i$ the purely external ones. The response solution of eqn (4.18) is assumed to be the equilibrium $\bar{x}$ and the perturbation Δx as in

$$x = \bar{x} + \Delta x \tag{4.19}$$

and the equilibrium state $\bar{x}$ can be approximated by the solution of the deterministic problem:

$$\ddot{x} + g(\bar{x}, \dot{x}) = f_1(t) \tag{4.20}$$

in which all random terms of eqn (4.18) are dropped. The search for the possible equilibrium will follow the harmonic balance method or any other methods so far as they can yield the admissible solutions that should be at least unbounded in the time-wise trend.

Once the equilibrium $\bar{x}$ is known, substitute eqn (4.19) into eqn (4.18) and expand g and F^i around $\bar{x}$, $\dot{\bar{x}}$ as follows.

$$g(\bar{x} + \Delta x, \dot{\bar{x}} + \Delta \dot{x}) = g(\bar{x}, \dot{\bar{x}}) + \frac{\partial}{\partial \dot{x}} g(x, \dot{x}) \bigg|_{\substack{x=\bar{x} \\ \dot{x}=\dot{\bar{x}}}} \Delta \dot{x}$$

$$+ \frac{\partial}{\partial x} g(x, \dot{x}) \bigg|_{\substack{x=\bar{x} \\ \dot{x}=\dot{\bar{x}}}} \Delta x + \text{higher order terms} \tag{4.21}$$

$$F^i(\bar{x} + \Delta x, \dot{\bar{x}} + \Delta \dot{x}) = F^i(\bar{x}, \dot{\bar{x}}) + \frac{\partial}{\partial \dot{x}} F^i(x, \dot{x}) \bigg|_{\substack{x=\bar{x} \\ \dot{x}=\dot{\bar{x}}}} \Delta \dot{x}$$

$$+ \frac{\partial}{\partial x} F^i(x, \dot{x}) \bigg|_{\substack{x=\bar{x} \\ \dot{x}=\dot{\bar{x}}}} \Delta x + \text{higher order terms} \tag{4.22}$$

Thus the differential equation for the perturbation Δx reads,

$$\Delta \ddot{x} + \frac{\partial}{\partial \dot{x}} \{g(x, \dot{x}) - R_i F^i(x, \dot{x})\} \bigg|_{\substack{x=\bar{x} \\ \dot{x}=\dot{\bar{x}}}} \Delta \dot{x} + \frac{\partial}{\partial x} \{g(x, \dot{x})$$

$$- R_i F^i(x, \dot{x})\} \bigg|_{\substack{x=\bar{x} \\ \dot{x}=\dot{\bar{x}}}} \Delta x = \{f_2^i(t) + F^i(\bar{x}, \dot{\bar{x}})\} R_i \tag{4.23}$$

Obviously this governing equation in Δx is a linear differential equation with random parametric excitations and also it is noticeable that both the equilibrium solution of the deterministic case and random inputs play a dual role of the purely external force and the parametric excitation to the motion. Equation (4.23) can be regarded as the most generalized expression which governs the behaviour of the perturbation Δx and is valid both for a scalar type and a vector type

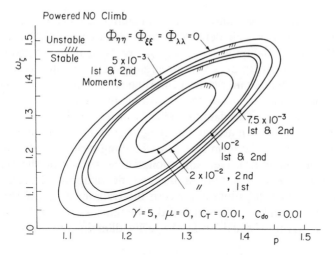

Fig. 8. First and second moment stability boundaries of coupled flap–leadlag motion.

Δx. Therefore it serves as a basic building block for the present application.

A numerical example is given in Fig. 8 where powered-no-climb flight is considered. It is seen that the instability domain will decrease under the presence of the turbulence. However, the turbulence level necessary to change instability region significantly seems unrealistically high, the conclusion drawn from this diagram could be a tentative statement that the turbulence might not aggravate the stability characteristics of this particular motion. Since the vertical turbulence velocity component appears also in the homogeneous terms, three-dimensional isotropic turbulence is considered in this problem.

The absolute response values under the high level turbulence would be a different matter which can be treated by a mathematically similar approach when the system is stable.

5 MODAL DAMPING COEFFICIENT OF A FLEXIBLE SYSTEM

This section deals with the derivation of the damping coefficient for a plate and its dispersion analysis. Also brief comments on the application to flexible spacecraft will be presented.

The important role played by a damping coefficient in flexible spacecraft design is well delineated in a few treatises.[16-18] Emphatically, the modal damping coefficient in terms of its mean value and dispersion turns out to be a very programmatic parameter as well as a technical one through all phases of flexible spacecraft development.

It is appropriate to make some introductory remarks on the basic nature of damping. The damping effect can be roughly classified into three categories: (i) material damping, (ii) interface damping, (iii) ambient damping. The term structural damping includes both material and interface dampings. Here attention will be focused on the material damping somewhat inherent in the specific material. Theorems of viscoelasticity which primarily concern the transverse thermal current in structural members allow us to derive the loss factor ζ from the complex modulus of elasticity E_r or specific damping energy density D_d as

$$\zeta = E_2/E_1 = ED_d/\pi\sigma_d^2 = \delta/\pi = 2\eta \qquad (5.1)$$

where

E = modulus of elasticity
$E_r = E_1 + iE_2$, complex modulus of elasticity
δ = logarithmic decrement
η = damping ratio, c_e/c_c
c_e = equivalent dashpot coefficient
c_c = critical dashpot coefficient, $2\sqrt{mk}$
σ_d = peak stress

Thus it is essential to obtain E_r and D_d by theoretical derivation.

The problem formulation[17] begins with the equation of the plate motion in the transverse direction as

$$D\Delta\Delta w + \frac{E\alpha}{1-v}\int_{-h}^{h}\Delta Tz\,dz + 2\rho h\frac{\partial^2 w}{\partial z^2} = 0 \qquad (5.2)$$

where

D = plate rigidity
w = deflection in the thickness direction
E = Young's modulus
α = thermal expansion coefficient
T = temperature
$\Delta = \dfrac{\partial^2}{\partial x^2} + \dfrac{\partial^2}{\partial y^2}$
$2h$ = plate thickness
ρ = plate material density

The energy equation is written as

$$kT_{,ii} = \rho C_E(\dot{T} + \tau_0 \ddot{T}) + (3\lambda + 2\mu)\alpha T_0(\dot{\varepsilon}_{jj} + \tau_0 \ddot{\varepsilon}_{jj}) \qquad (5.3)$$

where

C_E = specific heat
ε_{jj} = volume strain
k = thermal conductivity
λ, μ = Lamé elastic constants
T_0 = stress-free (equilibrium) temperature
τ_0 = relaxation time

Assuming that $k \, \partial^2 T/\partial x^2$ and $k \, \partial^2 T/\partial y^2$ are negligibly small compared to $k \, \partial^2 T/\partial z^2$, using the plane strain condition and introduction of normalized quantities: $\bar{x} = x/a$, $\bar{y} = y/a$, $\bar{z} = z/h$, $\bar{w} = w/h$, $\bar{t} = t/\tau$, $\bar{\tau} = a^2 \alpha T/h^2$, lead to the following modified forms of the motion and energy equations

$$\bar{\Delta}\bar{\Delta}\bar{w} + \frac{3(1+\nu)}{2} \int_{-1}^{1} \bar{\Delta}\bar{T}\bar{z} \, d\bar{z} + \frac{1}{\bar{\tau}^2}\frac{\partial^2 \bar{w}}{\partial \bar{t}^2} = 0 \qquad (5.4)$$

$$\frac{\partial^2 \bar{T}}{\partial \bar{z}^2} = \frac{\partial \bar{T}}{\partial \bar{t}} + \gamma \frac{\partial^2 \bar{T}}{\partial \bar{t}^2} - \frac{\beta}{1-2\nu}\left[\frac{\partial \bar{\Delta}\bar{w}}{\partial \bar{t}} + \gamma \frac{\partial^2 \bar{\Delta}\bar{w}}{\partial \bar{t}^2}\right]\bar{z} \qquad (5.5)$$

where

$$\beta = \frac{E\alpha^2 T_0}{\rho C_E}, \qquad \tau = \frac{\rho C_E h^2}{k}, \qquad \tau_0 = \frac{3k}{s^2 \rho C_E}$$

$$\bar{\tau} = \omega_0 \tau, \qquad \gamma = \tau_0/\tau$$

$$\bar{\Delta} = a^2 \Delta, \qquad 1/\omega_0 = a^2(2\rho h/D)^{1/2}$$

and

a = plate length
s = material sound velocity

Now the solutions of eqns (5.4) and (5.5) need to be in the following forms

$$\bar{w} = W(\bar{x}, \bar{y})e^{i\bar{\omega}\bar{t}} \qquad (5.6)$$

$$\bar{T} = \Theta(\bar{x}, \bar{y}, \bar{z})e^{i\bar{\omega}\bar{t}}$$

Then, the equations for W and Θ turn out to be

$$\bar{\Delta}\bar{\Delta}W + \frac{3(1+v)}{2}\int_{-1}^{1}\bar{\Delta}\Theta\bar{z}\,d\bar{z} - \left(\frac{\bar{\omega}}{\bar{\tau}}\right)^{2}W = 0 \tag{5.7}$$

$$\frac{\partial^{2}\Theta}{\partial\bar{z}^{2}} = \xi^{2}\Theta - \frac{\beta}{1-2v}\xi^{2}\bar{\Delta}W\bar{z} \tag{5.8}$$

with $\xi^{2} = i\bar{\omega} - \gamma\bar{\omega}^{2}$.

The thermal boundary conditions are given by

$$\frac{\partial\Theta}{\partial\bar{z}} \pm B_{i}\Theta = 0 \quad \text{at} \quad \bar{z} = \pm 1 \tag{5.9}$$

where $B_{i} = hH/k$, is the Biot number and H is the boundary conductance.

The solution of eqns (5.8) and (5.9) is obtained as

$$\Theta = \frac{\beta}{1-2v}\left[\bar{z} - (1+B_{i})\frac{\sinh\xi\bar{z}}{\xi\cosh\xi + B_{i}\sinh\xi}\right]\bar{\Delta}W \tag{5.10}$$

Substituting this expression and carrying out the integration yield the following equation for W

$$(1+g)\bar{\Delta}\bar{\Delta}W - \left(\frac{\bar{\omega}}{\bar{\tau}}\right)^{2}W = 0 \tag{5.11}$$

Therefore, the thermomechanical coupling function g is given by

$$g = \frac{1+v}{1-2v}\beta\left[1 - 3(1+B_{i})\frac{\xi - \tanh\xi}{\xi^{2}(\xi + B_{i}\tanh\xi)}\right] \tag{5.12}$$

The imaginary part of the above g gives the thermoelastic damping coefficient, i.e.

$$\eta = \text{Im}(g) \tag{5.13}$$

In the plane stress problem, the multiplier $(1+v)/(1-2v)$ in eqn (5.12) should be replaced by $(1+v)/(1-v)$.

It is appropriate to make some comments on the derivation and the result. First, the thermoelastic damping coefficient of a plate has the identical form as one for a beam apart from the multiplier determined by Poisson's ratio. Secondly, as is the case for the beam problem, the thermoelastic coupling arises from the temperature gradient across the plate thickness. If the temperature remains constant everywhere, the thermoelastic coupling never exists. Thus this type of modeling reflects little correlation between the shear deformation and the damping. In a very simplistic explanation the present analysis raises the damping factor by taking into account the bending type motion of the plate. The dispersion of the damping coefficient can be obtained by eqn (5.13) considering that η is a function of probabilistic parameters,

$$\eta = F(X_1, X_2, \ldots, X_N) \tag{5.14}$$

The Taylor expansion around the mean values of X_i yields

$$\eta - \bar{\eta} \cong \sum \frac{\partial F}{\partial X_i}\bigg|_{X_i = \bar{X}_i} (X - \bar{X}_i) \tag{5.15}$$

where the bar indicates the mean.

Denoting $\eta - \bar{\eta} = \eta'$ and $X_i - \bar{X}_i = X_i'$ as the perturbations around the mean and taking the expectation of the square of eqn (5.15) result in the following

$$E[\eta'^2] \cong \sum_i \sum_j \frac{\partial F}{\partial X_i}\bigg|_{X_i = \bar{X}_i} \frac{\partial F}{\partial X_j}\bigg|_{X_j = \bar{X}_j} E[X_i' X_j'] \tag{5.16}$$

If no correlation exists among the parameters, eqn (5.16) can be written simply as

$$E[\eta'^2] = \sum_i \left\{ \frac{\partial F}{\partial X_i}\bigg|_{X_i = \bar{X}_i} \right\}^2 E[X_i'^2] \tag{5.17}$$

Obviously X_i are considered here to be plate dimensions, physical and thermal properties of its material.

Next it will be seen how the theoretical result mentioned above can be applied to the engineering problem. Equation (5.13) exhibits the parabolic profile versus frequency which has the maximum peak at a certain frequency as exemplified in Fig. 9. The four data points are taken from the experiment with the honeycomb sandwich panel, called the substrate.[16] This substrate comprises a major part of the solar array paddle structure of a Marine Observation Satellite (MOS)-1. (This spacecraft was launched on February 19, 1987 and is in operation.) The data can be divided into two groups, thus two curve-fittings are done separately for them, which are drawn as solid lines. The dotted lines show the dispersion in the sense of the root mean square, i.e. the root of eqn (5.17). Since the analytical formula of eqn (5.13) does not take into account the inhomogeneity of the honeycomb plate, the curve fitting should be understood as one of approximation in the discussion.

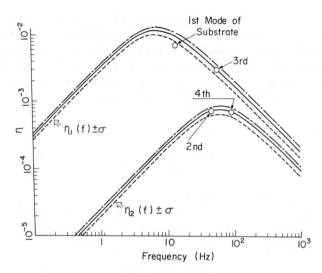

Fig. 9. Theoretical damping coefficient and its dispersion.

The two curves in Fig. 9 can be utilized to predict the modal damping coefficients of the entire solar array paddle structure.[16] The possible scatter band computed by eqn (5.17) may serve as a preliminary guideline to evaluate the capability of the attitude and orbit control subsystem of the spacecraft.[17] Regarding the computation of the system reliability the author proposed that the system reliability should be represented by just one probability that considers everything during the mission life. Otherwise many probabilities would become necessary to describe different aspects of various hazards and eventually it would become controversial to determine which one is the most important.

However, it remains an unsettled question how to establish a proper risk assessment model for the total system out of the many parameters involved, conventional system reliability calculus, the probability of a hit by a micrometeoroid or man-made debris, the survival probability after a collision and so forth.

ACKNOWLEDGEMENT

This article celebrates the sixty-fifth birthday of Professor Y. K. Lin who has achieved pioneering researches in stochastic structural dynamics and bestowed his warm discipline on so many students. His encouragement and guidance has made the author what the author is today. Being one of his students remains the author's utmost pride and honour for good. Also the author's sincere thanks are due to Professor S. T. Ariaratnam who kindly invited the author to write this article.

REFERENCES

1. Lin, Y. K., *Probabilistic Theory of Structural Dynamics*. McGraw-Hill, New York, 1967.
2. Tack, D. H. & Lambert, R. F., Response of bars and plates to boundary layer turbulence. *J. Aerospace Sci.*, March 1962, 311–22.
3. Fujimori, Y., Response analysis of the plate to boundary layer pressure fluctuations. *Proceedings of the 19th Japan National Congress for Applied Mechanics*. University of Tokyo Press, 1969, pp. 121–7.
4. Fujimori, Y., Response analysis of a simply-supported plate to boundary layer pressure fluctuations. NAL TR-207, September 1970.
5. Fujimori, Y., Response analysis of a clamped plate to boundary layer pressure fluctuations. NAL TR-401, January 1975.
6. Lin, Y. K. & Maekawa, S., Decomposition of turbulence forcing field and structural response. *AIAA J.*, **15** (5) (May 1977) 609–10.
7. Fujimori, Y., Spectrum analysis of the response of multimodal linear system to nonstationary input. *Proceedings of the 23rd Japan National Congress for Applied Mechanics*. University of Tokyo Press, 1973, pp. 467–75.
8. Priestly, M. B., Evolutionary spectra and nonstationary processes, *J. of the Royal Statistical Society, Series B,* **27** (1–3) (1965) 204–37.
9. Howell, L. J. & Lin, Y. K., Response of flight vehicles to nonstationary random atmospheric turbulence, *AIAA J.,* **9** (11) (1971) 2201–7.
10. Fujimori, Y. & Lin, Y. K., Analysis of airplane response to nonstationary turbulence including wing bending flexibility. *AIAA J.,* **11** (3) (March 1973) 334–9.
11. Fujimori, Y. & Lin, Y. K., Analysis of airplane response to nonstationary turbulence including wing bending flexibility II. *AIAA J.,* **11** (9) (September 1973) 1343–5.
12. Fujimori, Y., Shear and moment response of the airplane wing to nonstationary turbulence. *AIAA J.,* **12** (11) (November 1974) 1459–60.
13. Lin, Y. K., Fujimori, Y. & Ariaratnam, S. T., Rotor blade stability in turbulent flows Part I. *AIAA J.,* **17** (6) (June 1979), 545–52.
14. Fujimori, Y., Lin, Y. K. & Ariaratnam, S. T., Rotor blade stability in turbulent flows Part II. *AIAA J.,* **17** (7) (July 1979) 673–8.
15. Fujimori, Y., The effect of the atmospheric turbulence on the rotor blade flap-leadlag motion stability in hovering. AIAA paper 81-0610.

16. Fujimori, Y., Kato, J. & Toda, S., Modal damping estimates of MOS-1 Solar Array Paddle. *Acta Astronautica,* **11** (10/11) (1984) 633–40 (IAF-83-415).
17. Toda, S. & Fujimori, Y., Structural damping factors in flexible spacecraft design, Structural Safety and Reliability, Vol. 1. *Proceedings of ICOSSAR 85,* May 1985, pp. 357–64.
18. Fujimori, Y., Kato, J., Motohashi, S., Kuwao, F. & Sekimoto, S., Modal damping measurement of MOS-1 Solar Array Paddle. Preprint of Automatic Control in Space, 10th IFAC Symposium 1985, pp. 125–9.

6

Response of Linear Systems to Random Series and Filtered Poisson Processes

MIRCEA GRIGORIU

Department of Structural Engineering, Cornell University, Ithaca, New York, USA

ABSTRACT

Two methods are developed for calculating mean upcrossing rates and other probabilistic descriptors for the response of linear systems subject to random time series and filtered Poisson processes. The first method is based on the joint characteristic function of the response and its derivative. It is exact but complex. The second method is based on the assumption that the response can be represented by a memoryless nonlinear transformation of a Gaussian process. It involves only the marginal distribution and the covariance function of the response. Response characteristics by this method compare satisfactorily with simulation results.

1 INTRODUCTION

Wave, wind, seismic, and bridge traffic loads can be modeled by independent or correlated random series with constant time step and filtered Poisson processes.[5,6,7,9,12] The response of linear systems to such load models has been studied extensively and methods have been developed for estimating moments of the response.[7,9–12] Probabilistic characteristics of the peak response, such as the mean rate at which the response process crosses with positive slope (upcrosses) a limit level, have received little attention.

This paper presents an exact and an approximate method for

calculating mean crossing rates and other probabilistic characteristics for the response of linear systems subject to random series and filtered Poisson processes. The exact method is based on the characteristic function of the response vector $\{X(t), \dot{X}(t) = dX(t)/dt\}$. This function can be inverted by Fast Fourier Transform algorithms to obtain the joint density function $f_{X,\dot{X}}(x, \dot{x}; t)$ of the response at any time t. The mean upcrossing rate of level x of $X(t)$ at time t can be obtained from.[2]

$$v_X(x; t) = \int_0^\infty \dot{x} f_{X,\dot{X}}(x, \dot{x}; t) \, d\dot{x} \tag{1}$$

It is independent of time for stationary responses. The approximate method is based on the assumption that $X(t)$ has the same mean upcrossing rate as the translation process[4]

$$X_T(t) = F_X^{-1}(\Phi(Z(t))) \tag{2}$$

in which F_X = the first order distribution of $X(t)$, Φ = the distribution of the standard Gaussian variable, and $Z(t)$ = a stationary Gaussian process with zero mean and unit variance. The covariance function of $Z(t)$ is so chosen that $X(t)$ and $X_T(t)$ have the same covariance functions in the vicinity of the origin. Then, the variance of $\dot{Z}(t) = dZ(t)/dt$ is[4]

$$\sigma_{\dot{Z}}^2 = \sigma_{\dot{X}}^2 \frac{1}{\displaystyle\int_{-\infty}^\infty \frac{\phi(\alpha) \, d\alpha}{\{f_X(F_X^{-1}(\Phi(\alpha)))\}^2}} \tag{3}$$

in which $\sigma_{\dot{X}}^2$ = the variance of $\dot{X}(t)$ and $\phi(\alpha) = \exp(-\alpha^2/2)/\sqrt{2\pi}$. The mean upcrossing rate of level x of $X(t)$ can be approximated by[1,4]

$$v_X^*(x) = \frac{\sigma_{\dot{Z}}}{\sqrt{2\pi}} \phi(\Phi^{-1}(F_X(x))) \tag{4}$$

This approximation depends only on the marginal distribution of $X(t)$ and the variance of $\dot{X}(t)$.

Mean upcrossing rates are calculated exactly and approximately from eqns (1) and (4) for simple oscillators subject to Gaussian and non-Gaussian random series and filtered Poisson processes. Simulation studies are used to evaluate the approximate mean upcrossing rate in eqn (4). Some of the proposed techniques are also applied to determine moments of the response.

2 RESPONSE TO CORRELATED TIME SERIES

Consider a linear system with unit impulse response function $h(\alpha)$ which is subject to a stationary series $Y(t)$ with time step Δ, mean m, variance σ^2, first order density f, and first order distribution F. The series takes on constant values Y_k during each time step Δ. The values of the series in different time steps can be independent or dependent. It is assumed that the system is initially at rest so that it has the response

$$X(t) = \int_0^t h(t - \tau) Y(\tau)\, d\tau \tag{5}$$

or, for $t = n\Delta$,

$$X(t) = \sum_{k=1}^n h_k Y_k \tag{6}$$

in which

$$h_k = \int_{(k-1)\Delta}^{k\Delta} h(t - \tau)\, d\tau \tag{7}$$

The first derivative of the response is

$$\dot{X}(t) = \sum_{k=1}^n \dot{h}_k Y_k \tag{8}$$

in which

$$\dot{h}_k = \int_{(k-1)\Delta}^{k\Delta} \frac{d}{dt}[h(t - \tau)]\, dr \tag{9}$$

From eqns (6) and (8), the variances of $X(t)$ and $\dot{X}(t)$ and the correlation coefficient $\rho_{X,\dot{X}}$ between these variables at $t = n\Delta$ are

$$\sigma_X^2 = \sigma^2 \sum_{k,l=1}^n h_k h_l \rho_{kl} \tag{10}$$

$$\sigma_{\dot{X}}^2 = \sigma^2 \sum_{k,l=1}^n \dot{h}_k \dot{h}_l \rho_{kl} \tag{11}$$

and

$$\rho_{X,\dot{X}} = \frac{\sigma^2}{\sigma_X \sigma_{\dot{X}}} \sum_{k,l=1}^n h_k \dot{h}_l \rho_{kl} \tag{12}$$

in which ρ_{kl} = the correlation coefficient between Y_k and Y_l.

This section develops recurrence formulas for the characteristic functions of the response $X(t)$ and the response vector $\{X(t), \dot{X}(t)\}$ for several input series $\{Y_k\}$. These characteristic functions can be inverted by Fast Fourier Transform algorithms to obtain the densities of $X(t)$ and $\{X(t), \dot{X}(t)\}$. From eqns (1)–(4), these densities can be used to calculate mean upcrossing rates exactly and approximately for the response of linear systems. Resultant characteristic functions and the densities can also be used to determine moments of the response.

2.1 Gauss–Markov Excitations
Let $Y_k = m + \sigma \tilde{Y}_k$ with[5]

$$\tilde{Y}_k = \rho \tilde{Y}_{k-1} + \sqrt{1 - \rho^2}\, W_k \tag{13}$$

in which $\{W_k\}$ are independent standard Gaussian variables. It can be shown that $\{Y_k\}$ is stationary for large values of k and follows a Gaussian distribution with mean m, variance σ^2, and correlation ρ^p between Y_k and Y_{k+p}. From eqns (6) and (8), $X(t)$ and $\dot{X}(t)$ are independent Gaussian variables for large values of t. The mean upcrossing rate of the stationary response is [2]

$$\nu_X(x) = \frac{\sigma_{\dot{X}}}{2\pi\sigma_X} \exp\left[-\frac{1}{2}\left(\frac{x - m_X}{\sigma_X}\right)^2\right] \tag{14}$$

in which the mean

$$m_X = m \sum_{k=1}^{n} h_k$$

The standard deviations σ_X and $\sigma_{\dot{X}}$ can be obtained from eqns (10) and (11) for $\rho_{kl} = \rho\,|k - l|$.

2.2 Markov Excitation
Consider the series[8]

$$Y_k = V_k Y_{k-1} + (1 - V_k)W_k \tag{15}$$

in which $\{V_k\}$, $\{W_k\}$ and Y_0 are independent random variables. The variables $\{W_k\}$ and Y_0 follow a distribution F with mean m and variance σ^2 while the variables $\{V_k\}$ take on the values 1 and 0 with the probabilities ρ and $1 - \rho$. The series $\{Y_k\}$ is a Markov chain with marginal distribution F. It coincides with Y_0 and W_l, $l = 1, 2, \ldots, k$, with probabilities ρ^k and $(1 - \rho)\rho^{k-1}$, respectively. The correlation coefficient between Y_k and Y_{k+p} is equal to ρ^p, as for the Gauss–Markov series in eqn (13).

Let

$$\psi_n(u; h_1, h_2, \ldots, h_n) = E\left[\exp\left(iu \sum_{k=1}^{n} h_k Y_k\right)\right] \qquad (16)$$

be the characteristic function of $X(t)$ in eqn (6) and $i = \sqrt{-1}$. Since Y_n coincides with Y_{n-1} or W_n with probability ρ or $1 - \rho$,

$$\psi_n(u; h_1, h_2, \ldots, h_{n-1}, h_n) = \rho\psi_{n-1}(u; h_1, h_2, \ldots, h_{n-2}, h_{n-1} + h_n)$$
$$+ (1 - \rho)\psi_{n-1}(u; h_1, h_2, \ldots, h_{n-2}, h_{n-1})\psi(uh_n) \qquad (17)$$

for $n > 2$. In this equation, ψ denotes the characteristic function of f and F so that, from eqn (16), $\psi_1(u; h) = \psi(uh)$.

The determination of the characteristic function of the response at $t = n\Delta$ and any value of u involves a relatively large number of calculations. It requires n evaluations of ψ_1 $(u; \sum_{k=1}^{K} h_k)$ for $K = 1, 2, \ldots, n$ and $n(n - 1)/2$ applications of eqn (17). However, the computation is not excessive even for relatively large values of n because eqn (17) involves elementary operations.

A similar recurrence formula can be developed for the joint characteristic function of $\{X(t), \dot{X}(t)\}$. From eqns (6) and (8),

$$\psi_n(u, v; h_1, \ldots, h_n, \dot{h}_1, \ldots, \dot{h}_n)$$
$$= \rho\psi_{n-1}(u, v; h_1, \ldots, h_{n-2}, h_{n-1} + h_n, \dot{h}_1, \ldots, \dot{h}_{n-2}, \dot{h}_{n-1} + \dot{h}_n)$$
$$+ (1 - \rho)\psi_{n-1}(u, v; h_1, \ldots, h_{n-1}, \dot{h}_1, \ldots, \dot{h}_{n-1})\psi(uh_n + v\dot{h}_n) \qquad (18)$$

for $n \geq 2$, in which

$$\psi_n(u, v; h_1, \ldots, h_n, \dot{h}_1, \ldots, \dot{h}_n) = E\left\{\exp\left[i\sum_{k=1}^{n} (uh_k + v\dot{h}_k)Y_k\right]\right\} \qquad (19)$$

and $\psi_1(u, v; a, b) = \psi(ua + vb)$. However, the determination of the joint characteristic function of $X(t)$ and $\dot{X}(t)$ in eqn (19) and the corresponding joint distribution of the response involves numerous computations.

The analysis simplifies significantly when $\{Y_k\}$ is an independent series $(\rho = 0)$. From eqns (16) and (19), the characteristic functions of $X(t)$ and $\{X(t), \dot{X}(t)\}$ are in this case

$$\psi_n(u; h_1, \ldots, h_n) = \prod_{k=1}^{n} \psi(uh_k) \qquad (20)$$

and

$$\psi_n(u, v; h_1, \ldots, h_n, \dot{h}_1, \ldots, \dot{h}_n) = \prod_{k=1}^{n} \psi(uh_k + v\dot{h}_k) \qquad (21)$$

These results could have been obtained directly from eqns (6) and (8).

The characteristic functions in eqns (16) and (19) or eqns (20) and (21) can be used to obtain moments of $X(t)$ and $\{X(t), \dot{X}(t)\}$. These moments can be employed to develop approximations for the distribution and the mean upcrossing rate of the response.

2.3 Non-ergodic Excitation
Let

$$Y_k = U_0 + U_k \qquad (22)$$

in which $\{U_k\}$, $k = 0, 1, \ldots$ are independent random variables following the densities f_k, distributions F_k, and characteristic functions ψ_k. The series is not ergodic even when $\{U_k\}$, $k \geq 1$, are identically distributed. It can be used, for example, to model independent series with uncertain means.

The joint characteristic function of $\{X(t), \dot{X}(t)\}$ at $t = n\Delta$ is, from eqns (6) and (8),

$$\psi_n(u, v; h_1, \ldots, h_n, \dot{h}_1, \ldots, \dot{h}_n)$$
$$= \psi_0\left(u \sum_{k=1}^{n} h_k + v \sum_{k=1}^{n} \dot{h}_k\right)\prod_{k=1}^{n} \psi_k(uh_k + v\dot{h}_k) \qquad (23)$$

It gives the characteristic function $\psi_n(u; h_1, \ldots, h_n)$ for $v = 0$.

The densities of $\{X(t), \dot{X}(t)\}$ and $X(t)$ can be obtained by inverting the characteristic functions of these responses. From eqns (1)–(4), mean upcrossing rates of any level x of $X(t)$ can then be determined. These functions can also be obtained from eqn (14) when $\{Y_k\}$ is a Gaussian series.

3   RESPONSE TO FILTERED POISSON PROCESSES

Consider the filtered Poisson process

$$Y(t) = \sum_{k=1}^{N(t)} Y_k w(t - \tau_k) \qquad (24)$$

in which $\tau_k =$ the times of occurrence of inhomogeneous Poisson events with mean rate $\lambda(\alpha)$, $N(t) =$ the number of such events in $(0, t)$, $\{Y_k\} =$ independent identically distributed random variables with distribution F, density f, and characteristic function ψ, and $w(\alpha) =$ a shape function which is zero for $\alpha < 0$.

The response of a linear system to the excitation in eqn (24) is, from eqn (5),

$$X(t) = \sum_{k=1}^{N(t)} Y_k \int_0^t h(t - \tau)w(\tau - \tau_k)\, d\tau \qquad (25)$$

and has the derivative

$$\dot{X}(t) = \sum_{k=1}^{N(t)} Y_k \int_0^t \dot{h}(t - \tau)w(\tau - \tau_k)\, d\tau \qquad (26)$$

in which $\dot{h}(\alpha) = dh(\alpha)/d\alpha$. The joint characteristic function of $\{X(t), \dot{X}(t)\}$ is the expectation of $\exp\{i[uX(t) + v\dot{X}(t)]\}$ and has the expression

$$\psi_{X,\dot{X}}(u, v) = E\left[i \sum_{k=1}^{N(t)} g(t, \tau_k)Y_k \right] \qquad (27)$$

in which

$$g(t, \alpha) = u \int_0^t h(t - \tau)w(\tau - \alpha)\, d\tau + v \int_0^t \dot{h}(t - \tau)w(\tau - \alpha)\, d\tau \qquad (28)$$

Following similar arguments as in Ref. 11 (pp. 153–5), one finds that

$$\psi_{X,\dot{X}}(u, v) = \exp\left\{ \int_0^t [\psi(g(t, \alpha)) - 1]\lambda(\alpha)\, d\alpha \right\} \qquad (29)$$

Letting $v = 0$, eqn (29) yields the characteristic function of $X(t)$. This function has the expression

$$\Psi_X(u) = \exp\left\{ \int_0^t \left[\psi\left(u \int_0^t h(t - \tau)w(\tau - \alpha)\, d\tau \right) - 1 \right] \lambda(\alpha)\, d\alpha \right\} \qquad (30)$$

The characteristic functions in eqns (29) and (30) can be inverted to determine the densities of $X(t)$ and $\{X(t), \dot{X}(t)\}$ and, from eqns (1)–(4), calculate mean crossing rates for the response process in eqn (25). These functions can be used to determine moments of the response.[12,13] Response moments can also be obtained directly from eqns (25) and (26).

4 NUMERICAL RESULTS

Mean upcrossing rates and other descriptors are determined for the response of a simple oscillator to random series and filtered Poisson processes. The unit impulse response functions of the oscillator is

$$h(\alpha) = \begin{cases} 0, & \alpha \leq 0 \\ \dfrac{e^{-\zeta\omega_0\alpha}}{\omega_1}\sin(w_1\alpha), & \alpha \geq 0 \end{cases} \tag{31}$$

in which $\omega_0 =$ the natural frequency of the oscillator, $\zeta =$ the damping ratio, and $\omega_1 = \omega_0\sqrt{1 - \zeta^2}$.

4.1 Response to Random Series

Table 1 gives mean upcrossing rates of level $\bar{x} = x/\sigma_X = 2$ and 3 for the response of time $t = n\Delta$ of a simple oscillator with natural frequency $\omega_0 = 1$ rad/s and damping ratio $\zeta = 5\%$ subject to random series $\{Y_k\}$ in eqns (13), (15) and (22) for $n = 100$ and $\Delta = 2\pi/(10\omega_0)$. The excitations in eqns (13) and (22) are Gaussian series with zero mean and unit variance so that the mean upcrossing rate $v_X(x)$ can be obtained exactly from eqn (14). The Markov excitation in eqn (15) is not a Gaussian series even when $\{W_k\}$ follows a Gaussian distribution. Thus, eqn (14) cannot be applied to calculate mean upcrossing rates for $X(t)$. The mean upcrossing rate of the response to this excitation has been calculated approximately from eqns (4), (16) and (17) for variables W_k following standard Gaussian and bilateral exponential probabilities. The bilateral exponential model has the following density, distribution, and characteristic function:[3]

$$f(y) = \tfrac{1}{2}\exp(-|y|) \tag{32}$$

$$F(y) = \begin{cases} \tfrac{1}{2}e^{y}, & y < 0 \\ 1 - \tfrac{1}{2}e^{-y}, & y > 0 \end{cases} \tag{33}$$

and

$$\psi(u) = \frac{1}{1 + u^2} \tag{34}$$

The small differences between the mean upcrossing rates in the table should not be misleading. These mean rates correspond to the same standardized thresholds $\bar{x} = x/\sigma_X$. However, the corresponding values

Table 1
Response of simple oscillators to random series

Excitation		Mean upcrossing rates		
		$\rho = 0 \cdot 0$	$\rho = 0 \cdot 8$	$\rho = 0 \cdot 95$
$\bar{x} = 2 \cdot 0$				
Gauss–Markov series (eqns (13) and (14))		$2 \cdot 15 \times 10^{-2}$	$1 \cdot 90 \times 10^{-2}$	$1 \cdot 44 \times 10^{-2}$
Non-ergodic Gaussian series (eqns (14) and (22))		$2 \cdot 15 \times 10^{-2}$	$1 \cdot 45 \times 10^{-2}$	$8 \cdot 30 \times 10^{-3}$
Markov series with Gaussian marginal distribution	Approximation (eqns (4), (16) and (17))	$2 \cdot 10 \times 10^{-2}$	$2 \cdot 11 \times 10^{-2}$	$1 \cdot 67 \times 10^{-2}$
	Simulation	$2 \cdot 00 \times 10^{-2}$	$2 \cdot 18 \times 10^{-2}$	$1 \cdot 69 \times 10^{-2}$
Markov series with bilateral exponential marginal distribution	Approximation (eqns (4), (16) and (17))	$2 \cdot 20 \times 10^{-2}$	$2 \cdot 18 \times 10^{-2}$	$1 \cdot 71 \times 10^{-2}$
	Simulation	$2 \cdot 08 \times 10^{-2}$	$1 \cdot 96 \times 10^{-2}$	$1 \cdot 67 \times 10^{-2}$
$\bar{x} = 3 \cdot 0$				
Gauss–Markov series (eqns (13) and (14))		$1 \cdot 77 \times 10^{-3}$	$1 \cdot 56 \times 10^{-3}$	$1 \cdot 18 \times 10^{-3}$
Non-ergodic Gaussian series (eqns (14) and (22))		$1 \cdot 77 \times 10^{-3}$	$1 \cdot 19 \times 10^{-3}$	$6 \cdot 81 \times 10^{-4}$
Markov series with Gaussian marginal distribution	Approximation (eqns (4), (16) and (17))	$1 \cdot 77 \times 10^{-3}$	$3 \cdot 30 \times 10^{-3}$	$3 \cdot 85 \times 10^{-3}$
	Simulation	$9 \cdot 58 \times 10^{-4}$	$2 \cdot 41 \times 10^{-3}$	$3 \cdot 32 \times 10^{-3}$
Markov series with bilateral exponential marginal distribution	Approximation (eqns (4), (16) and (17))	$2 \cdot 28 \times 10^{-3}$	$5 \cdot 11 \times 10^{-3}$	$5 \cdot 65 \times 10^{-3}$
	Simulation	$1 \cdot 24 \times 10^{-3}$	$3 \cdot 20 \times 10^{-3}$	$4 \cdot 37 \times 10^{-3}$

Note: $\omega_0 = 1$ rad/s; $\zeta = 5\%$; $\Delta = \dfrac{2\pi}{10\omega_0}$; $n = 100$

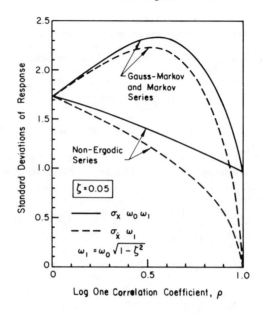

Fig. 1. Standard deviation of responses of simple oscillators to random time
series.

of x can differ significantly because of differences in the corresponding
standard deviations, as shown in Fig. 1. Note also that the approximate mean crossing rates of the response to Markovian inputs in eqn
(4) agree satisfactorily with simulation results.

4.2 Response to Filtered Poisson Processes

It is assumed for simplicity that $w(\alpha)$ in eqn (24) is the Dirac delta
function $\delta(\alpha)$. The characteristic functions in eqns (29) and (30)
become

$$\psi_{X,\dot{X}}(u, v) = \exp\left\{\int_0^t [\psi(uh(t - \alpha) + v\dot{h}(t - \alpha)) - 1]\lambda(\alpha)\,d\alpha\right\} \quad (35)$$

and

$$\psi_X(u) = \exp\left\{\int_0^t [\psi(uh(t - \alpha)) - 1]\lambda(\alpha)\,d\alpha\right\} \quad (36)$$

These functions can be used to calculate moments of the response. For example,

$$m_X = \frac{1}{i} \frac{d\psi_X(0)}{du} = m \int_0^t h(t - \alpha)\lambda(\alpha) \, d\alpha \qquad (37)$$

and

$$\alpha_{X^2} + m_{X^2} = \frac{1}{i^2} \frac{d^2\psi_X(0)}{du^2} = (m^2 + \sigma^2) \int_0^t h^2(t - \alpha)\lambda(\alpha) \, d\alpha \qquad (38)$$

If inverted, the characteristic functions in eqns (35) and (36) provide the densities $f_{X,\dot{X}}$ and f_X of the response. These densities can be used to determine mean upcrossing rates of $X(t)$ from eqns (1)–(4).

Figure 2 shows mean upcrossing rates of the response $X(t)$ for $\omega_0 = 1$ rad/s, $\zeta = 5\%$, $\lambda = 10(\omega_0/2\pi) = 1\cdot59$ occurrences/s, $t = 100/\lambda$, and impulses with standard Gaussian and Bilateral Exponential

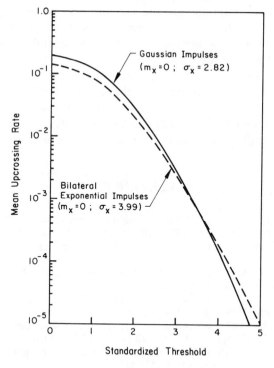

Fig. 2. Mean upcrossing rates of responses of simple oscillators to filtered Poisson processes.

distribution. The analysis has been based on eqns (4) and (36). The mean upcrossing rates of the response to exponential impulses resemble the ones to Gaussian impulses because the density of the impulses is high. There are on average 10 impulses arriving during any period $2\pi/\omega_0$ of the oscillator.

5 CONCLUSIONS

Two methods have been developed for the determination of mean upcrossing rates and other probabilistic descriptors for the response of linear systems subject to random time series and filtered Poisson processes. The first method is based on the joint characteristic function of the response and its derivative. It is exact but involves extensive calculations. The second method is an approximate solution. It is based on the assumption that the response can be obtained from a Gaussian process by a memoryless nonlinear transformation. The method involves only marginal probabilistic characteristics of the response and its derivative.

The approximate method has been applied to determine mean upcrossing rates for the response of simple oscillators to random time series and filtered Poisson processes with Gaussian and bilateral exponential impulses. The approximate mean upcrossing rates compare satisfactorily with mean upcrossing rates obtained by simulation.

REFERENCES

1. Buss, A. H., Crossings of non-Gaussian processes with reliability applications. Thesis presented to Cornell University, at Ithaca, NY, in 1987, in partial fulfillment of the requirements for the degree of Doctor of Philosophy.
2. Cramer, H. & Leadbetter, M. R., *Stationary and Related Stochastic Processes*. John Wiley, New York, 1967.
3. Feller, W., *An Introduction to Probability Theory and Its Application*, Vol. II. John Wiley, New York, 1966.
4. Grigoriu, M., Crossings of non-Gaussian translation processes. *Journal of Engineering Mechanics, ASCE*, **110** (4) (April 1984) 610–20.
5. Grigoriu, M., Extremes of correlated non-Gaussian series. *Journal of Structural Engineering, ASCE*, **110** (7) (July 1984) 1485–94.
6. Grigoriu, M., Load combination analysis by translation processes. *Journal of Structural Engineering, ASCE*, **110** (8) (August, 1984) 1725–34.

7. Iwankiewicz, R. & Sobczyk, K., Dynamic response of linear structures to correlated random impulses. *Journal of Sound and Vibration,* **86** (3) (1983) 303–17.
8. Jacobs, P. A. & Lewis, P. A. W., Discrete time series generated by mixtures. I: Correlation and runs properties. *Journal of the Royal Statistical Society, B,* **40** (1) (1978) 94–105.
9. Lin, Y. K., Nonstationary excursion and response in linear systems treated as sequences of random pulses. *Journal of the Acoustical Society of America,* **38** (1965) 453–60.
10. Lin, Y. K., First-excursion failure of randomly excited structures. *AIAA Journal,* **8** (4) (April 1970) 720–5.
11. Parzen, E., *Stochastic Processes.* Holden-Day, San Francisco, 1962.
12. Roberts, J. B., The response of linear vibratory systems to random impulses. *Journal of Sound and Vibration,* **2** (a4) (1965) 375–90.

7

Application of the Extended Kalman Filter–WGI Method in Dynamic System Identification

MASARU HOSHIYA

Department of Civil Engineering, Musashi Institute of Technology, Tokyo, Japan

ABSTRACT

In order to obtain stable solutions as well as their fast convergency to the optimal ones in dynamic system identification problems, a method of incorporating a weighted global iteration procedure with the extended Kalman filter (the EK–WGI method) was developed by the author and his co-researchers, and was clarified to be a powerful tool in extensive engineering problems. This paper intends to emphasize the latent usefulness of the method for a case in which the solutions may diverge due to insufficient observation data unless this method is used. Special attention is focused on parameter identification of linear and nonlinear structural systems.

1 INTRODUCTION

Recently, the extended Kalman filter,[1-3] has been applied on structural identification problems by many authors,[4-8] which is essentially a method of sequential least squares estimation, and the vast applicability of the filter was clarified not only for dynamic parameter identification but also for many other, static or dynamic system identification problems.[9-16]

In order to obtain stable solutions as well as their fast convergency to the optimal ones, a method of incorporating a weighted global iteration procedure with the extended Kalman filter was developed and applied on structural dynamic problems by the author and his co-researchers.[17-20] This method was named the EK–WGI method.

The purpose of this paper is to demonstrate the efficiency of the EK–WGI method and to discuss how effectively each identification problem is put into a set of state vector equation and an observation vector equation which are needed in conjunction with the method. Since this formulation is not necessarily monistic, it needs to be much more innovative dependent upon the nature of problems so that stability and fast convergency of estimated values to optimal ones could be effectively attained.

Special attention is focused on parameter identification of linear and nonlinear structural dynamic systems, and numerical analyses are performed with observed data which are numerically simulated on the responses of a structural model of known parameters, so that stability and covergency of estimated parameter values may be correctly identified.

2   EK–WGI METHOD

The extended Kalman filter algorithm,[1-3,18] is a recursive procedure to estimate the optimal state vector $\hat{\mathbf{X}}(tk/tk)$ and the corresponding error covariance matrix $\mathbf{P}(tk/tk)$ such that $\mathbf{P}(tk/tk)$ becomes minimum, on the basis of a nonlinear continuous state vector equation and a nonlinear discrete observation vector equation;

$$d\mathbf{X}_t/dt = f(\mathbf{X}_t, t), \qquad \mathbf{X}_{to} \sim N(\hat{\mathbf{X}}_{to}, \mathbf{P}_{to}) \qquad (1)$$

$$\mathbf{Y}_{tk} = h(\mathbf{X}_{tk}, t_k) + \mathbf{V}_{tk} \qquad (2)$$

in which $\mathbf{X}_{to} \sim N(\hat{\mathbf{X}}_{to}, \mathbf{P}_{to})$ implies that $\mathbf{X}_{to}$ is an initial state vector of Gaussian random variables with mean $\hat{\mathbf{X}}_{to}$ and an error covariance $\mathbf{P}_{to}$, $\mathbf{V}_{tk}$ = a vector of zero mean white noise Gaussian processes with $E[\mathbf{V}_{tk}\mathbf{V}_{tk}^T] = R(k)\delta_{kl}$, and δ_{kl} = the Kronecker delta.

If the initial state vector $\hat{\mathbf{X}}(to/to) = \hat{\mathbf{X}}_{to}$ and the error covariance matrix $\mathbf{P}(to/to) = \mathbf{P}_{to}$ are given, and then as the observation data $\mathbf{Y}_{tk}$ are processed, it is possible to estimate the state vector $\hat{\mathbf{X}}(tk/tk)$ and the error covariance $\mathbf{P}(tk/tk)$ from the following extended Kalman filter consisting of eqns (3)–(7).

$$\hat{\mathbf{X}}(tk + 1/tk) = \hat{\mathbf{X}}(tk/tk) + \int_{tk}^{tk+1} f[\hat{\mathbf{X}}(t/tk), t]\, dt \qquad (3)$$

$$\mathbf{P}(tk + 1/tk) = \mathbf{\Phi}[tk + 1, tk; \hat{\mathbf{X}}(tk/tk)]\mathbf{P}(tk/tk) * \mathbf{\Phi}^T[tk + 1, tk; \hat{\mathbf{X}}(tk/tk)] \tag{4}$$

$$\hat{\mathbf{X}}(tk + 1/tk + 1) = \hat{\mathbf{X}}(tk + 1/tk) + \mathbf{K}[tk + 1; \hat{\mathbf{X}}(tk + 1/tk)]$$
$$* [y_{tk+1} - \mathbf{h}[\hat{\mathbf{X}}(tk + 1/tk), tk + 1]] \tag{5}$$

$$\mathbf{P}(tk + 1/tk + 1) = [\mathbf{I} - \mathbf{K}[tk + 1; \hat{\mathbf{X}}(tk + 1/tk)]\mathbf{M}[tk + 1; \hat{\mathbf{X}}(tk + 1/tk)]]$$
$$* \mathbf{P}(tk + 1/tk)[\mathbf{I} - \mathbf{K}[tk + 1; \hat{\mathbf{X}}(tk + 1/tk)$$
$$* \mathbf{M}[tk + 1; \hat{\mathbf{X}}(tk + 1/tk)]]^T + \mathbf{K}[tk + 1; \hat{\mathbf{X}}(tk + 1/tk)]$$
$$* R(k + 1)\mathbf{K}^T[tk + 1; \hat{\mathbf{X}}(tk + 1/tk)]] \tag{6}$$

$$\mathbf{K}[tk + 1; \hat{\mathbf{X}}(tk + 1/tk)] = \mathbf{P}(tk + 1/tk)\mathbf{M}^T[tk + 1; \hat{\mathbf{X}}(tk + 1/tk)]$$
$$* [\mathbf{M}[tk + 1; \hat{\mathbf{X}}(tk + 1/tk)]\mathbf{P}(tk + 1/tk) * \mathbf{M}^T[tk + 1; \hat{\mathbf{X}}(tk + 1/tk)]$$
$$+ \mathbf{R}(k + 1)]^{-1} \tag{7}$$

in which $\hat{\mathbf{X}}(tk/tk) =$ state estimate at tk given $\mathbf{D}tk$ as follows: $\mathbf{P}(tk/tk) =$ covariance matrix of error in $\hat{\mathbf{X}}(tk/tk)$; $\mathbf{\Phi}[tk + 1, tk; \hat{\mathbf{X}}(tk/tk)] =$ state transfer matrix from tk to $tk + 1$; $\mathbf{K}[tk + 1; \hat{\mathbf{X}}(tk + 1, tk)] =$ Kalman gain matrix at $tk + 1$; $\mathbf{D}tk = [y_{to} \ldots y_{tk}]$; $y_{tk} =$ observation at tk; and

$$\mathbf{M}[tk; \hat{\mathbf{X}}(tk/tk)] = [\partial h_i(\mathbf{X}_{tk}, tk)/\partial x_j] \quad \text{at} \quad \mathbf{X}_{tk} = \hat{\mathbf{X}}(tk/tk)$$

in which $\mathbf{I} =$ a unit matrix; $h_i =$ the ith component of $\mathbf{h}(\mathbf{X}_{tk}, tk)$; and $x_j =$ the jth component of the vector, $\mathbf{X}_{tk}$. The $\mathbf{\Phi}$ matrix in this algorithm is given from Taylor's expansion of first order as follows:

$$\mathbf{F}[tk; \hat{\mathbf{X}}(tk/tk)] = [\partial f_i(\mathbf{X}_{tk}, tk)/\partial x_j] \quad \text{at} \quad \hat{\mathbf{X}}_{tk} = \hat{\mathbf{X}}(tk/tk)$$
$$\mathbf{\Phi}[tk + 1, tk; \hat{\mathbf{X}}(tk/tk)] = \mathbf{I} + \Delta\mathbf{F}[tk; \hat{\mathbf{X}}(tk/tk)] \tag{8}$$

in which $\Delta =$ sampling interval of observation waves.

In case of identification problems, parameters to be identified are included as additional state variables in the state vector[4,7,18–20] or a stationary state vector equation may be formulated consisting of only unknown constant parameters.[14,15] Thus, the extended Kalman filter for the state estimation problem can be also directly used for the identification.

However, it must be carefully examined to see whether or not the estimated parameters are stable and convergent to true ones, since only a finite duration time of observation data and arbitrarily assigned values $\hat{\mathbf{X}}_{to}$ and $\mathbf{P}_{to}$ for initial conditions are to be used in the analysis.

Therefore, it is necessary to assess the correct estimated parameters in terms of stability and convergency, making use of maximum information on parameters by repeatedly processing observation data, and in this direction a weighted global iteration procedure, including an objective function for the stability, has been recently developed.[18] This procedure of the EK–WGI method is shown in Fig. 1.

The extended Kalman filter algorithm is initially used with the initial conditions, $\hat{\mathbf{X}}(to/to)$ and $\mathbf{P}(to/to)$, to obtain $\hat{\mathbf{X}}(ts/ts)$ and $\mathbf{P}(ts/ts)$. Then, $\hat{\mathbf{X}}(to/to)$ and $\mathbf{P}(to/to)$ are replaced, respectively, by $\hat{\mathbf{X}}(ts/ts)$ and $\mathbf{P}(ts/ts)$, multiplied by a weight, $\mathbf{W}$, and, again, the extended Kalman filter algorithm is iterated. This same procedure is repeated until the initial values, $\hat{\mathbf{X}}(to/to)$, become essentially constant and almost equal to the final values of $\hat{\mathbf{X}}(ts/ts)$, in which ts = the discrete time for the last datum. The stability and convergency of the estimated parameters are also estimated by the nature of the objective function, $\bar{\theta}$. When the state vector (whose components are the parameters to be estimated) does not attain a reasonable value or tends to diverge, the proper value parameters are revealed on the basis of the minimum value of the objective function. In this method, the weight $\mathbf{W}$ is incorporated along with the initial covariance matrix in each iteration.

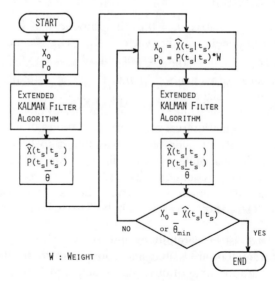

Fig. 1. EK–WGI method.

The reason why the weight **W** and the objective function $\bar{\theta}$ are introduced in this procedure is best described in the following paragraph. Since the error covariance, $\mathbf{P}(ts/ts)$, is the conditional expectation of the square value of the difference between the exact value $\mathbf{X}(ts/ts)$ and estimated value $\hat{\mathbf{X}}(ts/ts)$, the large initial covariance is favorable in order to accelerate the extended Kalman filter processing, while the stability might be sacrificed to some extent. Therefore, the weight used leads to a great fluctuation of $\hat{\mathbf{X}}(tk/tk)$ in the initial stage for the enlargement of the initial covariance in each iteration, and consequently, the weight seems to play an important role in the case of the promotion of convergency. The objective function, $\bar{\theta}$ is introduced as follows:

$$q_{ik} = y_{(i)tk} - h_i[\hat{\mathbf{X}}(t_k/t_k), t_k]; \qquad \gamma_i = \sum_{k=1}^{s} q_{ik}^2 \Big/ \sum_{k=1}^{s} y_{(i)tk}^2 \qquad (9)$$

$$\bar{\beta} = 1/\alpha' * \sum_{i=1}^{\alpha'} \gamma_i; \qquad \bar{\theta} = \left[\sum_{i=1}^{\alpha'} (\gamma_i - \bar{\beta})^2 \right]^{1/2} \qquad (10)$$

in which s = number of sampling points of observation waves; and α' = size of observation vector, $\mathbf{Y}_{tk}$. Here, γ_i $(i = 1, 2, \ldots)$ represents the normalized mean-square of the difference between the ith observation wave, $y_{(i)}$, and the corresponding estimate, $\hat{x}_i$. It is noted that $\hat{x}_i$ and $y_{(i)}$ are the ith elements of the vectors, $\hat{\mathbf{X}}$ and $\mathbf{y}_{tk}$, respectively. Since $\bar{\beta}$ is the average of γ_i, the objective function, $\bar{\theta}$, shows the scatterness in the root-mean-square of each γ_i from the central value of $\bar{\beta}$. Therefore, $\bar{\theta}_{\min}$ indicates that the difference between each observation and corresponding estimate becomes totally minimum, thus keeping their balance.

As to application of γ_i and $\bar{\theta}$ for the stability evaluation, numerical examples were carried out on parameter estimation of a linear system of shear type and the results were given in Ref. 18, where it was found that these indices are useful especially for a delicate stability problem.

3 IDENTIFICATION OF LINEAR SYSTEM

A multiple degree of freedom mass and spring system which is subject to a dynamic excitation at the support is expressed by the following equation:

$$M\ddot{z} + C\dot{z} + Kz = -MI\ddot{f}(t) \qquad (11)$$

where $\ddot{f}(t)$ = input excitation, z = response displacement vector, M = mass matrix, C = viscous damping coefficient matrix, K = elastic stiffness matrix and $I = [1, 1, \ldots, 1]^T$.

It is possible to identify directly either two of unknowns M, C and K by formulating a state vector equation on the basis of eqn (11). However, it was observed that the identification was not necessarily accurate, unless sufficient data sets were used as observation quantities.[7,18]

On the other hand, in engineering problems dynamic properties such as natural circular frequencies, modal damping coefficients and participation factors of each natural mode for existing structures are often required. Here, an efficient method of identifying such quantities is formulated in which even if only responses at one mass as well as input excitation are observed, these quantities are quite accurately identified.

Transforming Z into η by $Z = \Phi\eta$, where Φ is the modal matrix of the system of eqn (11) and has the components ϕ_{ij}, eqn (11) becomes a set of independent equations as follows.

For the jth mode,

$$\ddot{\eta}_j + 2\beta_j\omega_j\dot{\eta}_j + \omega_j^2\eta_j = -\delta_j\ddot{f}(t) \tag{12}$$

and

$$\delta_j = \sum_{k=1}^{n} m_k\phi_{kj} \tag{13}$$

where the viscous damping term is assumed to be diagonalized, and β_j = the jth modal damping coefficient, ω_j = the jth natural circular frequency, and ϕ_{kj} = the components of the jth mode.

In order to arrange eqn (12) into a state vector equation so that the EK–WGI method may be applied to the system parameter identification, eqn (12) is expressed as

$$\ddot{\zeta}_{ij} + 2\beta_j\omega_j\dot{\zeta}_{ij} + \omega_j^2\zeta_{ij} = -P_{ij}\ddot{f}(t) \tag{14}$$

where

$$\zeta_{ij} = \phi_{ij}\eta_j \quad \text{and} \quad P_{ij} = \phi_{ij}\delta_j$$

It is noted that ζ_{ij} is the displacement response of the ith mass in the jth mode, and thus, when all mode effects are taken into account, the actual displacement response of the ith mass is given by

$$Z_i = \sum_{j=1}^{n} \zeta_{ij} \tag{15}$$

Let $x_{1ij} = \zeta_{ij}$, $x_{2ij} = \dot{\zeta}_{ij}$, $x_{3ij} = \ddot{\zeta}_{ij}$, $x_{4j} = \beta_j$, $x_{5j} = \omega_j$, and $x_{6ij} = P_{ij}$. Then, by applying the linear acceleration method, eqn (14) becomes the following state vector equation in a discrete version of time $t = k \Delta t$ where k is an integer.[21,22]

$$\begin{bmatrix} \mathbf{X}_{i1}(k) \\ \hline \mathbf{X}_{i2}(k) \\ \hline \vdots \\ \hline \mathbf{X}_{in}(k) \end{bmatrix} = \begin{bmatrix} g_{i1}(k-1) \\ \hline g_{i2}(k-1) \\ \hline \vdots \\ \hline g_{in}(k-1) \end{bmatrix} \tag{16}$$

where

$\mathbf{X}_{ij}(k) =$

$$\begin{bmatrix} x_{1ij}(k) \\ x_{2ij}(k) \\ x_{3ij}(k) \\ x_{4j}(k) \\ x_{5j}(k) \\ x_{6ij}(k) \end{bmatrix} = \begin{bmatrix} D_{11}x_{1ij}(k-1) + D_{12}x_{2ij}(k-1) + D_{13}x_{3ij}(k-1) + D_{14}x_{6ij}(k-1)\ddot{f}(k) \\ D_{21}x_{1ij}(k-1) + D_{22}x_{2ij}(k-1) + D_{23}x_{3ij}(k-1) + D_{24}x_{6ij}(k-1)\ddot{f}(k) \\ D_{31}x_{1ij}(k-1) + D_{32}x_{2ij}(k-1) + D_{33}x_{3ij}(k-1) + D_{34}x_{6ij}(k-1)\ddot{f}(k) \\ x_{4j}(k-1) \\ x_{5j}(k-1) \\ x_{6ij}(k-1) \end{bmatrix}$$

$$= [\mathbf{g}_{ij}(k-1)] \tag{17}$$

$D_{11} = 1 + (\Delta t)^2 D_2/6$, $D_{12} = (\Delta t)(1 + (\Delta t)D_3/6)$, $D_{13} = ((\Delta t)^2/3)(1 + D_4/2)$

$D_{14} = (\Delta t)^2 D_1/6$, $D_{21} = (\Delta t)D_2/2$, $D_{22} = 1 + (\Delta t)D_3/2$

$D_{23} = (\Delta t)/2 + (\Delta t)D_4/2$, $D_{24} = (\Delta t)D_1/2$, $D_{31} = D_2$, $D_{32} = D_3$, $D_{33} = D_4$

$D_{34} = D_1$, $D_1 = -(1 + (\Delta t)x_{4j}(k-1)x_{5j}(k-1) + (\Delta t)^2 x_{5j}^2(k-1)/6)^{-1}$

$D_2 = D_1 x_{5j}^2(k-1)$, $D_3 = D_1(2x_{4j}(k-1)x_{5j}(k-1) + (\Delta t)x_{5j}^2(k-1))$ and

$D_4 = D_1((\Delta t)x_{4j}(k-1)x_{5j}(k-1) + (\Delta t)^2 x_{5j}^2(k-1)/3)$

The observation equation is dependent upon the choice of observed data. If the displacement and velocity responses at the ith mass are chosen as the observation data, then we have from eqn (15),

$$\mathbf{Y}_i(k) = \begin{bmatrix} Z_i(k) \\ \dot{Z}_i(k) \end{bmatrix} = [\mathbf{J}, \mathbf{J}, \dots, \mathbf{J}] \begin{bmatrix} \mathbf{X}_{i1}(k) \\ \hline \mathbf{X}_{i2}(k) \\ \hline \vdots \\ \hline \mathbf{X}_{in}(k) \end{bmatrix} + \mathbf{V}(k) \qquad (18)$$

where

$$\mathbf{J} = \begin{bmatrix} 1,0,0,0,0,0 \\ 0,1,0,0,0,0 \end{bmatrix}$$

It is noted that $\mathbf{V}$ is the noise vector involved in the observation and is treated as a vector of zero mean white noise Gaussian processes as indicated at the earlier stage. Note also that if acceleration response is observed as well, it is easy to reformulate eqn (18) such that the observation equation may be compatible with the state vector equation (16).

If only a most predominant mode, say the jth mode, is considered treating the other mode effects as noises, then the state vector equation becomes eqn (17), and the observation equation becomes simply

$$\mathbf{Y}_i(k) = \mathbf{J}\mathbf{X}_{ij}(k) + \mathbf{V} \qquad (19)$$

If observation data at each mass are available, then each mode shape is also identified in the following manner.

$$\begin{bmatrix} \phi_{1j} \\ \phi_{2j} \\ \vdots \\ \phi_{nj} \end{bmatrix} \propto \begin{bmatrix} P_{1j} \\ P_{2j} \\ \vdots \\ P_{nj} \end{bmatrix} \qquad (20)$$

Where $P_{ij}(= x_{5ij})$ is identified as one of the components of the state vector.

It is clear that eqn (16) is a nonlinear state vector equation, and with eqn (18) may be integrated directly into the EK–WGI procedure for identification.

The foregoing problem may also be assessed in the frequency domain. The advantage of using data which are Fourier-transformed as seen below is that one can reduce the computer time by utilizing only such a portion of data that center around the target natural frequency and consequently contain more information on the dynamic properties.

Fourier-transforming eqn (14), we have

$$\bar{\xi}_{ij}(\omega) = -P_{ij}/(-\omega^2 + \omega_j^2 + 2i\beta_j\omega_j\omega) * \bar{\bar{f}}(\omega) \tag{21}$$

If the most predominant mode and its properties are concerned, the state vector equation and observation equation may be formulated from eqn (21). Since parameters to be identified are considered as constant values, the state vector equation may be represented by putting $x_{1j} = \beta_j$, $x_{2j} = \omega_j$ and $x_{3ij} = P_{ij}$ as follows,

$$\mathbf{X}_{ij}(k) = \begin{bmatrix} x_{1j}(k) \\ x_{2j}(k) \\ x_{3ij}(k) \end{bmatrix} = \begin{bmatrix} 1, 0, 0 \\ 0, 1, 0 \\ 0, 0, 1 \end{bmatrix} \begin{bmatrix} x_{1j}(k-1) \\ x_{2j}(k-1) \\ x_{3ij}(k-1) \end{bmatrix} \tag{22}$$

The observation equation becomes from eqn (21)

$$\mathbf{Y}_i(k) = \begin{bmatrix} \mathrm{Re}[\bar{\xi}_{ij}(k)] \\ \mathrm{Im}[\bar{\xi}_{ij}(k)] \end{bmatrix} = \begin{bmatrix} \dfrac{-x_{3ij}[x_{2j}^2 - (k\Delta\omega)^2]\bar{\bar{f}}(k\Delta\omega)}{[[x_{2j}^2 - (k\Delta\omega)^2]^2 + 4x_{1j}^2 x_{2j}^2 (k\Delta\omega)^2]} \\ \dfrac{2x_{3ij}x_{1j}x_{2j}(k\Delta\omega)\bar{\bar{f}}(k\Delta\omega)}{[[x_{2j}^2 - (k\Delta\omega)^2]^2 + 4x_{1j}^2 x_{2j}^2 (k\Delta\omega)^2]} \end{bmatrix} + \mathbf{V}(k) \tag{23}$$

where the variable ω is discretized to $\omega = k\Delta\omega$.

If we take into account the top few predominant modes, then eqns (22) and (23) may be expanded into

$$\begin{bmatrix} \mathbf{X}_{ij}(k) \\ \mathbf{X}_{im}(k) \\ \vdots \end{bmatrix} = \begin{bmatrix} \mathbf{X}_{ij}(k-1) \\ \mathbf{X}_{im}(k-1) \\ \vdots \end{bmatrix} \tag{24}$$

and

$$\mathbf{Y}_i(k) = \begin{bmatrix} \mathrm{Re}[\bar{\xi}_{ij}(k)] + \mathrm{Re}[\bar{\xi}_{im}(k)] + \cdots \\ \mathrm{Im}[\bar{\xi}_{ij}(k)] + \mathrm{Im}[\bar{\xi}_{im}(k)] + \cdots \end{bmatrix} + \mathbf{V}(k) \tag{25}$$

An identification problem is numerically demonstrated based on the state vector equation (16) and the observation equation (18). As an example, a four-degree-of-freedom linear system of shear type[22] is employed. The input acceleration $\ddot{f}(t)$ has the maximum amplitude of 100 cm/s^2 and the time duration of 10 s with the discretized time interval of 0.01 s. The frequency characteristic is a white noise which is band limited in the range of $0.1–10 \text{ Hz}$.

Masaru Hoshiya

Table 1
Initial conditions

State variables	$x_{1ij}(to/to)$ 0.0	$x_{2ij}(to/to)$ 0·0	$x_{3ij}(to/to)$ 0·0	$x_{4j}(to/to)$ 1·0	$x_{5j}(to/to)$ 10·0	$x_{6ij}(to/to)$ 2·0
Covariances	$P_{11}(to/to)$ 1·0	$P_{22}(to/to)$ 1·0	$P_{33}(to/to)$ 1·0	$P_{44}(to/to)$ 100·0	$P_{55}(to/to)$ 100·0	$P_{66}(to/to)$ 100·0

$\mathbf{R} = 0·01^a$; $\mathbf{W} = 100·0^b$.
[a] Variances of $\mathbf{V}_k$ in eqn (18), [b] Weight in the EK–WGI method.

The displacement and velocity responses at the first mass (the top mass) of the known system were numerically simulated and taken up as the observation data so that stability and convergency of estimated parameter values may be correctly identified.

In the analysis, it is important how to set the initial conditions of the state vector $\hat{\mathbf{X}}_{ij}(to/to)$ and the error covariances $\mathbf{P}(to/to)$ in order to avoid divergency of the solutions. Here, the initial conditions in Table 1 were used, and the state vector equation based on a single mode (17), which was turned out to be the 1st mode as the dominant mode, was used for the first step of the EK–WGI method. The identified parameters are listed in the third column of Table 2.

For the next step, the state vector equation was expanded into one that corresponds to a two-degree-of-freedom system. In other words, eqns (16) and (18) were used, taking into account the unknown two predominant modes. As a result, they were found to be the first and second modes in this case. For the initial conditions for the first mode, the estimated parameters at the first step were used, whereas the initial conditions in Table 1 were used for the second mode. The identified parameters were shown in the fourth column of the Table 2.

In the same fashion, following the step by step expansion, all parameters of each mode were identified. They are shown in the Table 2. It was found that the estimated ones are almost identical to the exact ones in the second column. It is also interesting that the coefficients of errors in the mean square sense between the observed data and the corresponding estimated state variables (γ_1 for the displacement and γ_2 for the velocity) decreased as the expansion of the state vector equations proceeded.

As an example in the frequency domain analysis, a two-degree-of-freedom system was employed, and the first mode was estimated based on eqns (22) and (23), where a white noise input was used to evaluate

Table 2
Estimated system parameters (observed displacement and velocity at the first mass were used)

Mode j γ_1 and γ_2	Exact values	Single degree of freedom	Two degrees of freedom	Three degrees of freedom	Four degrees of freedom
1 $\hat{\omega}_1$ $\hat{\beta}_1$ $\hat{P}_{11}$	4·625 0·05781 1·350	4·621 0·05624 1·3170	4·624 0·05772 1·3470	4·625 0·05781 1·350	4·625 0·05781 1·350
2 $\hat{\omega}_2$ $\hat{\beta}_2$ $\hat{P}_{12}$	10·78 0·1348 −0·4182		10·76 0·1292 −0·4036	10·78 0·1348 −0·4181	10·78 0·1348 −0·4182
3 $\hat{\omega}_3$ $\hat{\beta}_3$ $\hat{P}_{13}$	17·40 0·2175 0·07013			17·38 0·2150 0·06927	17·38 0·2173 0·07001
4 $\hat{\omega}_4$ $\hat{\beta}_4$ $\hat{P}_{14}$	30·57 0·3821 −0·00159				30·81 0·3840 −0·00152
γ_1 γ_2		$3·67 \times 10^1$ $1·89 \times 10^0$	$1·32 \times 10^{-3}$ $9·35 \times 10^{-3}$	$2·72 \times 10^{-4}$ $2·58 \times 10^{-4}$	$1·25 \times 10^{-4}$ $1·18 \times 10^{-4}$

$\hat{\omega}$; rad/s, γ_1 and γ_2; per cent.

numerically the response displacement at the second mass. These input and response time histories were Fourier-transformed and were used as the input and observation data. The results are shown in Table 3. As to the second mode, the solutions were not stable and further research is needed for improvement.

Table 3
Estimated system parameters

Mode k		β_k	ω_k	P_{2k}
1	Exact value	0·05	8·822	0·621
	Initial value	0·01	8·8	0·5
	Estimated value	0·047	8·8 (fixed)	0·633
2	Exact value	0·05	28·463	0·379
	Initial value	0·01	26·4	0·5
	Estimated value	?	26·4 (fixed)	?

4 IDENTIFICATION OF NONLINEAR SYSTEMS

General nonlinear systems may be represented by a versatile hysteretic model investigated by Bouc[23] and Baber and Wen[24], which exhibits deterioration in strength or stiffness or both. Identification problems of a single-degree-of-freedom system by this model are investigated based on the following equations.

$$\ddot{u}(t) + 2h_0\omega_0\dot{u}(t) + \omega_0^2\phi(u(t)) = -\ddot{f}(t) \tag{26}$$

$$\dot{\phi}(u(t)) =$$
$$\frac{A(t)\dot{u}(t) - v(t)[\beta \, |\dot{u}(t)| \, |\phi(u(t))| \, |^{n-1}\phi(u(t)) - \gamma\dot{u}(t) \, |\phi(u(t))|^n]}{\eta(t)} \tag{27}$$

where $\ddot{f}(t)$ = input excitation, h_0 = fraction of critical viscous damping for small amplitudes, ω_0 = undamped natural circular frequency of small amplitude response (= pre-yielding natural circular frequency). The parameters β, γ, $A(t)$, $v(t)$, $\eta(t)$ and n control the hysteresis shape and degradation of the system.

The parameters $A(t)$, $v(t)$, and $\eta(t)$ are functions of the dissipated hysteretic energy, $\varepsilon(t)$, given by

$$\dot{\varepsilon}(t) = \omega_0^2\dot{u}(t)\phi(u(t)) \tag{28}$$

Then $\varepsilon(t)$ may be obtained by integration, provided $\dot{u}(t)$ and $\phi(u(t))$ are known.

The parameters $A(t)$, $v(t)$ and $\eta(t)$ may then be written as

$$\left.\begin{array}{l} A(t) = 1{\cdot}0 - \delta_A\varepsilon(t) \\ v(t) = 1{\cdot}0 + \delta_v\varepsilon(t) \\ \eta(t) = 1{\cdot}0 + \delta_\eta\varepsilon(t) \end{array}\right\} \tag{29}$$

where δ_A, δ_v and δ_η are constants specified for the desired rate of degradation.

Equations (26)–(29) are put into a continuous nonlinear state vector equation by introducing state variables $x_1 = u(t)$, $x_2 = \dot{u}(t)$, $x_3 = \phi(u(t))$, $x_4 = \varepsilon(t)$, $x_5 = h_0$, $x_6 = \omega_0$, $x_7 = \beta$, $x_8 = \gamma$, $x_9 = \delta_A$, $x_{10} = \delta_v$ and $x_{11} = \delta_\eta$.

$$
\begin{bmatrix} \dot{x}_1 \\ \dot{x}_2 \\ \dot{x}_3 \\ \dot{x}_4 \\ \dot{x}_5 \\ \dot{x}_6 \\ \dot{x}_7 \\ \dot{x}_8 \\ \dot{x}_9 \\ \dot{x}_{10} \\ \dot{x}_{11} \end{bmatrix} = \begin{bmatrix} x_2 \\ -2x_5 x_6 x_2 - x_6^2 x_3 - \ddot{f}(t) \\ \dfrac{(1 \cdot 0 - x_9 x_4)x_2 - (1 \cdot 0 + x_{10}x_4)(x_7\,|x_2|\,|x_3|^{n-1}x_3 + x_8 x_2\,|x_3|^n)}{(1 \cdot 0 + x_{11}x_4)} \\ x_6^2 x_3 x_2 \\ 0 \\ 0 \\ 0 \\ 0 \\ 0 \\ 0 \\ 0 \end{bmatrix}
$$

$$(30)$$

If observation data for the response displacement $u(t)$ and the response velocity $u(t)$ are available, the observation vector equation is given by

$$
\begin{bmatrix} y_1 \\ y_2 \end{bmatrix} = \begin{bmatrix} 1, 0, 0, 0, 0, 0, 0, 0, 0, 0, 0 \\ 0, 1, 0, 0, 0, 0, 0, 0, 0, 0, 0 \end{bmatrix} \mathbf{X} + \mathbf{V} \qquad (31)
$$

It is noted that the state variables x_5–x_{11} are the parameters to be identified in this study. Regarding the parameter n appearing in eqn (30), it is to be treated as a predetermined constant value for $n = 1$ in order to avoid divergency during the EK–WGI processing.

Equations (30) and (31) are incorporated directly into the EK–WGI method for parameter identification.[25,26]

As a numerical example, observation data were simulated on a single-degree-of-freedom nonlinear hysteretic system by the versatile hysteretic model with $7 \cdot 07$ rad/s for ω_0 and $0 \cdot 1$ for h_0, and the parameters on the hysteresis were $0 \cdot 05$ for β, $0 \cdot 05$ for γ, $0 \cdot 001$ for δ_A, $0 \cdot 002$ for δ_v, $0 \cdot 002$ for δ_η and 1 for n. The responses were simulated at a discrete time of $0 \cdot 01$ s due to an Imperial Valley earthquake acceleration record of the maximum value of $213 \cdot 1$ gal and of the time duration of 30 s.

Figure 2 shows the input and noise imposed response time histories as well as the hysteretic restoring force characteristic of the system. Figure 2(b) shows the known true hysteresis, whereas Fig. 2(c) shows

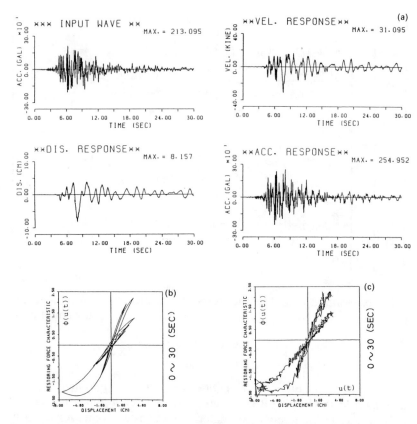

Fig. 2. Input and observation data: (a) input and noise imposed response time histories; (b) true hysteresis; and (c) noise imposed hysteresis.

Table 4
Initial condition for linear response

Initial conditions	x_1	x_2	x_3	x_4	$x_5(=h_0)$	$x_6(=\omega_0)$
$\mathbf{X}(to/to)$	0·0	0·0	0·0	0·0	0·5	5·0
$\mathbf{P}(to/to)$	1·0	1·0	1·0	1·0	100·0	100·0

Note: Covariance of $\mathbf{V} = 100·0$.

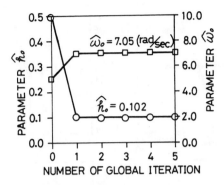

Fig. 3. Convergency process in WGI method.

a hysteresis calculated by using noise imposed response time histories with eqn (26). It is noted that the noise on the responses was uniformly random with the maximum value of 10% of the response amplitude.

In the identification analysis, the displacement and velocity responses were used as the observation data. The identification proce-

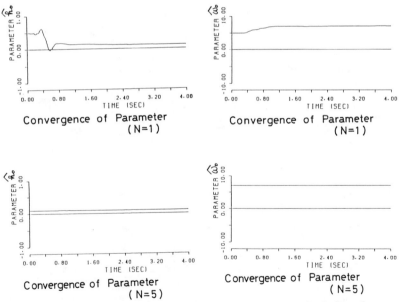

Fig. 4. Convergency process of a function of time in single iteration.

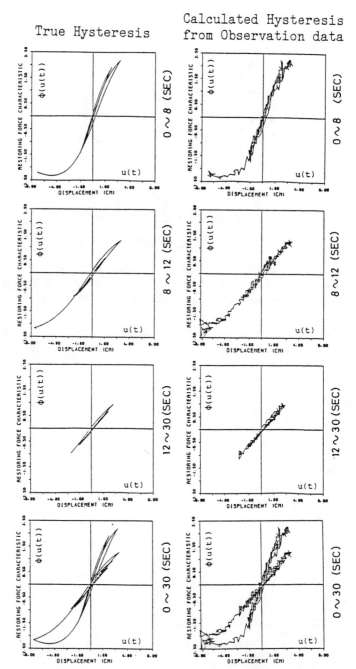

Fig. 5. Estimated hysteresis restoring force characteristics.

Estimated at 1st
Iteration

Estimated at 5th
Iteration

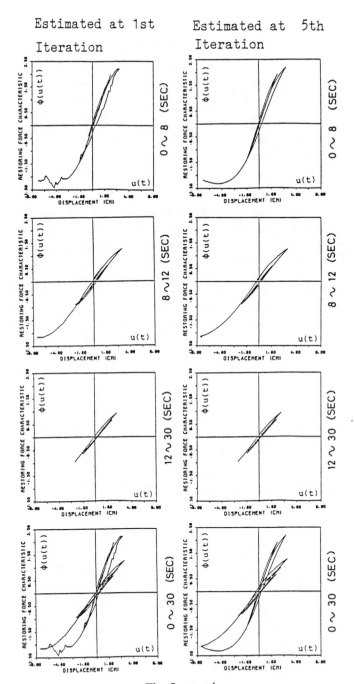

Fig. 5—*contd.*

Table 5
Initial condition for nonlinear response

Initial conditions	x_1	x_2	x_3	x_4	$x_5(=h_0)$	$x_6(=\omega_0)$	$x_7(=\beta)$	$x_8(=\gamma)$	$x_9(=\delta_A)$	$x_{10}(=\delta_v)$	$x_{11}(=\delta_\eta)$
$\mathbf{X}(to/to)$	0·0	0·0	0·0	0·0	0·1	7·05	0·0	0·0	0·0	0·0	0·0
$\mathbf{P}(to/to)$	1·0	1·0	1·0	1·0	0·0	0·0	100·0	100·0	1·0	1·0	1·0

Note: Covariance of $\mathbf{V} = 1000 \cdot 0$.

dure was divided into two stages. At the first stage, the parameters h_0 and ω_0 were identified using the first portion of 4·0 s duration of the observation data on the assumption that the responses were linear at the initial stage of oscillation. Thus, letting $x_7 = x_8 = x_9 = x_{10} = x_{11} = 0 \cdot 0$ as constants, only the parameters x_5 $(=h_0)$ and x_6 $(=\omega_0)$ were identified. The initial conditions are given in Table 4.

Figure 3 shows the convergency processes of the parameters h_0 and ω_o in terms of the number of the weighted global iteration, and Fig. 4

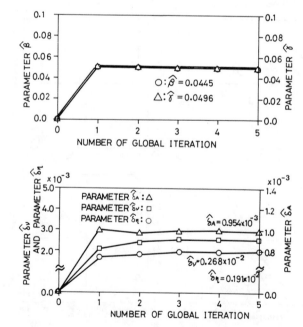

Fig. 6. Convergency process in number of iterations.

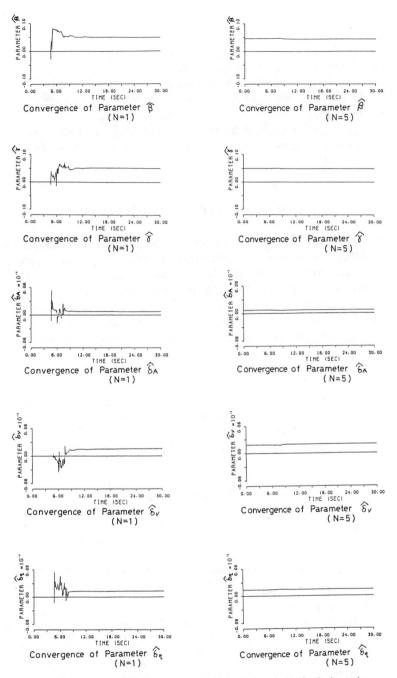

Fig. 7. Convergency process of a function of time in a single iteration.

shows how these parameters converge to true ones at the first and fifth iterations. The weight of 100 was used in the global iterations.

At the second stage of identification, the estimated values of x_5 and x_6 were frozen, and the other parameters x_7–x_{11} which govern the hysteresis were identified for the initial conditions in Table 5.

Figures 5–7 show the results. It is so clear that the identification was satisfactory.

5 CONCLUSION

A weighted global iteration method of the extended Kalman filter was described for structural identification problems, and was applied to linear and nonlinear systems. It was found by numerical examples that the method is a powerful tool for parameter identification. However, a remark should be made on the stability and convergency of solutions that the accuracy is totally dependent on how suitably the state vector equation and observation vector equation are formulated with reasonable initial conditions.

Some practical applications using actual observation data have been carried out successfully elsewhere.[14,17,27,28]

Finally, it should be noted that this paper was edited from technical papers published by the author and his colleagues at Musashi Institute of Technology, Tokyo, and the author thanks Dr E. Saito and Dr O. Maruyama for their cooperation in the preparation of this paper.

REFERENCES

1. Kalman, R. E., A new approach to linear filtering and prediction problems. *J. Basic Engng, Trans. ASME,* **82**(1) (1960) 35–45.
2. Kalman, R. E. & Bucy, R. S., New results in linear filtering and prediction theory. *J. Basic Engng, Trans. ASME,* **83**(1) (1961) 95–108.
3. Jazwinski, A. H., *Stochastic Processes and Filtering Theory,* Academic Press, New York, 1976.
4. Carmichael, D. G., The state estimation problem in experimental structural mechanics. *Proc. 3rd Inter. Conf. on Appl. of Statistics and Probability in Soil and Structural Engineering,* Sydney, Australia, 1979.
5. Carmichael, D. G., Identification of cyclic material constitutive relationships. Seventh Australian Conf. on the Mechanics of Structures and Materials, University of Western Australia, May, 1980.

6. Yun, C.-B. & Shinozuka, M., Identification of nonlinear structural dynamics systems, *J. Struct. Mech.*, **8**(2) (1980) 1371–90.
7. Shinozuka, M., Yun, C.-B. & Imai, M., Identification of linear structural dynamics systems. *J. Engng Mech. Div., ASCE* (1982).
8. Hoshiya, M. & Ishii, K., Deconvolution method between kinematic interaction and dynamic interaction of soil foundation systems based on observed data. *J. Soil Dynamics Earthquake Engng*, **3**(3) (July, 1984).
9. Bayless, J. W. & Brigham, E. O., Application of the Kalman filter to continuous signal restoration. *Geophysics*, **35**(1) (Feb., 1970) 2–23.
10. Hino, M., On-line prediction of hydrologic systems. *Proc. 15th Congress of IAHR*, Vol. 4, 1973.
11. Ogihara, K. & Tanaka, S., Application of Kalman filter theory to prediction of temperature in reservoir. 3rd Inter. World Congress of Water Resources, Mexico City, April, 1979.
12. Hoshiya, M. & Saito, E., Linearized liquefaction process by Kalman filter. *J. Geotech. Engng, ASCE*, **112**(2) (Feb. 1986).
13. Murakami, A. & Hasegawa, T., Observational prediction of settlement using Kalman filter theory. 5th Inter. Conf. on Numerical Methods in Geomechanics, Nagoya, April, 1985.
14. Hoshiya, M., Kodama, K. & Sakai, K., Identification of coefficient of horizontal subgrade reaction by EK–WGI method. 1st East Asian Conf. on Structural Engng and Construction, Bangkok, Jan., 1986.
15. Hoshiya, M. & Maruyama, O., Adaptive multiple nonlinear regression analysis and engineering applications. 1st East Asian Conf. on Structural Engng and Construction, Bangkok, Jan., 1986.
16. William, W-G. Yeh, Review of parameter identification procedures in ground water hydrology. *Water Resources Research*, **22**(2) (Feb. 1986) 95–108.
17. Hoshiya, M. & Saito, E., Identification of dynamic properties of a building by the EK–WGI method on microtremor records. *Proc. Japan Society of Civil Engineers*, No. 350/I-2, Oct., 1984 (in Japanese).
18. Hoshiya, M. & Saito, E., Structural identification by extended Kalman filter. *J. Engng Mech., ASCE*, **112**(12) (Dec. 1984).
19. Hoshiya, M., Saito, E. & Yamazaki, A., An equivalently linearized dynamic response analysis method for liquefaction of multi-layered sandy deposits. *Proc. Japan Society of Civil Engineers*, No. 356/I-3, April, 1985 (in Japanese).
20. Hoshiya, M. & Maruyama, O., Identification of running load and beam system. *J. Engng Mech., ASCE*, **113**(6) (June 1987).
21. Hoshiya, M. & Saito, E., Structural identification by extended Kalman filter. 1st East Asian Conf. on Structural Engng and Construction, Bangkok, Jan., 1986.
22. Hoshiya, M. & Maruyama, O., Dynamic parameters identification by EK–WGI Procedure. 8th European Conference of Earthquake Engineering, Lisbon, Portugal, Sept., 1986.
23. Bouc, R., Forced vibration of mechanical system with hysteresis. *Proc. Fourth Conf. on Nonlinear Oscillation*, Prague, Czechoslovakia, 1967.

24. Baber, T. T. & Wen, Y. K., Random vibration of hysteretic degrading system. *J. Engng Mech. Div., ASCE,* **107** (EM6) (1981) 1069–87.
25. Hoshiya, M. & Maruyama, O., Identification of nonlinear structural systems. *Proc. Fifth Inter. Conf. on Appl. of Statistics and Probability in Soil and Struct. Eng.,* Vancouver, Canada, June, 1987.
26. Hoshiya, M. & Maruyama, O., Identification of a restoring force model by EK–WGI procedure. 3rd Inter. Conf. on Soil Dyn. and Earth. Eng., Princeton, USA, June, 1987.
27. Sakai, K. & Hoshiya, M., Prediction and control of behaviors on driving shields using Kalman filter theory. *Proc. JSCE,* No. 385/VI-7, Sept. 1987 (in Japanese).
28. Hoshiya, M. & Maruyama, O., Kalman filter of versatile restoring systems. US.-Japan Scientific Cooperative Program Seminar, at Florida Atlantic University, Proceedings of Stochastic Approaches in Earthquake Engineering, May, 1987, pp. 68–86.

8

Nonlinear Response Phenomena in the Presence of Combination Internal Resonance

R. A. IBRAHIM* & W. LI

Department of Mechanical Engineering, Texas Tech University, Lubbock, Texas, USA

ABSTRACT

This paper examines the nonlinear interaction of a three-degree-of-freedom structural model subjected to a wide band random excitation. The nonlinearity of the system results in different critical regions of internal resonance which have significant effect on the response statistics. With reference to combination internal resonance of the summed type the system response is analyzed by using the Fokker–Planck equation approach together with a nonGaussian closure scheme. The nonGaussian closure is based on the cumulant properties of order greater than three. As a first order approximation the scheme yields 209 first order differential equations in first through fourth order joint moments of the response coordinates. The analysis is carried out with the aid of the computer algebra software MACSYMA. The response statistics are determined numerically in the time and frequency (internal detuning) domains. Contrary to the Gaussian closure scheme the nonGaussian closure solution yields a strictly stationary response in addition to a number of complex response characteristics to be first reported in the literature of the area of nonlinear random vibration. These include multiple solutions and jump phenomena (jump and collapse in the response mean squares) at internal detuning slightly shifted from the exact internal resonance condition. At exact internal resonance the system response possesses a unique limit cycle in a

* Present address: Department of Mechanical Engineering, Wayne State University, Detroit, Michigan 48202, USA.

stochastic sense. The regions of multiple solutions are defined in terms of system parameters (damping ratios and nonlinear coupling parameter) and excitation spectral density level.

1 INTRODUCTION

The linear modeling of any dynamical system is commonly acceptable as long as the actual response characteristics to various types of loading follow the linear solution. However, under certain situations the system may experience certain complex characteristics that cannot be justified by the linear solution. These complex response features owe their origin to the system inherent nonlinearities. In structural dynamics the nonlinearity may take one of three classes:[1,2] elastic, inertia, and damping nonlinearities. Elastic nonlinearity stems from nonlinear strain–displacement relations which are inevitable. Inertia nonlinearity is derived, in Lagrangian formulation, from the kinetic energy. In multi-degree-of-freedom systems the normal modes may involve nonlinear inertia coupling which may give rise to what are effectively parametric instability phenomena within the system. The parametric action is not due to the external loading, as in the case of parametric vibration, but to the motion of the system itself and, hence, is described as autoparametric.[3] The main feature of autoparametric coupling is that the motion of one normal mode gives rise to loading of other modes through time-independent coefficients in the corresponding equation of motion. The natural frequencies of the normal modes involved in the autoparametric interaction are usually related by a linear algebraic relationship known as 'internal resonance' condition.

According to the order of nonlinear coupling the system may exhibit certain types of response phenomena.[4,5] For example, systems with quadratic nonlinear inertia coupling may experience saturation phenomenon, amplitude jump, nonlinear resonance absorption effect, and multi-response behavior. For the case of two-degree-of-freedom systems which possess third order inertial resonance condition $\omega_2 = 2\omega_1$, it is found that if the second mode is externally excited it behaves, in the beginning, like a linear single-degree-of-freedom system, and the first mode remains dormant. As the excitation amplitude reaches a certain critical level the first mode becomes unstable, and the second mode reaches an upper level. This mode is said to be saturated, and energy is then transferred to the first mode. This feature has been predicted theoretically and observed experimentally by Haxton & Barr[6] in their autoparametric vibration absorber; Nayfeh *et al.*[7] and Mook *et al.*[8] in ship dynamics involving nonlinear coupling between pitching and rolling motion; and Haddow *et al.*[9] in nonlinear motion of coupled beams. For three-degree-of-freedom systems possessing internal resonance of the summed type $\omega_3 = \omega_1 + \omega_2$ similar features were reported by Ibrahim & Woodall,[10] Bux & Roberts[11] and Roberts & Zhang.[12]

The response behavior of nonlinear systems under harmonic excitation may be changed if the excitation is a random process. The theory of nonlinear random vibration is not as well developed as its deterministic counterpart. The theory of nonlinear random vibration requires advanced background in the theory of random processes and stochastic differential equations.[13–15] Few attempts have been made to predict the response statistics of nonlinear two-degree-of-freedom systems. These include the work of Ibrahim & Roberts,[16,17] Schmidt,[18] and Ibrahim & Heo[19,20] who examined the autoparametric interaction of two-degree-of freedom systems subjected to wide band random excitations. The response statistics of these systems share a number of nonlinear characteristics of deterministic results such as nonlinear resonance absorption effect. However, the saturation phenomenon did not take place because the excitation contains a wide range of frequencies which result in a continuous variation of the external detuning. Recently, Ibrahim & Hedayati[21] have examined the effect of quadratic nonlinear inertia coupling in a three-degree-of-freedom structure subjected to a wide band random excitation. They used the Fokker–Planck equation approach together with a Gaussian closure scheme. In the neighborhood of the combination internal resonance condition, $\omega_3 = \omega_1 + \omega_2$, the nonlinear interaction was found to take place between the second and third normal modes at an internal detuning parameter $r = \omega_3/(\omega_1 + \omega_2) = 1 \cdot 18$. At $r = 1 \cdot 0$ all attempts converged to the linear random response statistics. The purpose of the present paper is to employ a nonGaussian closure scheme which takes into account the effect of the response deviation from normal distribution with the purpose of exploring three modal nonlinear interaction. For the sake of completeness a brief review of the results of Ref. 21 will be given in Section 6. The effect of excitation spectral density level on the response characteristics will be examined in Section 7.

2 BASIC MODEL AND EQUATIONS OF MOTION

Figure 1 shows a schematic diagram of an analytical model of an aircraft subjected to random excitation $F(t)$. The fuselage is represented by the main mass m_3, linear spring k_3 and dashpot C_3. Attached to the main mass on each side are two coupled beams with tip masses m_1 and m_2, stiffnesses k_1 and k_2, and lengths l_1 and l_2. In the analysis of the shown system only the symmetric motions of the two sides of the model are considered. Under random excitation the system response will be described by the generalized coordinates q_1, q_2 and q_3 as shown in the figure. The equations of motion are derived by applying Lagrange's equation

$$\frac{d}{dt}\left\{\frac{\partial L}{\partial \dot{q}}\right\} - \left\{\frac{\partial L}{\partial q}\right\} = \{Q_{NC}\} \tag{1}$$

where $L = T - V$.

The kinetic energy T is given by the expression

$$T = \frac{1}{2}\left\{m_1 + m_2\left[1 + \left(\frac{3l_2}{2l_1}\right)^2\right]\right\}\dot{q}_1^2 + \frac{1}{2}m_2\dot{q}_2^2 + \frac{1}{2}(m_1 + m_2 + m_3)\dot{q}_3^2$$

$$+ \frac{3m_2l_2}{2l_1}\dot{q}_1\dot{q}_2 + (m_1 + m_2)\dot{q}_1\dot{q}_3 + \frac{9m_2l_2}{20l_1^2}(q_1\dot{q}_1^2 + 5q_1\dot{q}_1\dot{q}_3)$$

$$+ \frac{3m_2}{2l_1}\left(\frac{q_1\dot{q}_1\dot{q}_2}{5} + q_1\dot{q}_2\dot{q}_3 + \dot{q}_1q_2\dot{q}_3 + \dot{q}_1^2q_2\right) + \frac{6m_2}{5l_2}(q_2\dot{q}_2\dot{q}_3 + \dot{q}_1q_2\dot{q}_2) \tag{2}$$

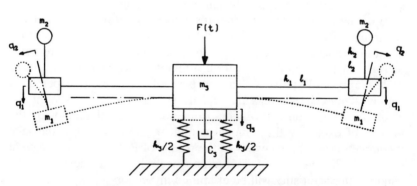

Fig. 1. Schematic diagram of the model.

where a dot denotes differentiation with respect to time t. Neglecting the gravitational effects, the potential energy V is given by the elastic energy

$$V = \tfrac{1}{2}(k_1 q_1^2 + k_2 q_2^2 + k_3 q_3^2) \tag{3}$$

Substituting for T and V in eqn (1) and bypassing energy dissipation due to damping (damping forces will be introduced later) yields the equations of motion in terms of the nondimensional coordinates $\bar{q}_i$

$$\omega_3^2 \begin{bmatrix} m_{11} & m_{12} & m_{13} \\ m_{12} & m_{22} & 0 \\ m_{13} & 0 & m_{33} \end{bmatrix} \begin{Bmatrix} \bar{q}_1'' \\ \bar{q}_2'' \\ \bar{q}_3'' \end{Bmatrix} + \begin{bmatrix} k_1 & 0 & 0 \\ 0 & k_2 & 0 \\ 0 & 0 & k_3 \end{bmatrix} \begin{Bmatrix} \bar{q}_1 \\ \bar{q}_2 \\ \bar{q}_3 \end{Bmatrix}$$
$$= \frac{1}{q_3^\circ} \begin{Bmatrix} 0 \\ 0 \\ F(\tau/\omega_3) \end{Bmatrix} - \frac{m_2 q_3^\circ \omega_3^2}{l_1} \begin{Bmatrix} \Psi_1 \\ \Psi_2 \\ \Psi_3 \end{Bmatrix} \tag{4}$$

where $\bar{q}_i = q_i/q_3^\circ$, $\tau = \omega_3 t$

q_3° is taken as the root-mean-square of the main mass when all other parts are locked under forced excitation, ω_3 is taken as the third eigenvalue of the system, and

$$m_{11} = m_1 + m_2[1 + 2\cdot25(l_2/l_1)^2], \qquad m_{22} = m_2$$
$$m_{33} = m_1 + m_2 + m_3, \qquad\qquad m_{12} = 1\cdot5m_2(l_2/l_1)$$
$$m_{13} = m_1 + m_2$$
$$\Psi_1 = 0\cdot45(l_2/l_1)(2\bar{q}_1\bar{q}_1'' + \bar{q}'^2 + 5\bar{q}_1\bar{q}_3'')$$
$$\quad + 1\cdot5(0\cdot2\bar{q}_1\bar{q}_2'' + \bar{q}_2\bar{q}_3'' + 2\bar{q}_2\bar{q}_1'' + 2\bar{q}_1'\bar{q}_2')$$
$$\quad + 1\cdot2(l_1/l_2)(\bar{q}_2\bar{q}_2'' + \bar{q}'^2)$$
$$\Psi_2 = 1\cdot5(0\cdot2\bar{q}_1\bar{q}_1'' - 0\cdot8\bar{q}_1'^2 + \bar{q}_1\bar{q}_3'')$$
$$\quad + 1\cdot2(l_1/l_2)(\bar{q}_2\bar{q}_3'' + \bar{q}_2\bar{q}_1'')$$
$$\Psi_3 = 2\cdot25(l_2/l_1)(\bar{q}_1\bar{q}_1'' + \bar{q}_1'^2)$$
$$\quad + 1\cdot5(\bar{q}_1\bar{q}_2'' + 2\bar{q}_1'\bar{q}_2' + \bar{q}_2\bar{q}_1'')$$
$$\quad + 1\cdot2(l_1/l_2)(\bar{q}_2\bar{q}_2'' + \bar{q}_2'^2) \tag{5}$$

where a prime denotes differentiation with respect to the dimensionless time τ.

3 EIGENVALUES AND MODE SHAPES

The system eigenvalues are determined from the conservative linear part of the equations of motion

$$[m]\{\ddot{q}\} + [k]\{q\} = \{0\} \tag{6}$$

the characteristic equation of (6) is

$$\text{Det }|[k] - \omega^2[m]| = 0 \tag{7}$$

where the roots of (7) give the eigenvalues ω_i of the system. Expanding the determinant gives the cubic equation

$$
\left(-1 + \frac{m_{12}^2}{m_{11}m_{22}} + \frac{m_{13}^2}{m_{11}m_{33}}\right)\left(\frac{\omega}{\omega_{33}}\right)^6 + \left[\left(\frac{\omega_{11}}{\omega_{33}}\right)^2\right.
$$
$$
+ \left(\frac{\omega_{22}}{\omega_{33}}\right)^2\left(1 - \frac{m_{13}^2}{m_{11}m_{33}}\right) + \left(1 - \frac{m_{12}^2}{m_{11}m_{22}}\right)\left]\left(\frac{\omega}{\omega_{33}}\right)^4\right.
$$
$$
- \left[\left(\frac{\omega_{11}}{\omega_{33}}\right)^2\left(\frac{\omega_{22}}{\omega_{33}}\right)^2 + \left(\frac{\omega_{11}}{\omega_{33}}\right)^2 + \left(\frac{\omega_{22}}{\omega_{33}}\right)^2\right]\left(\frac{\omega}{\omega_{33}}\right)^2 + \left(\frac{\omega_{11}}{\omega_{33}}\right)^2\left(\frac{\omega_{22}}{\omega_{33}}\right)^2 = 0 \tag{8}
$$

where the frequency parameters $\omega_{ii} = \sqrt{k_i/m_{ii}}$ are the natural frequencies of the individual components of the model. The IMSL (International Mathematical and Statistical Library) subroutine ZPOLR (Zeros of a Polynomial with Real coefficients) is used to find the roots of eqn (8). Figure 2 shows a sample of the dependence of the natural frequency ratio $r = \omega_3/(\omega_1 + \omega_2)$ on the ratios ω_{11}/ω_{33} and ω_{22}/ω_{33} for beams of length ratio $l_2/l_1 = 0.25$, and mass ratios $m_2/m_1 = 0.5$, and $m_3/m_1 = 5.0$. The importance of these curves is to define the critical points where the structure possesses internal combination resonance $r = 1.0$. It is seen that the most critical region is located on the curves belonging to the values of ω_{22}/ω_{33} ranging from 1 to 2. For the present analysis the curve corresponding to $\omega_{22}/\omega_{33} = 1.4$ will be adopted. The mode shapes of the model corresponding to the three eigenvalues which satisfy the internal resonance condition $r = 1.0$ are evaluated by the method of matrix decomposition[22] and are shown in Fig. 3. The eigenvectors of the system will be used in Section 4 to construct the modal matrix $[R]$.

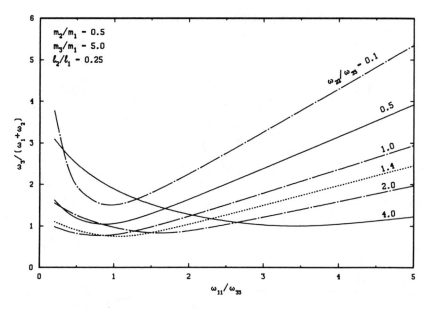

Fig. 2. Normal mode frequency ratios $r = \omega_3/(\omega_1 + \omega_2)$ versus system parameters ω_{11}/ω_{33} for various of ω_{22}/ω_{33}.

4 TRANSFORMATION INTO NORMAL COORDINATES

Equations (4) include linear and nonlinear dynamic coupling. It is convenient to eliminate the linear dynamic coupling by transforming eqn (4) into principal coordinates Y_i, by using the transformation

$$\{q\} = [R]\{Y\} \qquad (9)$$

where $[R]$ is the modal matrix whose columns are the eigenvectors,

$$[R] = \begin{vmatrix} 1 & 1 & 1 \\ n_1 & n_2 & n_3 \\ \bar{n}_1 & \bar{n}_2 & \bar{n}_3 \end{vmatrix} \qquad (10)$$

Rewriting eqns (4) in terms of the principal coordinates in the matrix form

$$[m][R]\{Y''\} + [k][R]\{Y\} = \{F(\tau)\} - \{\psi(\mathbf{Y}, \mathbf{Y}', \mathbf{Y}'')\} \qquad (11)$$

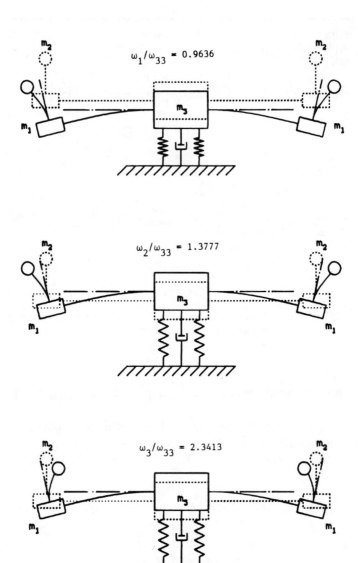

Fig. 3. Mode shapes corresponding to the combination internal resonance $\omega_3 = \omega_1 + \omega_2$.

Premultiplying eqn (11) by the transpose of the modal matrix results in diagonalizing the mass and stiffness matrices. The resulting equations involve nonlinear coupling and have the form

$$m_1\omega_3^2 \begin{bmatrix} M_{11} & 0 & 0 \\ 0 & M_{22} & 0 \\ 0 & 0 & M_{33} \end{bmatrix} \begin{Bmatrix} Y_1'' \\ Y_2'' \\ Y_3'' \end{Bmatrix} + k_1 \begin{bmatrix} k_{11} & 0 & 0 \\ 0 & k_{22} & 0 \\ 0 & 0 & k_{33} \end{bmatrix} \begin{Bmatrix} Y_1 \\ Y_2 \\ Y_3 \end{Bmatrix}$$

$$= \frac{1}{q_3^\circ} \begin{Bmatrix} \bar{n}_1 F(\tau/\omega_3) \\ \bar{n}_2 F(\tau/\omega_3) \\ \bar{n}_3 F(\tau/\omega_3) \end{Bmatrix} - \frac{m_2 q_3^\circ \omega_3^2}{l_1} \begin{Bmatrix} \psi_1 \\ \psi_2 \\ \psi_3 \end{Bmatrix} \quad (12)$$

where

$$M_{ii} = 1 + \alpha(1 + 2 \cdot 25\beta^2 + 3\beta n_i + 2\bar{n}_i + n_i^2 + \bar{n}_i^2) + [1 + m_3/m_1)]\bar{n}_i^2 + 2\bar{n}_i$$

$$k_{ii} = 1 + (k_2/k_1)n_i^2 + (k_3/k_1)\bar{n}_i^2$$

$$\psi_i = Y_1''(L_{i11}Y_1 + L_{i21}Y_2 + L_{i31}Y_3) + Y_2''(L_{i12}Y_1 + L_{i22}Y_2 + L_{i32}Y_3)$$
$$\quad + Y_3''(L_{i13}Y_1 + L_{i23}Y_2 + L_{i33}Y_3) + M_{i11}Y_1'^2 + M_{i22}Y_2'^2 + M_{i33}Y_3'^2$$
$$\quad + M_{i12}Y_1'Y_2' + M_{i13}Y_1'Y_3' + M_{i23}Y_2'Y_3'$$

$$\alpha = m_2/m_1, \qquad \beta = l_2/l_1$$

$$L_{ijk} = 0 \cdot 9\beta + 2 \cdot 25\beta\bar{n}_k + 0 \cdot 3n_k + 1 \cdot 5n_j\bar{n}_k + 3n_j + (1 \cdot 2/\beta)n_j n_k$$
$$\quad + n_i[0 \cdot 3 + 1 \cdot 5\bar{n}_k + (1 \cdot 2/\beta)n_j(1 + \bar{n}_k)]$$
$$\quad + \bar{n}_i[2 \cdot 25\beta + 1 \cdot 5(n_j + n_k) + (1 \cdot 2/\beta)n_j n_k]$$

$$M_{ijk} = 0 \cdot 9\beta + 3(n_j + n_k) + (2 \cdot 4/\beta)n_j n_k$$
$$\quad - 2 \cdot 4n_i + \bar{n}_i[4 \cdot 5\beta + 3(n_j + n_k) + (2 \cdot 4/\beta)n_j n_k] \qquad (j \neq k)$$

$$M_{ill} = 0 \cdot 45\beta + 3n_l + (1 \cdot 2/\beta)n_l^2$$
$$\quad - 1 \cdot 2n_i + \bar{n}_i[2 \cdot 25\beta + 3n_l + (1 \cdot 2/\beta)n_l^2] \qquad (j = k = l) \qquad (13)$$

5 DIFFERENTIAL EQUATIONS OF RESPONSE MOMENTS

The response coordinates can be approximated as a Markov vector if the random excitation is represented by a zero mean physical white noise $W(\tau)$ having the autocorrelation function

$$R_w(\Delta\tau) = E[W(\tau)W(\tau + \Delta\tau)] = 2D\delta(\Delta\tau) \qquad (14)$$

where $2D$ is the spectral density and $\delta(\ )$ is the Dirac delta function.

This modeling is justified as long as the relevant Wong–Zakai[14] correction term is preserved. In order to construct the response Markov vector the acceleration terms involved in the nonlinear functions Ψ_i must be removed by successive elimination. In view of the complexity of the equations of motion the elimination process is performed by using the symbolics manipulation software MACSYMA. Equation (12) takes the new form

$$Y_i'' + 2\zeta_i r_{i3} Y_i' + r_{i3}^2 Y_i = f_i W(\tau) + \varepsilon g_i(\mathbf{Y}, \mathbf{Y}'), \qquad i = 1, 2, 3 \quad (15)$$

where linear viscous damping terms have been introduced to account for energy dissipation, and

$$\omega_i^2 = (k_{ii}/M_{ii})(k_1/m_1) \qquad\qquad \varepsilon = q_3^\circ / l_1$$

$$r_{i3} = \omega_i / \omega_3 \qquad\qquad W(\tau) = \frac{1}{q_3^\circ \omega_3^2 m_1} F(\tau/\omega_3)$$

$$f_i = \bar{n}_i / M_{ii}$$

Introducing the transformation into the Markov state vector $\mathbf{X}$

$$\{Y_1, Y_1', Y_2, Y_2', Y_3, Y_3'\} = \{X_1, X_2, X_3, X_4, X_5, X_6\} \quad (16)$$

eqn (15) may be written in the standard form of a Stratonovich differential equation

$$dX_i = F_i(\mathbf{X}, \tau)\, d\tau + \sum_{j=1}^{6} G_{ij}(\mathbf{X}, \tau)\, dB_j(\tau) \quad (17)$$

where the white noise $W(\tau)$ has been replaced by the formal derivative of the Brownian motion process $B(\tau)$, i.e.

$$W(\tau) = \sigma\, dB(\tau)/d\tau, \qquad \sigma^2 = 2D$$

Alternatively, eqn (17) may in turn be transformed into the Ito type equation

$$dX_i = \left[F_i(\mathbf{X}, \tau) + \tfrac{1}{2} \sum_{k=1}^{6} \sum_{j=1}^{6} G_{kj}(\mathbf{X}, \tau) \frac{\partial G_{ij}(\mathbf{X}, \tau)}{\partial X_k} \right] d\tau$$

$$+ \sum_{j=1}^{6} G_{ij}(\mathbf{X}, \tau)\, dB_j(\tau) \quad (18)$$

where the double summation expression is the called the Wong–Zakai (or Ito) correction term.[14]

The system stochastic Ito equations are

$$dx_1 = x_2\,d\tau, \qquad dx_3 = x_4\,d\tau, \qquad dx_5 = x_6\,d\tau$$

$$
\begin{aligned}
dx_2 = \{&2\xi_1 r_{13}x_2 - r_{13}^2 x_1 + (2b_1\xi_1 r_{13}x_2 + r_{13}^2 x_1)(L_{111}x_1 + L_{121}x_3 + L_{131}x_5) \\
&+ (2b_1\xi_2 r_{23}x_4 + b_1 r_{23}^2 x_3)(L_{112}x_1 + L_{122}x_3 + L_{132}x_5) \\
&+ (2b_1\xi_3 x_6 + b_1 x_5)(L_{113}x_1 + L_{123}x_3 + L_{133}x_5) \\
&+ b_1[-M_{111}x_2^2 - M_{122}x_4^2 - M_{133}x_6^2 - M_{112}x_2 x_4 \\
&- M_{123}x_4 x_6 - M_{113}x_2 x_6]\}\,d\tau \\
&+ \{f_1 + c_{11}x_1^2 + c_{12}x_3^2 + c_{13}x_5^2 + c_{14}x_1 x_3 + c_{15}x_1 x_5 + c_{16}x_3 x_5 \\
&+ c_{17}x_1 + c_{18}x_3 + c_{19}x_5\}\,dB
\end{aligned}
$$

$$
\begin{aligned}
dx_4 = \{&2\xi_2 r_{23}x_4 - r_{23}^2 x_3 + (2b_2\xi_1 r_{13}x_2 + r_{13}^2 x_1)(L_{211}x_1 + L_{221}x_3 + L_{231}x_5) \\
&+ (2b_2\xi_2 r_{23}x_4 + b_1 r_{23}^2 x_3)(L_{212}x_1 + L_{222}x_3 + L_{232}x_5) \\
&+ (2b_2\xi_3 x_6 + b_2 x_5)(L_{213}x_1 + L_{223}x_3 + L_{233}x_5) \\
&+ b_2[-M_{211}x_2^2 - M_{222}x_4^2 - M_{233}x_6^2 - M_{212}x_2 x_4 \\
&- M_{223}x_4 x_6 - M_{213}x_2 x_6]\}\,d\tau \\
&+ \{f_2 + c_{21}x_1^2 + c_{22}x_3^2 + c_{23}x_5^2 + c_{24}x_1 x_3 + c_{25}x_1 x_5 + c_{26}x_3 x_5 \\
&+ c_{27}x_1 + c_{28}x_3 + c_{29}x_5\}\,dB
\end{aligned}
$$

$$
\begin{aligned}
dx_6 = \{&2\xi_3 x_6 - x_5 + (2b_3\xi_1 r_{13}x_2 + r_{13}^2 x_1)(L_{311}x_1 + L_{321}x_3 + L_{331}x_5) \\
&+ (2b_3\xi_2 r_{23}x_4 + b_3 r_{23}^2 x_3)(L_{312}x_1 + L_{322}x_3 + L_{332}x_5) \\
&+ (2b_3\xi_3 x_6 + b_3 x_5)(L_{313}x_1 + L_{323}x_3 + L_{333}x_5) \\
&+ b_3[-M_{311}x_2^2 - M_{322}x_4^2 - M_{333}x_6^2 - M_{312}x_2 x_4 \\
&- M_{323}x_4 x_6 - M_{313}x_2 x_6]\}\,d\tau \\
&+ \{f_3 + c_{31}x_1^2 + c_{32}x_3^2 + c_{33}x_5^2 + c_{34}x_1 x_3 + c_{35}x_1 x_5 + c_{36}x_3 x_5 \\
&+ c_{37}x_1 + c_{38}x_3 + c_{39}x_5\}\,dB
\end{aligned} \tag{19}
$$

where

$$b_i = \varepsilon\alpha/M_{ii}$$

$$c_{i1} = b_i \sum_{k=1}^{3}\left(b_k \sum_{j=1}^{3} f_j L_{i1k}L_{k1j}\right)$$

$$c_{i2} = b_i \sum_{k=1}^{3}\left(b_k \sum_{j=1}^{3} f_j L_{i2k}L_{k2j}\right)$$

$$c_{i3} = b_i \sum_{k=1}^{3} \left(b_k \sum_{j=1}^{3} f_j L_{i3k} L_{k3j} \right)$$

$$c_{i4} = b_i \left[\sum_{k=1}^{3} \left(b_k \sum_{j=1}^{3} f_j L_{i1k} L_{k2j} \right) + \sum_{k=1}^{3} \left(b_k \sum_{j=1}^{3} f_j L_{i2k} L_{k1j} \right) \right]$$

$$c_{i5} = b_i \left[\sum_{k=1}^{3} \left(b_k \sum_{j=1}^{3} f_j L_{i1k} L_{k3j} \right) + \sum_{k=1}^{3} \left(b_k \sum_{j=1}^{3} f_j L_{i3k} L_{k1j} \right) \right]$$

$$c_{i6} = b_i \left[\sum_{k=1}^{3} \left(b_k \sum_{j=1}^{3} f_j L_{i2k} L_{k3j} \right) + \sum_{k=1}^{3} \left(b_k \sum_{j=1}^{3} f_j L_{i3k} L_{k2j} \right) \right]$$

$$c_{i7} = -b_i \sum_{j=1}^{3} f_j L_{i1j}$$

$$c_{i8} = -b_i \sum_{j=1}^{3} f_j L_{i2j}$$

$$c_{i9} = -b_i \sum_{j=1}^{3} f_j L_{i3j}$$

The evolution of the probability density of the joint response coordinates $\mathbf{X}$ is described by the Fokker–Planck equation

$$\frac{\partial p(\mathbf{X}, \tau)}{\partial \tau} = -\sum_{i=1}^{6} \frac{\partial}{\partial X_i} [a_i(\mathbf{X}, \tau) p(\mathbf{X}, \tau)]$$
$$+ \frac{1}{2} \sum_{i=1}^{6} \sum_{j=1}^{6} \frac{\partial^2}{\partial X_i \partial X_j} [b_{ij}(\mathbf{X}, \tau) p(\mathbf{X}, \tau)] \qquad (20)$$

where $p(\mathbf{X}, \tau)$ is the response joint probability density function, and $a_i(\mathbf{X}, \tau)$ and $b_{ij}(\mathbf{X}, \tau)$ are the first and second incremental moments evaluated as follows

$$a_i(\mathbf{X}, \tau) = \lim_{\Delta\tau \to 0} \frac{1}{\Delta\tau} E[X_i(\tau + \Delta\tau) - X_i(\tau)] \qquad (21)$$

$$b_{ij}(\mathbf{X}, \tau) = \lim_{\Delta\tau \to 0} \frac{1}{\Delta\tau} E[\{X_i(\tau + \Delta\tau) - X_i(\tau)\}\{X_j(\tau + \Delta\tau) - X_j(\tau)\}]$$

In order to construct the Fokker–Planck equation of the present system the coefficients a_i and b_{ij} are evaluated. In view of the complicated analytical manipulations involved in this section and subsequent sections the MACSYMA programming is used throughout the analysis of this paper. It is obvious that the system Fokker–Planck equation cannot be solved even for the stationary case. Instead, one may generate a general first order differential equation describing the evolution of response joint moments of any order. This equation is obtained by multiplying both sides of the Fokker–Planck equation by the scalar function $\Phi(\mathbf{X})$

$$\Phi(\mathbf{X}) = X_1^{k1} X_2^{k2} \ldots X_6^{k6} \tag{22}$$

and integrating by parts over the entire state space $-\infty < \mathbf{X} < +\infty$. The boundary conditions are used

$$p(\mathbf{X} \rightarrow -\infty) = p(\mathbf{X} \rightarrow \infty) = 0 \tag{23}$$

The resulting moment equation is very long and it will not be listed in this paper. However, the general form of this equation is

$$m'_N = F_N(m_1, m_2, \ldots, m_N, m_{N+1}) \tag{24}$$

In deriving the moment equation the following notation is adopted

$$m_{k1,k2,\ldots,k6} = \int_{-\infty}^{\infty} \ldots \int p(\mathbf{X}, \tau) \Phi(\mathbf{X}) \, dX_1 \, dX_2 \ldots dX_6 \tag{25}$$

Equation (24) constitutes a set of infinite coupled equations. In other words, the differential equation of order N contains moment terms of order N and $N + 1$. In Ref. 21, these equations were closed via a closure scheme based on the assumption that the response process is very close to a normal process. However, the system mean square responses revealed that the nonlinear interaction took place only between two normal modes, instead of three, although the system was tuned to the combination internal resonance. In order to clarify this deficiency the system response will be further examined by a nonGaussian closure based on the concept of cumulant-neglect.

6 GAUSSIAN CLOSURE SOLUTION

This section briefly reviews the main results obtained in Ref. 21 for the sake of completeness. The moment equations were closed by setting all third order cumulants to zero, i.e.

$$\lambda_3[X_i X_j X_k] = E[X_i X_j X_k] - \sum^{3} E[X_i]E[X_j X_k] + 2E[X_i]E[X_j]E[X_k] = 0 \tag{26}$$

where the number over summation sign refers to the number of terms generated in the form of the indicated expression without allowing permutation of indices. Relation (26) is used to obtain expressions for third order moments in terms of first and second order moments. These expressions are then used to close the second order moment equations generated from the general eqn (24). In this case one can generate 27 coupled equations in the first (6 equations) and second (21 equations) order moments. The solution of the closed 27 coupled moment equations is obtained numerically by using the IMSL DVERK Subroutine (Runge–Kutta–Verner fifth and sixth numerical integration method). Depending on the value of internal detuning parameter r the system response may be reduced to the linear response or may become quasi-stationary which deviates significantly from the linear solution. The numerical integration is carried out on the IBM-3081 computer which takes 61·08 s CPU time with accuracy 0·1D-06 for the case of a quasi-stationary solution ($r = 1·18$). Figure 4 presents the transient and steady-state responses for $\zeta_i = 0·01$, $\varepsilon = 0·05$, and $r = 1·12$. The steady state solution converges to the stationary linear solution derived in Ref. 21. For $r = 1·18$ the response significantly deviates from the linear solution. Figures 5 and 6 show the

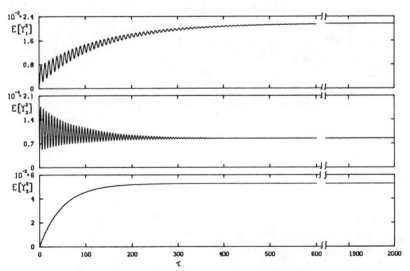

Fig. 4. Transient and steady mean square responses according to Gaussian closure solutions for $\zeta_i = 0·01$, $\varepsilon = 0·05$, and $r = 1·12$. The steady state response converges to the stationary linear solution at $r = 1·12$.

transient and steady state responses indicated by the dotted curves (G) for excitation spectral density $D/2\zeta_3 = 1\cdot0$, damping ratios $\zeta_i = 0\cdot01$, and nonlinear coupling parameter $\varepsilon = 0\cdot025$ and $0\cdot05$, respectively. The transient response shows that the autoparametric coupling takes place between the second and third normal modes in a form of energy exchange. It is seen that the steady state response does not achieve a stationary value but fluctuates between two limits. The values of the two limits are divided by the linear solution and the ratios are plotted against the detuning parameter r as shown in Figs 7 and 8 for two different values of the nonlinear coupling parameter ε. The region of autoparametric interaction is seen to become wider as the nonlinear coupling parameter increases and as the damping ratios ζ_i decrease. These two figures reveal the fact that the nonlinear interaction takes place within a small range of internal detuning parameter around $r = 1\cdot18$ which is well remote from the exact internal resonance $r = 1\cdot0$. The authors have made several attempts to determine the response statistics under the condition of exact internal resonance $r = 1\cdot0$. However, all numerical solutions converge to the linear response and the Gaussian closure fails to predict any nonlinear interaction between the three modes at $r = 1\cdot0$. Inspection of the frequency ratios ω_3/ω_2, ω_3/ω_1, and ω_2/ω_1, shown in Fig. 9, reveal that at $r = 1\cdot18$ the second and third modes are in exact internal resonance, i.e. $\omega_3 = 2\omega_2$. If one considers only the equations of motion which govern the nonlinear coupling of the second and third modes with the condition $\omega_3 = 2\omega_2$, the system moment equations are then reduced to 69 equations whose numerical solutions coincides with the response presented in Figs 7 and 8. It is obvious that the Gaussian closure scheme is not adequate to predict the nonlinear three modal interaction and this is the main reason for considering the nonGaussian closure approach in the next section.

7 NONGAUSSIAN CLOSURE SOLUTION

The nonGaussian closure scheme takes into account the effect of nonnormality of the response probability density and thus is expected to provide adequate modeling for the system nonlinear random response. As a first order approximation the third and fourth order cumulants will be considered in the analysis and all higher order cumulants are set to zero. In this case one has to generate moment

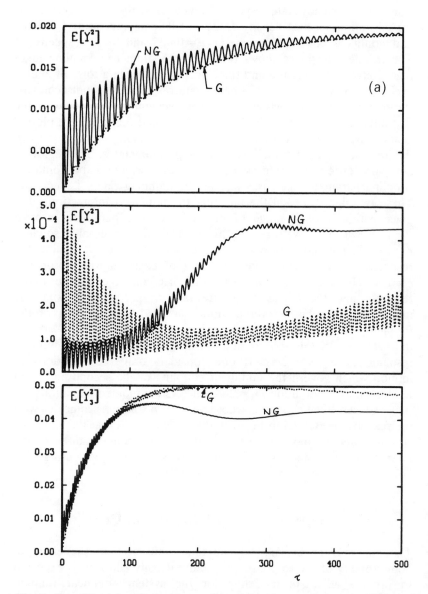

Fig. 5. Transient (a) and steady state (b) responses according to Gaussian (G) and nonGaussian (NG) closure solutions, for $\zeta_i = 0.01$, $\varepsilon = 0.025$, $\omega_3/(\omega_1 + \omega_2) = 1.18$, $D/2\zeta_3 = 1.0$.

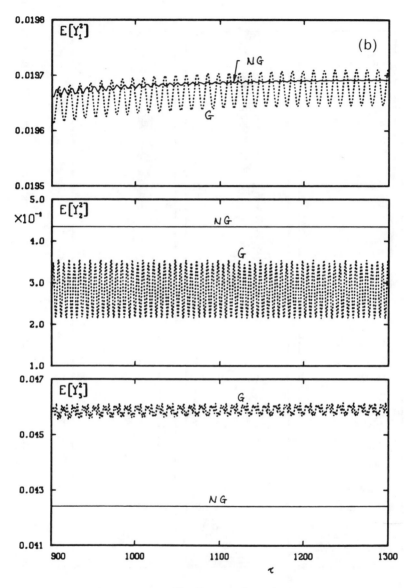

Fig. 5.—*contd.*

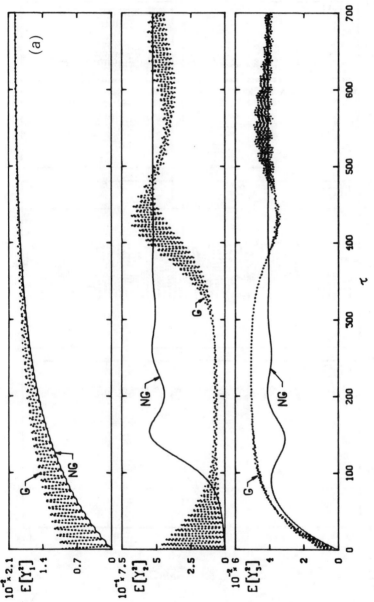

Fig. 6. Transient (a) and steady state (b) responses according to Gaussian (G) and nonGaussian (NG) closure solutions, for $\zeta_i = 0.01$, $\varepsilon = 0.05$, $\omega_3/(\omega_1 + \omega_2) = 1.18$, $D/2\zeta_3 = 1.0$.

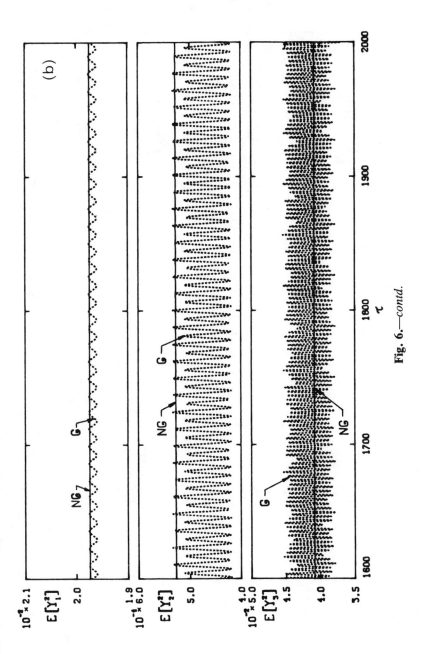

Fig. 6.—*contd.*

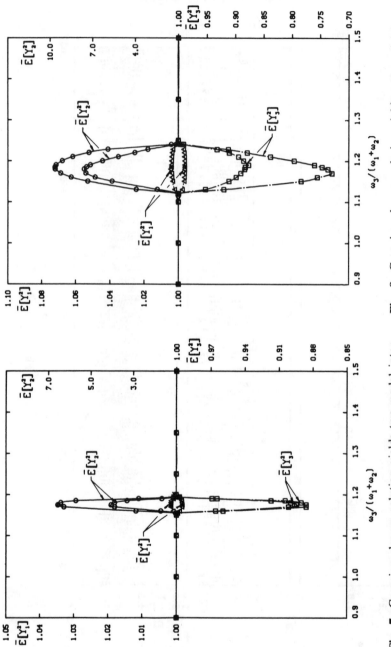

Fig. 7. Gaussian closure solution yields two-modal interaction between second and third normal modes ($\varepsilon = 0.025$, $\zeta_i = 0.01$, $D/2\zeta_3 = 1.0$).

Fig. 8. Gaussian closure solution yields two-modal interaction between second and third normal modes ($\varepsilon = 0.05$, $\zeta_i = 0.01$, $D/2\zeta_3 = 1.0$).

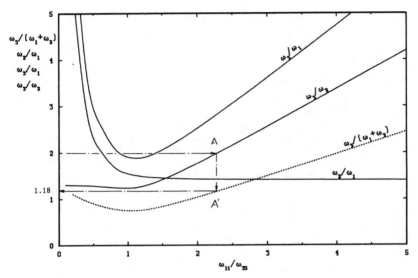

Fig. 9. Frequency ratios of two and three normal modes versus system parameters for $\omega_{22}/\omega_{33} = 1\cdot4$, $m_2/m_1 = 0\cdot5$, $m_3/m_1 = 5\cdot0$, $l_2/l_1 = 0\cdot25$.

equations up to fourth order. Fifth order moments which appear in the fourth order moment equations will be replaced by fourth and lower order moments by using the relationship

$$\lambda_5[X_iX_jX_kX_lX_m] = E[X_iX_jX_kX_lX_m] - \sum^5 E[X_i]E[X_jX_kX_lX_m]$$

$$+ 2\sum^{10} E[X_i]E[X_j]E[X_kX_lX_m]$$

$$- 6\sum^{10} E[X_i]E[X_j]E[X_k]E[X_lX_m]$$

$$+ 2\sum^{15} E[X_i]E[X_jX_k]E[X_lX_m]$$

$$- \sum^{10} E[X_iX_j]E[X_kX_lX_m]$$

$$+ 24E[X_i]E[X_j]E[X_k]E[X_l]E[X_m] = 0 \qquad (27)$$

The number N of moment equations of order K is given by the relationship[14]

$$N = n(n+1)(n+2)\ldots(n+K-1)/K! \qquad (28)$$

where n is the number of state coordinates $\mathbf{X}$. For the present problem one needs to generate 6 equations for first order moments, 21 equations for second order, 56 equations for third order and 126 equations for fourth order, with a total of 209 first order differential equations which are closed by using relation (27).

The 209 differential equations are solved by numerical integration using the DVERK subroutine on the IBM3081 computer. For one complete time history response the numerical integration took 748·48 s CPU time (i.e. over 12 times the CPU time of the Gaussian closure solution) with accuracy of 0·1D-06. Figures 5 and 6 show the transient and steady state responses indicated by solid curves (NG) for $r = 1·18$ which corresponds to two-modal interaction between second and third modes. Again the transient response shows nonlinear interaction in a form of energy exchange between second and third normal modes. Contrary to the Gaussian closure solution, it is seen that the steady state response achieves a strictly stationary solution. The numerical integration is carried out for the 209 equations at $r = 1·0$ to find out if the nonGaussian closure predicts nonlinear three modal interaction. Figure 10 shows the transient and steady state mean square response for $\zeta_i = 0·01$, $\varepsilon = 0·05$. The CPU time taken for one complete time

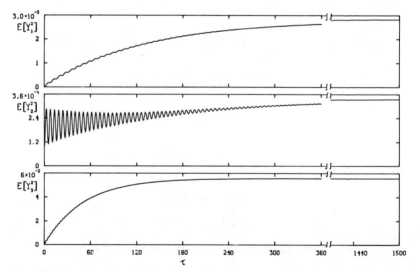

Fig. 10. NonGaussian closure of nonlinear three-modal interaction for $\varepsilon = 0·05$, $\zeta_i = 0·01$, $D/2\zeta_3 = 1·0$.

history record is 1414 s which is much longer than the CPU time for Gaussian closure solution. The fluctuations observed in the transient response are entirely dependent on the initial conditions introduced in the numerical algorithm. For example all response records obtained with zero initial conditions do not show any fluctuations in the transient period.

Since the nonGaussian closure scheme yields a stationary solution it is possible to solve only for the steady state by setting the left hand sides of the 209 equations to zero and solving the resulting nonlinear algebraic equations numerically. The numerical solution is carried out by using the ZSPOW (Solve a System of Nonlinear Equations) subroutine on the IBM-3081 computer. The solution is obtained by assuming initial guessing (approximate) values for the roots of the equations. Convergence of the solution is reached when the initial roots are closed to the exact solution. The decision of accepting valid roots is based on two main criteria. The first is the nonnegativeness of the even order moments, and the second is to satisfy Schwarz's inequality. Another important criterion is the positiveness of the joint probability density of the response coordinates especially at the tails. However, in view of the problem complexity the authors did not inspect this criterion. It is noteworthy to mention that once the solution is obtained for one point, the solution of all subsequent points is generated with less effort. The CPU time for one point solution varies between 40 and 60 s depending on the initial guessing values, with accuracy of $0·1D-06$.

The steady state solution is plotted against the internal detuning parameter r for various values of system parameters and excitation spectral density level. Figures 11–15 show three- and two-modal nonlinear interactions which occur at $r = 1·0$ and $1·18$, respectively. It is seen that the regions of autoparametric interaction become wider as the nonlinear coupling parameter ε and excitation level $D/2\zeta_3$ increase and as the damping ratios decrease. For most system and excitation parameters used in Figs 11–14 the two modal interaction is stronger than the three modal interaction. Significant three modal interaction arises for relatively larger values of the nonlinear coupling parameter ε and small ratios ζ_i as shown in Fig. 15.

Since the main objective of this study is to examine the random response characteristics of three modal interaction, attention is focused on the response characteristics in the neighborhood of exact internal resonance $r = 1·0$. The mean square responses of the three

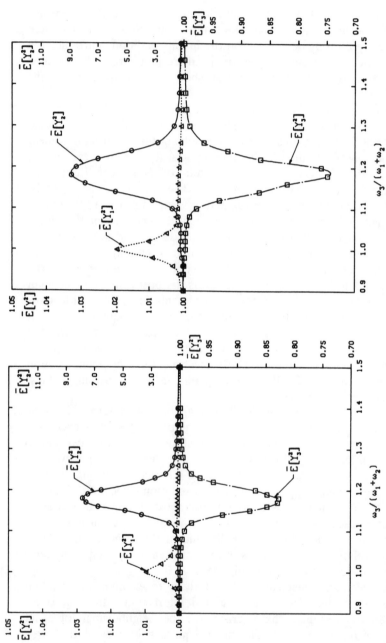

Fig. 11. NonGaussian closure solution showing two-modal interaction at $r = 1.18$, and weak three-modal interaction at $r = 1.0$, for $\varepsilon = 0.025$, $\zeta_i = 0.01$, $D/2\zeta_3 = 1.0$.

Fig. 12. NonGaussian closure solution showing two-modal interaction at $r = 1.18$, and weak three-modal interaction at $r = 1.0$, for $\varepsilon = 0.025$, $\zeta_i = 0.01$, $D/2\zeta_3 = 2.0$.

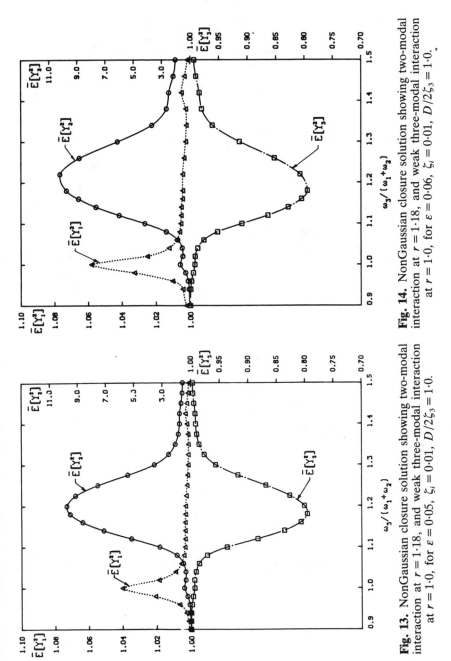

Fig. 13. NonGaussian closure solution showing two-modal interaction at $r = 1 \cdot 18$, and weak three-modal interaction at $r = 1 \cdot 0$, for $\varepsilon = 0 \cdot 05$, $\zeta_i = 0 \cdot 01$, $D/2\zeta_3 = 1 \cdot 0$.

Fig. 14. NonGaussian closure solution showing two-modal interaction at $r = 1 \cdot 18$, and weak three-modal interaction at $r = 1 \cdot 0$, for $\varepsilon = 0 \cdot 06$, $\zeta_i = 0 \cdot 01$, $D/2\zeta_3 = 1 \cdot 0$.

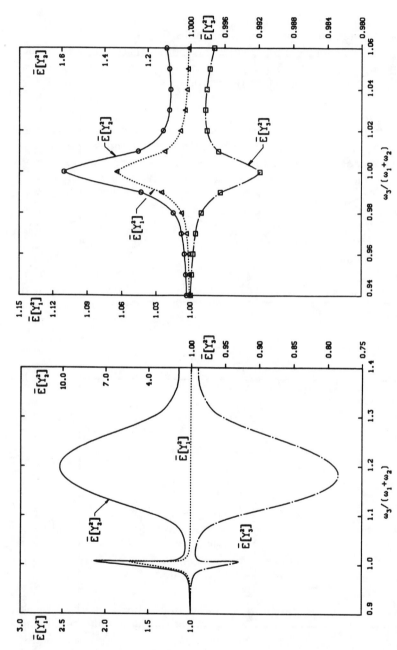

Fig. 15. NonGaussian closure solution showing two-modal interaction at $r = 1.18$, and weak three-modal interaction at $r = 1.0$, for $\varepsilon = 0.05$, $\zeta_i = 0.003$, $D/2\zeta_3 = 1.0$.

Fig. 16. NonGaussian closure solution showing strong three-modal interaction for $\varepsilon = 0.025$, $\zeta_i = 0.004$, $D/2\zeta_3 = 1.0$.

normal modes are plotted in Figs 16–21 for various values of nonlinear coupling parameter ε and damping ratios ζ_i. These figures demonstrate the development of complex response characteristics as the nonlinear coupling parameter increases and as the damping ratios decrease. The autoparametric interaction occurs between the three modes in such a manner that the mean square responses of the first two normal modes is always greater than the linear solution ($>1\cdot0$) while it is less for the third normal mode. This means that the nonlinear interaction takes place between the first and second modes on one hand and the third mode on the other hand. A new feature of considerable interest is the contrast in the form of the mean square response curves above the exact detuning ratio $r > 1\cdot0$ for a certain combination of system parameters and excitation level as shown in Figs 18–21. This is indicated by the multiple solutions over a finite portion of the internal detuning parameter. At points of vertical tangency the response mean squares exhibit the jump and collapse phenomena indicated by the arrows AB and CD, respectively, see Fig. 18. Those solutions shown by the upper and lower branches BC and AD are verified by numerical integration. However, all numerical integration attempts made at points very close to the middle branch AC converge to either the upper or lower branches. This implies that the middle solution is always unstable which is analogous to a great extent to deterministic solutions of nonlinear systems. For systems with quadratic nonlinearity, the deterministic theory of nonlinear vibrations predicts similar phenomena.

The well-known saturation phenomenon reported by Nayfeh & Mook[4] does not take place for dynamic systems with quadratic nonlinearity subjected to wide band random excitation since the excitation includes a wide range of frequencies which always excite the system normal modes. The influence of the excitation spectral density level $D/2\zeta_3$ upon the response mean squares is shown in Fig. 22 damping ratios $\zeta_i = 0\cdot002$ and coupling parameter $\varepsilon = 0\cdot05$. This figure shows that the system has three possible solutions for the same excitation level only if the internal detuning is slightly shifted from the exact internal resonance $r = 1\cdot0$. At points of vertical tangency the response mean square will experience the jump and collapse phenomena as shown by the arrows AB and CD, respectively.

Figures 18–21 reveal that the region of internal detuning over which multiple solutions take place is dependent on the nonlinear coupling parameter and damping ratios of the three normal modes. In order to

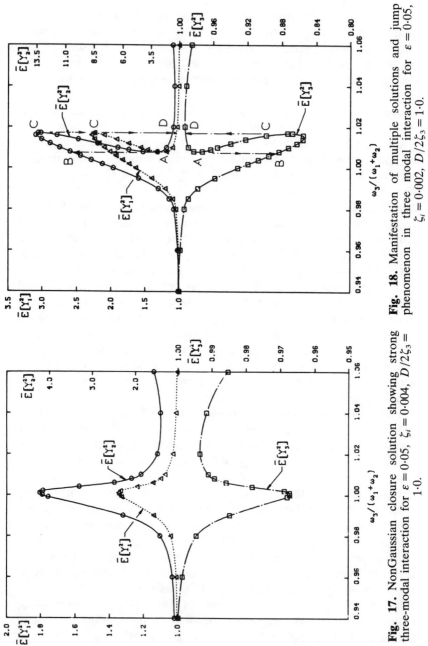

Fig. 17. NonGaussian closure solution showing strong three-modal interaction for $\varepsilon = 0.05$, $\zeta_i = 0.004$, $D/2\zeta_3 = 1.0$.

Fig. 18. Manifestation of multiple solutions and jump phenomenon in three modal interaction for $\varepsilon = 0.05$, $\zeta_i = 0.002$, $D/2\zeta_3 = 1.0$.

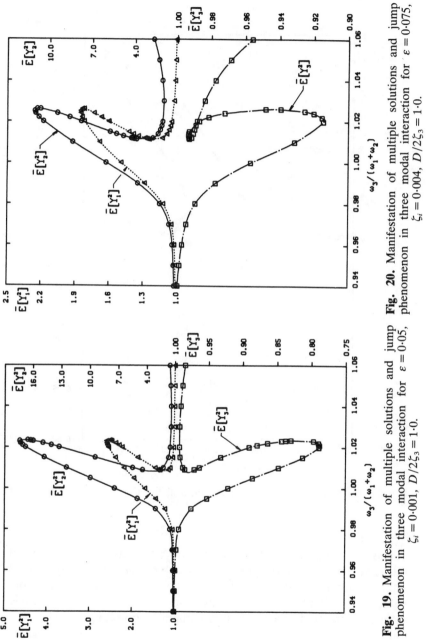

Fig. 19. Manifestation of multiple solutions and jump phenomenon in three modal interaction for $\varepsilon = 0.05$, $\zeta_i = 0.001$, $D/2\zeta_3 = 1.0$.

Fig. 20. Manifestation of multiple solutions and jump phenomenon in three modal interaction for $\varepsilon = 0.075$, $\zeta_i = 0.004$, $D/2\zeta_3 = 1.0$.

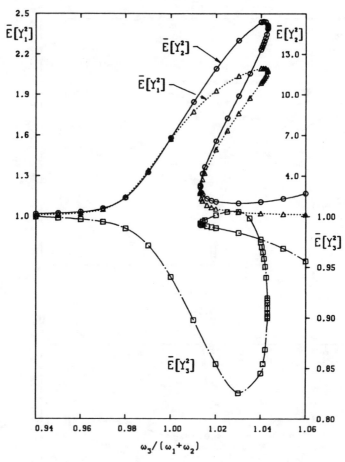

Fig. 21. Development of complex response in the third normal mode mean square for $\varepsilon = 0{\cdot}075$, $\zeta_i = 0{\cdot}002$, $D/2\zeta_3 = 1{\cdot}0$.

define the region of multiple solutions a parametric study is carried out. The results are shown in Fig. 23 which displays a set of regions of multiple solutions for various values of damping ratios. The threshold value of ε^* where the mean square responses bifurcate into multiple solutions is plotted as a function of damping ratios ζ_i in Fig. 24.

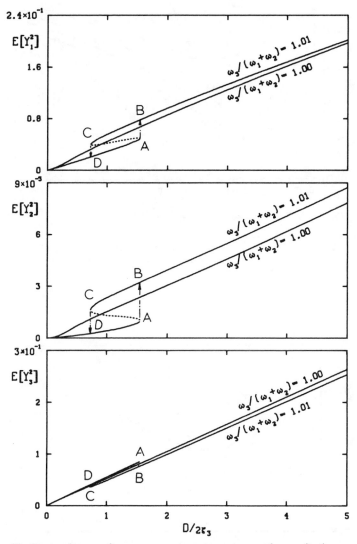

Fig. 22. Dependence of mean square responses on the excitation spectral density, a region of multiple solutions for $\varepsilon = 0.05$, $\zeta_i = 0.002$.

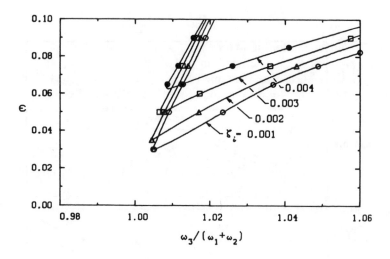

Fig. 23. Region of multiple solutions for various values of ζ_i.

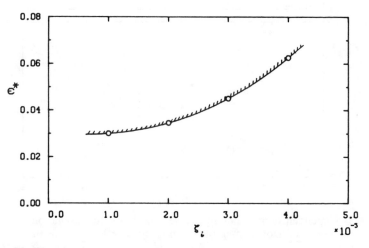

Fig. 24. Threshold value of ε above where the mean square responses have multiple solutions.

8 CONCLUSIONS

The nonlinear random modal interaction of a three-degree-of-freedom structural model is examined in the neighborhood of combination internal resonance of the summed type. The Fokker–Planck equation approach together with a nonGaussian closure scheme are used to determine the response statistics. Contrary to the Gaussian closure scheme results, the nonGaussian closure yields several new features of response characteristics. These include weak and strong three modal interaction, multiple solutions, and jump phenomena. Multiple solutions only occur over a finite region of internal detuning parameter which is slightly greater than the exact internal resonance condition. At exact internal combination resonance the system possesses a unique stable limit cycle.

ACKNOWLEDGEMENT

This research is supported by a grant from the Air Force Office of Scientific Research under grant no. AFOSR-85-0008. Dr A. Amos is the program director.

REFERENCES

1. Bolotin, V. V., *The Dynamic Stability of Elastic Systems*. Holden-Day, Inc., San Francisco, 1964.
2. Barr, A. D. S., Some developments in parametric stability and nonlinear vibration. *Proceedings of International Conference of Recent Advances in Structural Dynamics*, Southampton, England, 1980, pp. 545–68.
3. Minorsky, N., *Nonlinear Oscillations*. Van Nostrand, New York, 1962.
4. Nayfeh, A. H. & Mook, D. T., *Nonlinear Oscillations*. Wiley-Interscience, New York, 1979.
5. Schmidt, G. & Tondl, A., *Nonlinear Vibrations*. Cambridge University Press, Cambridge, 1986.
6. Haxton, R. S. & Barr, A. D. S., The autoparametric vibration absorber. *Journal of Engineering for Industry*, **94** (1972) 119–25.
7. Nayfeh, A. H., Mook, D. T. & Marshall, L. R., Nonlinear coupling of pitch and roll modes in ship motions. *Journal of Hydronautics*, **7** (1973) 145–52.
8. Mook, D. T., Marshall, L. R. & Nayfeh, A. H., Subharmonic and superharmonic resonances in the pitch and roll modes of ship motions. *Journal of Hydronautics*, **8** (1974) 32–40.

9. Haddow, A. G., Barr, A. D. S. & Mook, D. T., Theoretical and experimental study of modal interaction in a two-degree-of-freedom structure. *Journal of Sound and Vibration*, **97**(3) (1984) 451–73.
10. Ibrahim, R. A. & Woodall, T. D., Linear and nonlinear modal analysis of aeroelastic structural systems. *Computer & Structures*, **22**(4) (1986) 699–707.
11. Bux, S. L. & Roberts, J. W., Nonlinear vibratory interactions in systems of coupled beams. *Journal of Sound and Vibration*, **104**(3) (1986) 497–520.
12. Roberts, J. W. & Zhang, J. Z., Nonlinear resonance absorption in vibratory systems. *Proceedings of the 5th International Modal Analysis Conference*, Imperial College, London, 1987, pp. 427–34.
13. Bolotin, V. V., *Random Vibration of Elastic Systems*. Martinus Nijhoff Publishers, The Netherlands, 1984.
14. Ibrahim, R. A., *Parametric Random Vibration*. John Wiley, New York, 1985.
15. Piszczek, K. & Niziot, J., *Random Vibration of Mechanical Systems*. John Wiley, New York, 1986.
16. Ibrahim, R. A. & Roberts, J. W., Broad band random excitation of a two degree-of-freedom system with autoparametric coupling. *Journal of Sound and Vibration*, **44**(3) (1976) 335–48.
17. Ibrahim, R. A. & Roberts, J. W., Stochastic stability of the stationary response of a system with autoparametric coupling. *Zeitschrift für Angewandte Mathematik und Mechanik (ZAMM)*, **57**(9) (1977) 643–9.
18. Schmidt, G., Probability densities of parametrically excited random vibration. *Stochastic Problems in Dynamics*, ed. B. L. Clarkson, Pitman, London, 1977, pp. 197–213.
19. Ibrahim, R. A. & Heo, H., Autoparametric vibration of coupled beams under random support motion. *Journal of Vibration, Acoustics, Stress, and Reliability in Design*, **108**(4) (1986) 421–6.
20. Ibrahim, R. A. & Heo, H., Stochastic response of nonlinear structures with parameter random fluctuations. *American Institute of Aeronautics and Astronautics Journal*, **25**(2) (1987) 331–8.
21. Ibrahim, R. A. & Hedayati, Z., Stochastic modal interaction in linear and nonlinear aeroelastic structures. *Probabilistic Engineering Mechanics*, **1**(4) (1986).
22. Craig, R. R., *Structural Dynamics: An Introduction to Computer Methods*. John Wiley, New York, 1981.

9

Multiple Response Moments and Stochastic Stability of a Nonlinear System

R. N. IYENGAR

Department of Civil Engineering, Indian Institute of Science, Bangalore, India

ABSTRACT

It is known that under a periodic input, the cubic oscillator or the Duffing equation can have two stable response amplitudes. This leads to the well investigated jump phenomenon. It is of interest to study how the above system responds under a narrow band excitation. Similarly when the excitation contains both a periodic term and noise, it would be interesting to investigate the mean and mean square responses. For the cubic oscillator the response moments can be obtained using the equivalent linearization approximation. It has been observed that the response variance shows multiple values for a certain range of parameters. If a stability analysis is done on these steady state moment values it is found that two solutions are stable out of the possible three. Similarly when the excitation is a noise plus periodic function, both the mean and variance show multiple values. However, these results have to be interpreted carefully. While the approximate moment equations may exhibit multiple stable solutions, this cannot be taken to mean that the original nonlinear equation also has multiple steady random solutions. It is essential to verify whether the corresponding solutions are asymptotically stable or not. In the present paper both the above problems are discussed along with a stochastic stability analysis. It is shown that even though the moment equations show multiple solutions, only one of these is asymptotically stable and hence realizable in practice.

159

1 INTRODUCTION

Nonlinear systems exhibit many interesting phenomena under deterministic periodic excitations. For example, it is known that under a periodic input, the cubic oscillator or the Duffing equation can have two stable response amplitudes. This leads to the well investigated jump phenomenon. Thus, it is of interest to study how the above system responds under a narrow band excitation. Similarly when the excitation contains both a periodic term and noise, it would be interesting to investigate the mean and mean square responses. Lyon et al.[1] in experiments and Lennox & Kuak[2] in numerical simulation have observed that there could be amplitude jumps in the sample response of the hardening cubic oscillator under narrow band excitation. In theoretical investigations, one generally finds the response moments instead of the sample responses. For the cubic oscillator the response moments using the equivalent linearization approximation have been obtained by Iyengar[3] and Davies & Nandlall.[4] Recently Jia & Fang[5] have obtained results on the response moments of coupled cubic oscillators. In these studies it has been observed that the response variance shows multiple values for a certain range of parameters. If a stability analysis is done on these steady state moment values it is found that two solutions are stable out of the possible three. Similarly when the excitation is a noise plus periodic function, both the mean and variance show multiple values. However, these results have to be interpreted carefully. While the approximate moment equations may exhibit multiple stable solutions, this cannot be taken to mean that the original nonlinear equation also has multiple steady random solutions. It is essential to verify whether the corresponding solutions are asymptotically stable or not. In the present paper both the above problems are discussed along with a stochastic stability analysis. It is shown that even though the moment equations show multiple solutions, only one of these is asymptotically stable and hence realizable in practice.

2 COMBINED NOISE AND PERIODIC EXCITATION

The cubic oscillator is governed by the equation

$$\ddot{z} + 2\eta\omega\dot{z} + \omega^2(z + \alpha z^3) = w(t) + Q\lambda^2 \sin \lambda t \qquad (1)$$

Here $w(t)$ is a Gaussian white noise with autocorrelation

$$\langle w(t_1)w(t_2)\rangle = I\delta(t_2 - t_1) \tag{2}$$

The response variable is now transformed in terms of the linear variance σ_L^2 as

$$x = z/\sigma_L; \qquad \sigma_L^2 = I/(4\eta\omega^3) \tag{3}$$

This leads to

$$\ddot{x} + 2\eta\omega\dot{x} + \omega^2(x + \alpha\sigma_L^2 x^3) = (w/\sigma_L) + (\lambda^2 Q/\sigma_L)\sin\lambda t \tag{4}$$

The process $x(t)$ can be expressed as

$$x(t) = m(t) + y(t) \tag{5}$$

where $m(t)$ is the mean response. Upon substituting this in eqn (4), and taking ensemble average, under the assumption of gaussianness of $y(t)$, it follows

$$\ddot{m} + 2\eta\omega\dot{m} + \omega^2 m + \varepsilon\sigma_L^2(m^3 + 3\sigma^2 m) = (\lambda^2 Q/\sigma_L)\sin\lambda t \tag{6}$$

Further, one gets for the process, y, the equivalent linear equation

$$\ddot{y} + 2\eta\omega\dot{y} + \omega^2 y + 3\varepsilon\sigma_L^2(m^2 + \sigma^2)y = w/\sigma_L \tag{7}$$

This equation has $m(t)$ and $\sigma(t)$ as time varying coefficients. However, for large time t, if $y(t)$ attains stationarity, σ^2 will be nearly a constant and hence $m(t)$ of eqn (6) can be taken as

$$m = R\sin(\lambda t - \phi) \tag{8}$$

Now, from harmonic balance one gets

$$R^2 = \bar{Q}^2\bar{\lambda}^4/[(1 + 3\varepsilon\sigma^2 + 0{\cdot}75\varepsilon R^2 - \bar{\lambda}^2)^2 + (2\eta\bar{\lambda})^2]$$
$$\tan\phi = 2\eta\bar{\lambda}/[1 + 3\varepsilon\sigma^2 + 0{\cdot}75\varepsilon R^2 - \bar{\lambda}^2] \tag{9}$$
$$\bar{Q} = Q/\sigma_L; \qquad \bar{\lambda} = \lambda/\omega, \qquad \varepsilon = \alpha\sigma_L^2$$

These contain σ^2 as an unknown for which an expression can be derived from eqn (7). The differential equations for the moments

$$S_1 = \langle y^2\rangle = \sigma^2; \qquad S_2 = \langle\dot{y}^2\rangle, \qquad S_3 = \langle y\dot{y}\rangle \tag{10}$$

are

$$\dot{S}_1 = 2S_3$$
$$\dot{S}_2 = I\sigma_L^{-2} - 2\omega^2[1 + 3\varepsilon(m^2 + \sigma^2)]S_3 - 4\eta\omega S_2 \tag{11}$$
$$\dot{S}_3 = S_2 - \omega^2[1 + 3\varepsilon(m^2 + \sigma^2)]S_1 - 2\eta\omega S_3$$

Here m is a time varying term. If in the first approximation, this is estimated as $m^2 \simeq 0 \cdot 5R^2$, in the steady state the moment derivatives vanish leading to

$$6\varepsilon\sigma^2 = [\{(1 + 1 \cdot 5\varepsilon R^2)^2 + 12\varepsilon\}^{1/2} - (1 + 1 \cdot 5\varepsilon R^2)] \qquad (12)$$

Equations (9) and (12) have to be solved simultaneously to find R, ϕ and σ.

3 STOCHASTIC STABILITY

In the presence of only the periodic excitation the amplitude R can have three solutions of which two are stable. Thus, it is expected that in the present case also, with added noise, R and hence σ exhibit three solutions. However, these will be realizable only if the sample solutions are stable. This stability study is complicated since eqn (4) contains a stochastic term also. An approximate analysis is possible by representing the sample solution of eqn (4) in the form

$$x_0(t) = R \sin(\lambda t - \phi) + a \sin(\omega_e t - \theta) \qquad (13)$$

Here $a(t)$ and $\theta(t)$ are slowly varying envelope and phase of the narrow band, approximately ergodic, Gaussian process $y(t)$. ω_e is the effective frequency of the oscillator given by

$$\omega_e = \omega[1 + 3\varepsilon(\sigma^2 + 0 \cdot 5R^2)]^{1/2} \qquad (14)$$

The above solution will be stable provided the solution of the variational equation of eqn (4), namely,

$$\ddot{v} + 2\eta\omega\dot{v} + \omega^2(1 + 3\varepsilon x_0^2)v = 0 \qquad (15)$$

is almost sure asymptotically stable. The details of the stability analysis are presented elsewhere (Iyengar[6]). The analysis shows that out of the possible three mean values and variances only one value is realizable. As an example the numerical results for a system with $\eta = 0 \cdot 08$, $\varepsilon = 0 \cdot 5$, $\bar{Q} = 0 \cdot 5$ is shown in Figs 1 and 2. In Fig. 1 the deterministic amplitude R for the case of $f = 0$ is also shown. The same problem has been studied by Bulsara et al.[7] by a combination of harmonic and statistical averaging without any kind of stability analysis. Their results are also shown in these figures. The theoretical results are also compared with numerical simulation considering 100 samples. It is observed that the numerical simulation results compare favourably well with the theoretical predictions.

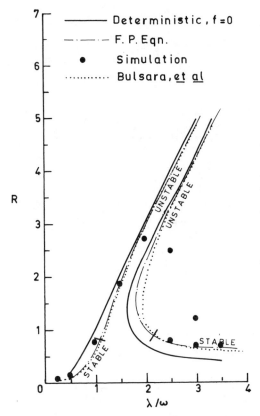

Fig. 1. Steady state amplitude. $\varepsilon = 0\cdot5$, $\eta = 0\cdot08$, $\bar{Q} = 0\cdot5$.

4 NARROW BAND EXCITATION

For studying the effect of a narrow band excitation on the system of eqn (1) it is convenient to assume that $f(t)$ is a filtered white noise. Hence the governing equations after linearization (Lin[8]) are

$$\ddot{x} + 2\eta\omega\dot{x} + \omega^2(1 + 3\alpha\sigma^2)x = f(t) \tag{16}$$

$$\ddot{f} + 2\xi\lambda\dot{f} + \lambda^2\dot{f} = w(t) \tag{17}$$

Here $w(t)$ is a white noise of strength I as in eqn (2). λ and ξ control the centre frequency and band width of the excitation $f(t)$. It is

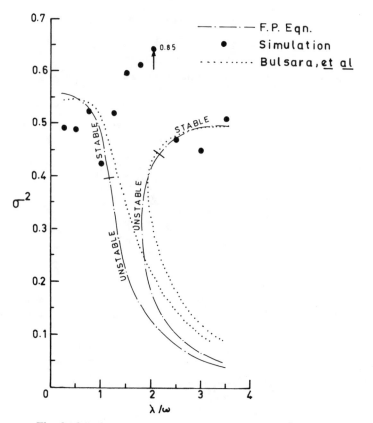

Fig. 2. Steady state variance. $\varepsilon = 0.5$, $\eta = 0.08$, $\bar{Q} = 0.5$.

convenient to nondimensionalize the response with respect to the linear system variance σ_L^2, corresponding to $\alpha = 0$, which is given by

$$\sigma_L^2 = (\sigma_f^2/\omega^4)\frac{[1 - \bar{\lambda}^2 + (1 + \bar{\xi}\bar{\lambda})(\bar{\lambda}^2 + 4\xi\eta\bar{\lambda})]}{[(1 - \bar{\lambda}^2)^2 + 4\eta\bar{\lambda}(1 + \bar{\xi}\bar{\lambda})(\xi + \eta\bar{\lambda})]} \quad (18)$$

The variance of (x/σ_L) is given by

$$\mu^2(1 + 3\varepsilon\mu^2)[\{D + 4\xi\eta\bar{\lambda}(1 + \bar{\xi}\bar{\lambda})\}(1 + 3\varepsilon\mu^2) + \bar{\lambda}^2\{4\eta^2(1 + \bar{\xi}\bar{\lambda}) - D\}]$$
$$= \beta^2[D + (1 + \bar{\xi}\bar{\lambda})(\bar{\lambda}^2 + 4\xi\eta\bar{\lambda})] \quad (19)$$

Here $\mu^2 = \sigma^2/\sigma_L^2$

$$D = (1 + 3\varepsilon\mu^2 - \bar{\lambda}^2); \qquad \beta^2 = \sigma_f^2/(\omega^4\sigma_L^2)$$
$$\bar{\lambda} = \lambda/\omega, \; \bar{\xi} = \xi/\eta; \qquad \varepsilon = \alpha\sigma_L^2 \tag{20}$$

The solution of the above equation gives three values for μ^2, in a certain range of $\bar{\lambda}$ values.

A typical solution of eqn (4) is shown in Fig. 3 for a system with $\varepsilon = 0.3$, $\eta = 0.08$ and $\bar{\xi} = 0.02$. A stability analysis is done along the lines mentioned previously. The variational equation is the same as

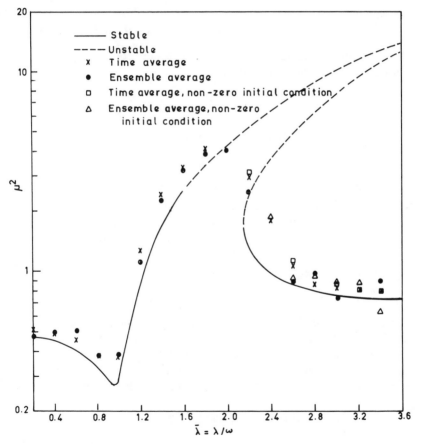

Fig. 3. Steady state variance. $\varepsilon = 0.3$, $\bar{\xi} = 0.02$, $\eta = 0.08$.

eqn (15) with

$$x_0 = R \sin(\lambda_e t - \theta) \tag{21}$$

The steady state centre frequency λ_e is taken as the average rate of upward zero crossing of x_0

$$\lambda_e = \lambda[(1 + \xi\bar{\lambda})(1 + 3\varepsilon\mu^2)]^{1/2}/[(1 + 3\varepsilon\mu^2) + \xi\bar{\lambda}(4\eta + \bar{\lambda}^2 + 4\xi\bar{\lambda})]^{1/2} \tag{22}$$

After introducing the transformations

$$\omega t = \tau, \qquad v = u \exp(-\eta\omega_e\tau) \tag{23}$$

$$\omega_e = \lambda_e/\omega; \qquad r = R/\sigma_L; \qquad C_1 = 1.5\varepsilon r^2 \omega_e^2; \qquad C_2 = C_1 + \omega_e^2(1 - \eta^2)$$

one gets

$$u'' + [C_0 - C_1 \cos 2(\tau - \theta)]u = 0 \tag{24}$$

This can be converted into two first order equations using the quasistatic approximation of Stratonovich[9]

$$\begin{aligned}
A' &= -C_{11}A - C_{12}B \\
B' &= C_{21}A + C_{22}B \\
C_{11} &= 0.25C_1 \sin 2\theta; \qquad C_{12} = 0.5(1 - C_0) - 0.25C_1 \cos 2\theta \\
C_{21} &= 0.5(1 - C_0) + 0.25C_1 \cos 2\theta; \qquad C_{22} = 0.05C_1 \sin 2\theta
\end{aligned} \tag{25}$$

These equations contain r and θ as slowly varying processes. The solution for A and B can be taken as

$$A = a_0 \exp\left(\int_o^\tau \zeta \, d\tau\right); \qquad B = b_0 \exp\left(\int_o^\tau \zeta \, d\tau\right) \tag{26}$$

For this to satisfy eqn (25) it is required that

$$\zeta = (C_{11}C_{22} - C_{12}C_{21})^{1/2} = 0.25[C_1^2 - 4(1 - C_0)^2]^{1/2} \tag{27}$$

The condition for the almost sure asymptotic stability of the solution x_0 is that every sample of v goes to zero with time. For this to happen the real part of $(-\eta\omega_e\tau + \int_0^\tau \zeta \, d\tau)$ should remain negative as $\tau \to \infty$. If the processes r and θ are assumed to be ergodic then ξ will also be ergodic and hence the above stability condition reduces to

$$\left[-\eta\omega_e + 0.25\int_{Re} [C_1^2 - 4(1 - C_0^2)]^{1/2} p(r) \, dr\right] < 0 \tag{28}$$

where the integration is to be done over the range of r such that the integral remains real. However to use this condition the probability density function of r is required. In the equivalent linearization since the process x_0 is Gaussian, r follows the Rayleigh distribution

$$p(r) = (r/\mu^2)e^{-(0\cdot5r^2/\mu^2)} \tag{29}$$

This stability analysis shows that in the region where μ^2 has three possible values in Fig. 3, the solution process $x(t)$ corresponding to the lowest value of μ^2 is the only one which is stable. These results have also been verified through extensive numerical simulation as shown in Fig. 3. It is seen that there is overall agreement between the theoretical predictions and the numerical simulation results. However, comparison is not very good in the intermediate frequency range of $\bar{\lambda}$ between 1·6 and 2·4. The linearization approximation which leads to the amplitude density function of eqn (29) and the consequent stability analysis are also very approximate in the above range of $\bar{\lambda}$. This can also be seen in another comparison shown Fig. 4 between the theoretical and simulated average rate of zero crossings. The linearization predicts a value of nearly unity for λ/λ_e up to about $\bar{\lambda} = 2\cdot0$

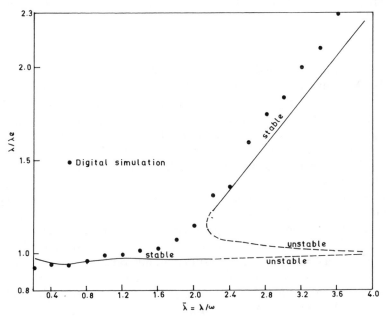

Fig. 4. Mean frequency of response. $\varepsilon = 0\cdot3$, $\eta = 0\cdot08$, $\xi = 0\cdot02$.

whereas the simulation maintains this behaviour only up to about $\bar{\lambda} = 1 \cdot 2$. Beyond about $\bar{\lambda} = 2 \cdot 4$ the theory and the simulation again compare favourably. Thus, in the transition region the linearization solution is not very good. Figure 4 can also be interpreted to mean that the response contains essentially the external frequency up to $\bar{\lambda} = 1 \cdot 0$. Beyond this, other frequencies are present in the response process. This observation is important in that it indicates the region of validity of the response representation used by Lennox & Kuak,[2] namely

$$x(t) = R \sin(\lambda_e t - \theta); \qquad \lambda_e / \lambda \doteq 1 \qquad (30)$$

to be $\bar{\lambda} \leqslant 1$. This fact can be made use of to find a new approximation to the amplitude distribution, instead of assuming it to be Rayleigh distributed. Now, following Lennox & Kuak,[2] if R and θ are taken to be slowly varying random variables attaining steady values such that $\dot{R} = \dot{\theta} = 0$, it can be shown that

$$E^2 = \omega^4 \sigma_L^2 [\{0 \cdot 75 \varepsilon r^3 - r(\bar{\lambda}^2 - 1)\}^2 + (2\eta \bar{\lambda} r)^2] \qquad (31)$$

Here E is the amplitude of the narrow band input, the density function of which is

$$p(E) = (E/\sigma_f^2) \exp(-0 \cdot 5 E^2 / \sigma_f^2) \qquad (32)$$

$r = (R/\sigma_L)$ is the nondimensional response amplitude. The above relation remains valid, as discussed previously, for values of $\bar{\lambda} \leqslant 1$. This is also the region in which eqn (31) does not show multiple solutions. As $\varepsilon \to 0$ eqn (31) degenerates to the linear case and yields a Rayleigh distribution for r. The nonRayleigh density function of r for $\varepsilon \neq 0$ can be easily found from eqn (31). A typical result is shown in Fig. 5 along with a numerical simulated histogram. The theoretical result refers to $p(r) \Delta r$ which is the probability of the response amplitude being in the interval $(r, r + \Delta r)$. For values of $\bar{\lambda}$ in the transition region or beyond, the representation of eqns (30)–(32) are not valid and hence cannot be used. It is interesting however to observe the digital simulation histograms shown in Fig. 6 for $\bar{\lambda} = 2$ and $\bar{\lambda} = 3$. The amplitude in the transition zone is bimodal indicating that the sample amplitudes would cluster around two values with transitions in between, which is the stochastic counterpart of the deterministic jump phenomenon. As $\bar{\lambda}$ increases, this behaviour vanishes leading again to a unimodal density function.

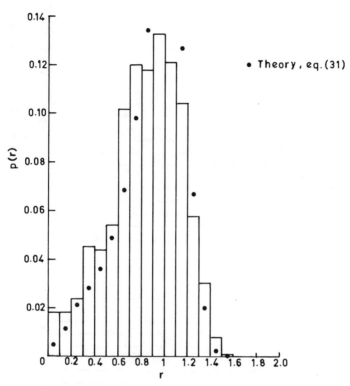

Fig. 5. Probability histogram of amplitude $r = R/\sigma_L$. $\bar{\lambda} = 1$, $\varepsilon = 3$, $\eta = 0.08$, $\xi = 0.02$.

5 SUMMARY AND CONCLUSION

The application of the equivalent linearization method to study the random response of the hardening Duffing oscillator under two kinds of excitation has been considered in this study. First the case of the input containing a periodic term and a white noise is considered. This is followed by the case of a narrow band excitation. In both the cases the equivalent linearization leads to multiple values for the moments.

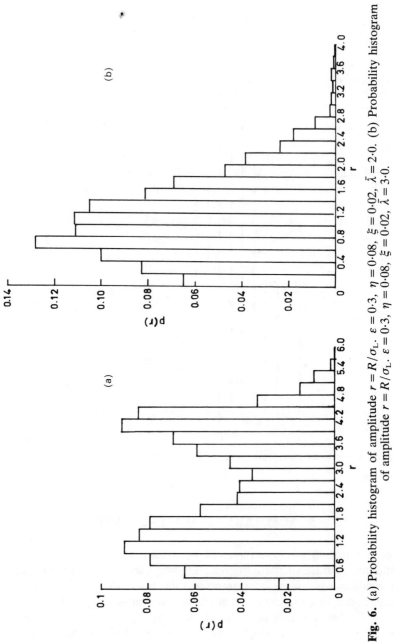

Fig. 6. (a) Probability histogram of amplitude $r = R/\sigma_L$. $\varepsilon = 0.3$, $\eta = 0.08$, $\xi = 0.02$, $\bar{\lambda} = 2.0$. (b) Probability histogram of amplitude $r = R/\sigma_L$. $\varepsilon = 0.3$, $\eta = 0.08$, $\xi = 0.02$, $\bar{\lambda} = 3.0$.

For practical application of this result it is necessary to conduct a stochastic stability analysis. Even though an exact stability analysis is perhaps not possible, an approximation is possible along the lines of the quasistatic approach of Stratonovich. It is found that whenever multiple roots arise for the moments, only one of them will be acceptable as shown by the stability analysis.

It is interesting to note that the introduction of a small noise into a sinusoidal excitation can alter the behaviour of the hardening cubic oscillator. The response variance of a hardening oscillator ($\varepsilon > 0$) is less than the linear value under a white noise input ($\lambda = 0$). With this in the background here the variance is represented as a fraction of the steady state linear variance. In the case of the combined excitation, the first and the second moments nonlinearly interact to reduce the variance further as the external frequency λ increases. As λ/ω_e approaches unity, the sinusoidal term controls the system and hence the mean amplitude increases with a reduction in the randomness represented in terms of the variance. On the other hand, far away from resonance the noise influences the system considerably and hence the variance increases with corresponding reduction in the mean amplitude. However in between there is a zone in which the solution is unstable indicating a breakdown of the steady state. This means the random part of the response is no more a stationary process and hence the variance fluctuates widely instead of being nearly constant. When the excitation is a narrow band process, there is again a dominant external frequency along with randomness. However, the mean of the response is zero, unlike in the previous case. Still the response variance shows multiple values as the external frequency is increased. Again a stability analysis is possible to show that only the lowest value of the variance is realizable. The numerical simulations closely compare with the theoretical values except in an intermediate region of frequencies. The analysis further indicates that the response is predominantly at the input centre frequency for values of $\bar{\lambda} \leqslant 1$. This helps in arriving at the nonRayleigh probability density function of the amplitude. However, the assumption of Rayleigh distribution for r in the stability analysis is not vitiated since the digital simulations which show unimodal histograms (Figs 5 and 6(b)) can be well approximated by the Rayleigh density function. But the interesting bimodal density function of the amplitude in the transition region of the frequency ratio $\bar{\lambda}$, as shown by simulation studies cannot be properly accounted

R. N. Iyengar

by the theory presented herein. This aspect remains unsolved and needs further investigation.

REFERENCES

1. Lyon, R. H., Heckl, M. & Hazelgrove, C. B., Narrow-band excitation of the hard spring oscillator. *J. Acoust. Soc. Amer.,* **33** (1961) 1404–11.
2. Lennox, W. C. & Kuak, Y. C., Narrow-band excitation of a nonlinear oscillator. *J. Appl. Mech.,* **43** (1976) 340–4.
3. Iyengar, R. N., Random vibration of a second order nonlinear elastic system. *J. Sound Vibration,* **40** (1975) 155–65.
4. Davies, H. G. & Nandlall, D., Phase plane for narrow-band random excitation of a Duffing oscillator. *J. Sound Vibration,* **104** (1986) 277–83.
5. Jia, W. & Fang, T., Jump phenomena in coupled Duffing oscillators under random excitation. *J. Acoust. Soc. Amer.* **81**(4) (1987) 961–5.
6. Iyengar, R. N., A nonlinear system under combined periodic and random excitation. *J. Statist. Phys.,* **44** (1986) 907–20.
7. Bulsara, A. R., Lindenberg, K. & Shuller, K. E., Spectral analysis of a nonlinear oscillator driven by random and periodic forces. *J. Statist. Phys.,* **27** (1982) 787–808.
8. Lin, Y. K., *Probabilistic Theory of Structural Dynamics.* McGraw-Hill, New York, 1967.
9. Stratonovich, R. L., *Topics in the Theory of Random Noise.* Gordon and Breach, New York, 1963.

10

Stability of Flexible Structures with Random Parameters

F. KOZIN

Polytechnic University, Farmingdale, New York, USA

ABSTRACT

As the stability of stochastic ordinary differential equations has been highly developed over the past three decades, there has also been research directed towards the questions of stability of stochastic partial differential equations. Recently, there has been an increased interest in this topic motivated by the problems of dynamics of flexible structures in space. In this paper will be presented a brief discussion of some of the directions these studies have taken in recent years.

1 INTRODUCTION

Typical of the types of problems that have been studied in the past are lateral displacement of strings subjected to random transverse loads, vibrations of panels in supersonic flows subjected to random end loads, columns subjected to random follower forces, and finally cylindrical shells subjected to radial and axial loads have been considered. These examples have appeared in a number of early papers (see for example, Refs 1–4) with further results in Refs 5 and 6. There has been continued development in recent years, through more general mathematical approaches based upon semi-group properties.[7,8] For stochastic partial differential equations with white noise coefficients, the fundamental development in Ref. 9, and the general concepts of stability of stochastic evolution equations,[10,11] are typical of this approach.

Recent directed applications have appeared in Ref. 12, and a series of papers.[13-15] Clearly other problems concerning control as well as parameter estimation for stochastic partial differential equations have also undergone developments in recent years (see for example Refs 16 and 17).

These references are but a small representation of the many studies that have appeared on this important topic. We shall here indicate only a few of the methods that have been applied to study this stochastic stability problem.

In order to focus our ideas we consider the transverse vibrations of the column.

$$\left.\begin{aligned}
&v_{tt} + 2\beta v_t + (f_0 + f(t))v_{xx} + v_{xxxx} = 0 \\
&0 \leq x \leq 1, \qquad 0 \leq t < \infty \\
&v(0, t) = v(1, t) = 0, \qquad v_{xx}(0, t) = v_{xx}(1, t) = 0
\end{aligned}\right\} \tag{1.1}$$

where $f(t)$ is a stationary, ergodic, zero mean process.

The traditional approach to this process is to separate variables and study the asymptotic behavior of the modes.

Thus, we can write

$$v(x, t) = \sum_{n=1}^{\infty} T_n(t) \sin n\pi x \tag{1.2}$$

where $T_n(t)$ is the solution to the stochastic differential equation.

$$\ddot{T}_n(t) + 2\beta \dot{T}_n(t) + [(n\pi)^4 - (n\pi)^2 f_0 - (n\pi)^2 f(t)]T_n(t) = 0 \tag{1.3}$$

The stochastic stability of stochastic ordinary differential equations is well documented and applying the 'best' condition obtained first by Infante,[18] through Lyapunov function ideas, it follows that,

$$E\{f^2(t)\} < 4\beta^2\left[1 - \frac{f_0}{(n\pi)^2}\right] \tag{1.4}$$

is sufficient for almost sure asymptotic sample stability of $T_n(t)$, for $n \geq 1$. Hence,

$$E\{f^2(t)\} < 4\beta^2\left[1 - \frac{f_0}{\pi^2}\right] \tag{1.5}$$

guarantees sample asymptotic stability for all modes, and thus stability for the column. However, it is clear that the sufficient condition (1.5) is quite conservative, and it would certainly be desirable to relax it as much as possible. One potential approach would be to study the qualitative properties of the deflection $v(x, t)$ directly without reference to the modes. However, we would then require a Lyapunov norm which would be a quadratic functional of the form

$$V(t) = \int_0^1 Q(v, v_t, v_x, v_{xx}, \ldots) \, dx \qquad (1.6)$$

The V functional can be considered as a norm defined upon the appropriate Hilbert space containing the solutions of the system (1.1). In Ref. 6, it was shown that the optimal quadratic function $V(t)$ can be obtained as,

$$V(t) = \int_0^1 [(v_t)^2 + 2\beta v v_t + 2\beta^2 v^2 + (v_{xx})^2 - f_0(v_x)^2] \, dx \qquad (1.7)$$

The form (1.7) was obtained through the optimal Lyapunov function for the individual modes, applying the given boundary conditions and finally identifying the constant $n\pi$ with the operator $\partial/\partial x$.

It was established, further, that

$$\dot{V}(t) = -2\beta V(t) + U(t) \leq 2[-\beta + \lambda(t)]V(t) \qquad (1.8)$$

where $\lambda(t)$ was obtained as a polygonal envelope

$$\lambda(t) = \begin{cases} |f(t)| & f(t) \leq f_{01} \\ \lambda_1(t) & f_{01} < f(t) \leq f_{12} \\ \lambda_2(t) & f_{12} < f(t) \leq f_{23} \\ \vdots & \vdots \\ \lambda_n(t) & f_{n-1,n} < f(t) \leq f_{n,n+1} \\ \vdots & \vdots \end{cases} \qquad (1.9)$$

where $\lambda_n(t)$ is given as

$$\lambda_n(t) = \frac{|2\beta^2 + (n\pi)^2 f(t)|}{[(n\pi)^4 - f_0(n\pi)^2 + \beta^2]^{1/2}}, \qquad n = 1, 2, \ldots \qquad (1.10)$$

and

$$f_{k-1,k} = \frac{2\beta^2}{\pi^2} \frac{1 - \sqrt{\dfrac{((k-1)\pi)^4 + \beta^2}{(k\pi)^4 + \beta^2}}}{k^2 \sqrt{\dfrac{((k-1)\pi)^4 + \beta^2}{(k\pi)^4 + \beta^2}} - (k-1)^2} \qquad (1.11)$$

The regions of sufficiency must then be obtained through computer calculation.

This approach has been applied to beams and cylindrical shells in Refs 14 and 15.

2 RECENT STUDIES

Although in the Introduction we studied a Lyapunov functional defined on a Hilbert space, we did not apply the principle of functional analysis on function spaces to solve the stability problem.

An early investigation in this direction appears in Ref. 4. In particular a system

$$\vec{u}_t(t, \vec{x}) = [\mathscr{L}_0 + G(t, \vec{x})]\vec{u}(t, \vec{x})$$
$$\beta_0\vec{u}(t, \vec{x}) = 0, \qquad \vec{x} \in \partial D \tag{2.1}$$

where $G(t, \vec{x})$ is a matrix whose non-zero components are stationary, ergodic processes in t, and ∂D denotes the boundary of the domain D.

It is assumed that the operator $\mathscr{L}_0$ is the infinitesimal generator of the strongly continuous semi-group $\{\Phi(t_2, t_1)\}$, $0 \leqslant t_1$, $t_2 \leqslant \infty$ of bounded operators on a Banach function space $\Gamma(\bar{D})$ into itself, which satisfy, for the associated norm, $\| \ \|$,

$$\|\Phi(t_2, t_1)\| \leqslant C \exp[-\gamma(t_2 - t_1)] \tag{2.2}$$

where C, γ are positive constants.

The operators $\{\Phi(t_2, t_1)\}$ are said to form a semi-group if for $t_0 \leqslant t_1 \leqslant t$ we have

$$\Phi(t, t_1)\Phi(t_1, t_0) = \Phi(t, t_0) \qquad \Phi(t, t) = I, \tag{2.3}$$

where I denotes identity.

In particular, if $\mathscr{L}_0$, with the given boundary conditions, possesses a Green's function, K, then we can write the operator $\Phi(t_2, t_1)$ as,

$$\Phi(t_2, t_1) = \int_D K(t_2, t_1, \vec{x}, \vec{x}^0)(\cdot)\, d\vec{x}^0 \tag{2.4}$$

It is established that under the condition (2.2), if

$$E\{\|G(t, \vec{x})\|\} < \gamma/C \tag{2.5}$$

then the trivial solution (2.1) is almost surely asymptotically stable.

The level of generality, however, has been significantly lifted recently in order to study the stability problem for partial differential equations with white noise[7] coefficients. In this case the concept of an infinite dimensional Ito integral is introduced, to yield an analysis equivalent to ordinary Ito differential equations, but through a semi-group interpretation of the stochastic partial differential equation. As in Ref. 7, we rewrite the equation,

$$v_{tt} + \alpha v_t + \mathscr{L}_0 v + \mathscr{L}_1 v \dot{B}(t) = 0$$
$$v(0) = v_0, \quad v_t(0) = v_1 \tag{2.6}$$

for $z' = (v, v_t)$, as the differential operator equality,

$$dz(t) = \mathscr{A}z(t)\,dt + Dz(t)\,dB(t), \quad z(0) = z_0 \tag{2.7}$$

Here, the B-process is the Brownian Motion, with $E\{B^2(t)\} = \sigma^2 t$. Then, if $\langle, \rangle$ denotes inner product in the associated Hilbert space, the Ito differential lemma becomes[19]

$$
y(t) = y(0) + \int_0^t \langle \Phi_z(z(s)), \mathscr{A}z(s) \rangle\,ds
$$
$$
+ \int_0^t \tfrac{1}{2} t_R [DZ(s)W(DZ(s))^* \Phi_{zz}(z(s))]\,ds
$$
$$
+ \int_0^t \langle \Phi_z(z(s)), Dz(s)\,dB(s) \rangle \tag{2.8}
$$

where $y(t) = \Phi(z(t))$, and Φ_z, Φ_{zz} denote Fréchét derivatives.

Assuming that $\mathscr{A}$ generates the semi-group S_t, we can write the so-called mild solution to (2.6) in the form

$$Z(t) = S_t z_0 + \int_0^t S_{t-s} Dz(s)\,dB(s) \tag{2.9}$$

For $\alpha > 0$, if $\omega(-\mathscr{A})$ denotes $\sup\{\operatorname{Re}(\lambda); \lambda\varepsilon \text{ spectrum } (-\mathscr{A})\}$, then the semi-group is exponentially stable with

$$\|S_t\| \leq e^{-\omega t} \tag{2.10}$$

where

$$\omega \geq \frac{2\alpha\,|\omega(-\mathscr{A})|}{4\,|\omega(-\mathscr{A})| + \alpha(\alpha + \sqrt{\alpha^2 + 4\,|\omega(-\mathscr{A})|})}. \tag{2.11}$$

For the damped, stretched string

$$\mathscr{L}_0 = -\frac{\partial^2}{\partial x^2}, \quad |\omega(-\mathscr{A})| = \pi^2$$

thus

$$\omega \geq \frac{2\alpha\pi^2}{4\pi^2 + \alpha(\alpha + \sqrt{\alpha^2 + 4\pi^2})} \tag{2.12}$$

yields the exponentially stable semi-group.

For the vibrating panel,

$$\mathscr{L}_0 = \frac{\partial^4}{\partial x^4} + f\frac{\partial^2}{\partial x^2}, \quad |\omega(-\mathscr{A})| = \pi^2(\pi^2 - f^2), \quad f^2 < \pi^2$$

and

$$\omega \geq \frac{2\alpha\pi^2(\pi^2 - f^2)}{4\pi^2(\pi^2 - f^2) + \alpha\sqrt{\alpha^2 + 4\pi^2(\pi^2 - f^2)}} \tag{2.13}$$

yields the exponentially stable semi-group.

The solution process of (2.6) or equivalently (2.9), is said to be mean square, exponentially stable if there exists constants C, $\gamma > 0$ such that

$$E\{\|z(t)\|^2\} \leq Ce^{-\gamma t}E\{\|z_0\|^2\} \tag{2.14}$$

Mean square exponential stability is shown to follow for each of the two examples above if $\sigma^2 < 2\omega$, for ω given respectively by (2.12) and (2.13).

Finally, upon application of the Ito differential formula (2.8) to second moments, it is established that mean square exponential stability implies almost sure sample exponential stability. That is, there exist positive constants, K, b and a random time $T(\omega) < \infty$ (depending upon the sample), such that for $t > T(\omega)$,

$$\|z(t)\|^2 \leq KE\{\|z_0\|^2\}e^{-bt} \tag{2.15}$$

The proof in function space is similar to the proof found in Ref. 20. This approach has been applied, in Ref. 21, to thin-walled beams.

3 CONCLUSIONS

Although we have presented only a brief view of the problem, there is no question that there have been many significant developments on the stability of continuous parameter systems during the past decade. However, interesting problems remain to be studied. For example it would be desirable to investigate the possibility of determining exact stability boundaries for the white noise coefficient case, and characterize the exact sample stability region, in terms of the moment stability regions. Associated problems related to identification, estimation and control are of great importance and also under intensive study currently.

We hope that this brief review of the topic will motivate the interested reader to look further into the many tools that have been developed and results that have been obtained.

REFERENCES

1. Ariaratnam, S. T., Dynamic stability of a column under random loading, *Proc. Inter. Conference Dynamic Stability of Structures.* Pergamon Press, New York, 1967, p. 267.
2. Samuels, J. C. & Eringen, A. C., On stochastic linear systems. *J. Math. Phys.*, **38** (1959) 83–103.
3. Caughey, T. K. & Gray, A. H., On the almost sure stability of linear dynamic systems with stochastic coefficients. *ASME J. Appl. Mech.*, **32** (1965) 365–72.
4. Wang, P. K. C., On the almost sure stability of linear stochastic distributed parameter dynamical systems. *ASME J. Appl. Mech.*, **33** (1966) 182–6.
5. Plaut, R. H. & Infante, E. F., On the stability of some continuous systems subjected to random excitations. *ASME J. Appl. Mech.*, **37** (1970) 623–7.
6. Kozin, F., Stability of the linear stochastic systems. *Lect. Notes in Math.*, **294** (1972) 186–229.
7. Curtain, R. F., Stability of stochastic partial differential equation. *J. Math. Anal. Appl.*, **79** (1981) 352–69.
8. Haussman, U. G., Asymptotic stability of the linear Ito equation in infinite-dimensions. *J. Math. Anal. Appl.*, **65** (1978) 219–35.
9. Pardoux, E., Doctoral thesis, L'Université de Paris Sud, Centre d'Orsay, 1975.
10. Chow, P. L., Stability of nonlinear stochastic-evolution equations. *J. Math. Anal. Appl.*, **89** (1982) 400–19.
11. Ichikawa, A., Stability of semilinear stochastic evolution equations. *J. Math. Anal. Appl.*, **90** (1) (1982) 12–44.
12. Sobczyk, K., *Stochastic Wave Propagation.* Elsevier, Amsterdam, 1985.
13. Tylikowski, A., Dynamic stability of non-linear rectangular plates. *ASME J. Appl. Mech.*, **45** (1978) 583–5.
14. Tylikowski, A. & Kurnik, W., Stochastic stability and nonstability of a linear cylindrical shell. *Ingen.-Archiv*, **53** (1983) 363–9.

15. Tylikowski, A., Stochastic stability of beams with impulse parametric excitation. *The 9th International Conference on Nonlinear Oscillations, Vol. 1, Analytical methods of the nonlinear oscillation theory,* ed. Yu. A. Mitropolsky. Kiev, Naukova Dumka, 1984, pp. 367–9.
16. Sunahara, Y. *et al.,* On the optimal control for distributed parameter systems with white noise coefficients. *Proc. Inter. Symp. on Stoch. Diff. Eqns,* Kinokuniya, Co. Ltd, Tokyo, 1976.
17. Curtain, R. F., A semi-group approach to the LQG problem for infinite dimensional systems. *IEEE Trans. Circuits and Systems,* **25** (1978) 713–20.
18. Infante, E. F., On the stability of some linear nonautonomous random systems. *ASME J. Appl. Mech.,* **35** (1968) 7–12.
19. Curtain, R. F. & Falb, P. L., Ito's lemma in infinite dimensions. *J. Math. Anal. Appl.,* **31** (1970) 434–48.
20. Kozin, F., On almost sure sample asymptotic properties of diffusion processes defined by stochastic differential equations. *J. Math., Kyoto University,* **4** (1965) 515–28.
21. Tylikowski, A., Stochastic stability of a thin-walled beam subjected to a time and space dependent loading. *ZAMM,* **66** (1986) 97–8.

11

An Algorithm for Moments of Response from Non-normal Excitation of Linear Systems

S. KRENK & H. GLUVER

Department of Structural Engineering, Technical University of Denmark, Lyngby, Denmark

ABSTRACT

A recursive algorithm is developed for the moments of the response of a linear system with input in the form of a polynomial of a normal process with rational power spectrum. The algorithm consists of a recursive system of linear equations for moments of increasing order. Examples illustrate how the kurtosis of the response is influenced by the bandwidth of the excitation and the damping of the system as well as by the relative magnitude of the dominating frequencies. Application of the method to structural fatigue analysis is discussed.

1 INTRODUCTION

Wind and waves can often be described well by normal stochastic processes. However, the corresponding forces on structures are generally determined by nonlinear functions, and therefore lead to non-normal excitation and non-normal response. A different type of non-normal excitation that is useful in connection with traffic loads is associated with the arrival of pulses with intensity governed by a probabilistic law. While the non-normal properties of response to pulse excitation can be treated in terms of characteristic functions—see for example Parzen[1] for the case of a filtered Poisson process—different mathematical tools are needed for the efficient solution of problems involving excitation in the form of a non-linear function of a normal process.

181

In the description of loads such as those from wind and waves it is important to be able to represent the power spectral density as well as the deviation from normality of the load process. This has been done in two different ways. Lutes & Hu[2] start out with a non-normal white noise process, and then apply a linear filter to introduce the desired spectral density in the form of a rational function. While this approach is operational, it has two disadvantages: the non-normal white noise is an artifice introduced to produce some measure of non-normality, and this deviation from normality is influenced by the subsequent filtering. The mathematical difficulties associated with non-normal white noise processes have been discussed by Grigoriu.[3]

Alternatively a normal process $Y(t)$ with suitable spectral density can be subjected to a memoryless transformation $f(Y(t))$, and $f(Y(t))$ used as input to the system. This approach has the advantage that $Y(t)$ is often associated directly with a physical process such as particle velocities or accelerations, while the function $f(\ )$ describes the non-linear force mechanism, e.g. stagnation pressure or drag force. For linear systems the response $X(t)$ can be determined in the form of a convolution integral, and thus all higher moments of $X(t)$ can in principle be determined by multiple convolutions (Kotulski & Sobczyk[4]; Lutes & Hu[2]). However, analytical evaluation of these integrals is only feasible for very simple systems with simple spectral density of the load, and in general numerical integration is necessary. If the spectral density of the process $Y(t)$ can be represented by a rational function and $f(\ )$ by a polynomial, a closed recursive set of equations can be obtained for the moments of increasing order. This was accomplished by Kotulski & Sobczyk[4] by integration of the Fokker–Planck equation, while the present paper and Grigoriu & Ariaratnam[5] arrive at these equations via a reformulation of the problem in terms of Ito differential equations.

2 SYSTEM DESCRIPTION

The system is described by a set of linear differential equations in a state space determined by the vector $\mathbf{X}(t)$. These equations can be given in incremental form as

$$d\mathbf{X}(t) = \mathbf{S}^x \mathbf{X}(t)\,dt + \mathbf{f}(\mathbf{Y}(t))\,dt + \mathbf{D}^x\,d\mathbf{W}^x(t) \tag{1}$$

$\mathbf{S}^x$ is a constant matrix describing the system, $\mathbf{f}(\mathbf{Y}(t))$ is a function describing the excitation by a stochastic vector process $\mathbf{Y}(t)$, and $\mathbf{D}^x$ is a matrix defining an additional excitation by a normal white noise vector process with the increment $d\mathbf{W}^x(t)$. Thus $\mathbf{W}^x(t)$ is a Wiener process. In the common case of a system governed by second order differential equations the vector $\mathbf{X}(t)$ contains the position and velocity of each degree of freedom of the system.

The stochastic vector process $\mathbf{Y}(t)$ is generated from a normal white noise vector process with increment $d\mathbf{W}^y(t)$ by use of a linear filter. Thus

$$d\mathbf{Y}(t) = \mathbf{S}^y\mathbf{Y}(t)\,dt + \mathbf{D}^y\,d\mathbf{W}^y(t) \tag{2}$$

The spectral density of the process $\mathbf{Y}(t)$ is a rational function, the coefficients of which are determined by the matrices $\mathbf{S}^y$ and $\mathbf{D}^y$.

Equations (1) and (2) are special cases of Ito stochastic differential equations. They can be combined into a single Ito stochastic differential equation in standard form

$$d\mathbf{Z}(t) = \mathbf{A}(\mathbf{Z})\,dt + \mathbf{B}(\mathbf{Z})\,d\mathbf{W}(t) \tag{3}$$

with the notation

$$\mathbf{Z}(t) = \begin{bmatrix} \mathbf{X}(t) \\ \mathbf{Y}(t) \end{bmatrix} \tag{4}$$

$$\mathbf{W}(t) = \begin{bmatrix} \mathbf{W}^x(t) \\ \mathbf{W}^y(t) \end{bmatrix} \tag{5}$$

$$\mathbf{A}(\mathbf{Z}) = \begin{bmatrix} \mathbf{S}^x & \mathbf{0} \\ \mathbf{0} & \mathbf{S}^y \end{bmatrix}\begin{bmatrix} \mathbf{X}(t) \\ \mathbf{Y}(t) \end{bmatrix} + \begin{bmatrix} \mathbf{f}(\mathbf{Y}) \\ \mathbf{0} \end{bmatrix} \tag{6}$$

$$\mathbf{B}(\mathbf{Z}) = \begin{bmatrix} \mathbf{D}^x & \mathbf{0} \\ \mathbf{0} & \mathbf{D}^y \end{bmatrix} \tag{7}$$

The present formulation is a special case in which the equations for $d\mathbf{Y}$ are linear, and the equations for $d\mathbf{X}$ are linear in $\mathbf{X}$ and non-linear in $\mathbf{Y}$. When the function $\mathbf{f}(\mathbf{Y})$ is a polynomial these equations lead to a closed recurrence scheme for the moments of increasing order. A similar procedure can be used for non-linear systems where the non-linear terms are given by $\mathbf{f}(\mathbf{X}, \mathbf{Y})$. In that case the moment equations do not form a closed set. However, approximate solutions can be obtained by cumulant-neglect closure as discussed by Wu & Lin.[6]

3 MOMENT EQUATIONS

Equations for the moments of the components of the vector $\mathbf{Z}(t)$ are found in a systematic way by use of Ito's formula for the increment of a function of a stochastic process, see for example Gardiner,[7] p. 95. The key to this formula lies in the fact that $(dW)^2 = dt$, i.e. the second order increment $(dW)^2$ of the Wiener process is identical with the first order time increment dt. Thus an expansion that is linear in dt must include quadratic terms in dW. For a function $g(\mathbf{Z})$ of the stochastic vector process $\mathbf{Z}(t)$ from (3) this implies that

$$
dg(\mathbf{Z}) = \frac{\partial}{\partial z_i} g(\mathbf{Z})\, dZ_i + \frac{1}{2}\frac{\partial^2}{\partial z_i\, \partial z_j} g(\mathbf{Z})\, dZ_i\, dZ_j
$$

$$
= \left[A_i(Z)\frac{\partial}{\partial z_i} g(\mathbf{Z}) + \tfrac{1}{2}B_{ik}B_{jk}\frac{\partial^2}{\partial z_i\, \partial z_j} g(\mathbf{Z}) \right] dt + B_{ij}\frac{\partial}{\partial z_i} g(\mathbf{Z})\, dW_j
$$

$$(8)$$

For stationary response the expected value of the increments $dg(\mathbf{Z})$ and dW_j vanish, leaving the equation

$$
E\left[A_i(\mathbf{Z})\frac{\partial}{\partial z_i} g(\mathbf{Z}) + \tfrac{1}{2}B_{ik}B_{jk}\frac{\partial^2}{\partial z_i\, \partial z_j} g(\mathbf{Z}) \right] = 0 \qquad (9)
$$

By introducing the special forms (6) and (7) of $\mathbf{A}(\mathbf{Z})$ and $\mathbf{B}$ appropriate for the response of linear systems eqn (9) takes the form

$$
E\left[S_{ik}^x x_k \frac{\partial}{\partial x_i} g(\mathbf{X}, \mathbf{Y}) + S_{ik}^y y_k \frac{\partial}{\partial y_i} g(\mathbf{X}, \mathbf{Y}) + \tfrac{1}{2}D_{ik}^x D_{jk}^x \frac{\partial^2}{\partial x_i\, \partial x_j} g(\mathbf{X}, \mathbf{Y}) \right.
$$

$$
\left. + \tfrac{1}{2}D_{ik}^y D_{jk}^y \frac{\partial^2}{\partial y_i\, \partial y_j} g(\mathbf{X}, \mathbf{Y}) + f_i(\mathbf{Y})\frac{\partial}{\partial x_i} g(\mathbf{X}, \mathbf{Y}) \right] = 0 \quad (10)
$$

When $\mathbf{f}(\mathbf{Y})$ and $g(\mathbf{X}, \mathbf{Y})$ are polynomials, (10) is an equation connecting moments of the components of the vectors $\mathbf{X}$ and $\mathbf{Y}$. It is seen that the two first terms are polynomials of the same degree as $g(\mathbf{X}, \mathbf{Y})$, while the two next terms are polynomials of smaller degree in $\mathbf{X}$ and $\mathbf{Y}$, respectively. When $f_i(\mathbf{Y})$ is a polynomial of degree greater than one the last term is a polynomial of degree greater than the polynomial $g(\mathbf{X}, \mathbf{Y})$.

To make the general structure clear we consider functions of the form

$$
g(\mathbf{X}, \mathbf{Y}) = X_1^{n_1} X_2^{n_2} \ldots X_N^{n_N} Y_1^{m_1} Y_2^{m_2} \ldots Y_M^{m_M} \qquad (11)
$$

The expectation is the moment

$$m(n_1, n_2, \ldots, n_N; m_1, m_2, \ldots, m_M) = E[g(\mathbf{X}, \mathbf{Y})] \qquad (12)$$

The moments are classified according to their degree in $\mathbf{X}$ and $\mathbf{Y}$, respectively. Thus a moment belongs to the class $M(r, l)$ when

$$r = \sum_1^N n_j, \qquad l = \sum_1^M m_j \qquad (13)$$

In this notation eqn (10) can be rewritten as

$$\sum_i \sum_k \{ S_{ik}^x n_i m(\ldots, n_i - 1, \ldots, n_k + 1, \ldots ; \ldots)$$

$$+ S_{ik}^y m_i m(\ldots ; \ldots, m_i - 1, \ldots, m_k + 1, \ldots) \}$$

$$= -\sum_i \sum_j \sum_k \{ D_{ik}^x D_{jk}^x n_i n_j m(\ldots, n_i - 1, \ldots, n_j - 1, \ldots ; \ldots)$$

$$+ D_{ik}^y D_{jk}^y m_i m_j m(\ldots ; \ldots, m_i - 1, \ldots, m_j - 1, \ldots) \Big\}$$

$$- \sum_i E\left[f_i(\mathbf{Y}) \frac{\partial}{\partial x_i} g(\mathbf{X}, \mathbf{Y}) \right] \qquad (14)$$

Now, let the forcing function $\mathbf{f}(\mathbf{Y})$ be represented by a polynomial of degree s, i.e. $\mathbf{f}(\mathbf{Y})$ is of the form

$$f_i(\mathbf{Y}) = \sum_j a_{ij} Y_1^{k_1} Y_2^{k_2} \ldots Y_M^{k_M} \qquad (15)$$

where

$$\sum_1^M k_j \le s \qquad (16)$$

for all terms.

Equation (14) has the following structure. All the terms on the left hand side are moments of the same class, $M(r, l)$. The right hand side contains three groups of terms. The first two contain moments of class $M(r - 2, l)$ and $M(r, l - 2)$, respectively. The moments in the last group are associated with the forcing function and all belong to the classes $M(r - 1, l)$, $M(r - 1, l + 1)$, \ldots, $M(r - 1, l + s)$.

The solution strategy is sketched in Table 1. The solution starts with the initial value

$$m(0, 0, \ldots, 0; 0, 0, \ldots, 0) = 1 \qquad (17)$$

Table 1
Solution procedure for fourth order response moments

$M(0, 0)$	$M(0, 1)$	$M(0, 2)$	$M(0, 3)$	$M(0, 4) \ldots M(0, 4s)$	
$M(1, 0)$	$M(1, 1)$	$M(1, 2)$	$M(1, 3) \ldots M(1, 3s)$		
$M(2, 0)$	$M(2, 1)$	$M(2, 2) \ldots M(2, 2s)$			
$M(3, 0)$	$M(3, 1) \ldots M(3, s)$				
$M(4, 0)$					

and the convention that the moment vanishes if any index is negative. $m(0, \ldots ; 0, \ldots)$ is the only member of the class $M(0, 0)$. The moments of the class $M(0, 1)$ can now be found from equations of the type (14), because the last term vanishes for $r = 0$. The moments of classes $M(0, 2)$, $M(0, 3)$, etc., are found by continuing this procedure. When a sufficient number of classes $M(0, l)$, $l = 1, 2, \ldots$ have been determined, the first index is increased to 1 and the procedure continued. Table 1 illustrates the solution procedure for response moments up to fourth order. The solution proceeds row by row, and the length of the rows decreases with the degree s of the polynomial $\mathbf{f(Y)}$.

All the moments in the first row are independent of $\mathbf{X}$. In the absence of any constant terms in the input the odd moments, i.e. the classes $M(0, 1)$, $M(0, 3)$, $\ldots$, vanish, and the even moments can be generated explicitly from the covariance matrix of $\mathbf{Y}$ by recurrence, see for example Madsen et al.,[8] p. 223. The equations for the covariance matrix of $\mathbf{Y}$, i.e. the moments of class $M(0, 2)$, are similar to those derived by Spanos[9] for spectral density integrals. In the present context it is hardly feasible to use explicit recurrence for the moment classes $M(0, 4)$, $M(0, 6)$, $\ldots$, as the solution of the corresponding equations generally constitutes only a small part of the computational effort.

The moment classes may contain one or more elements that are trivially zero such as for example $E(X^j \dot{X})$. In a general computational procedure it is not advisable to remove these moments from the systems of equations, as this will influence the simple structure of the equations.

4   EXAMPLES

The following examples are concerned with a linear oscillator with one degree of freedom excited by a non-linear function of a stochastic process produced from normal white noise by a second order filter. This formulation is sufficiently general to permit investigation of how deviations from normal response are influenced by the excitation bandwidth, the structural damping, and the magnitude of the structural resonance frequency relative to any dominant excitation frequency. The influence of the forcing function $f(Y)$ is also investigated, and particular attention is given to forms that can represent wave forces of Morison-type on slender structural members. This problem has recently been demonstrated to be of relevance to fatigue of offshore structures (Brouwers & Verbeek[10]).

The oscillator and its forcing function are described by the equation

$$\ddot{X} + 2\zeta_0\omega_0\dot{X} + \omega_0^2 X = \alpha_1 Y + \alpha_3 Y^3 + \beta_1 \dot{Y} + \beta_3 \dot{Y}^3 \tag{18}$$

where the stochastic process $Y(t)$ is generated by the second order filter

$$\ddot{Y} + 2\zeta_g\omega_g\dot{Y} + \omega_g^2 Y = 2\omega_g\sqrt{\zeta_g\omega_g}\,\xi(t) \tag{19}$$

$\xi(t) = dW/dt$ is a normal white noise process with covariance function

$$E[\xi(t)\xi(t+\tau)] = \delta(\tau) \tag{20}$$

Equation (19) has been normalized such that (Madsen *et al.*[8])

$$E[Y^2] = 1, \qquad E[\dot{Y}^2] = \omega_g^2 \tag{21}$$

By introduction of the state variables

$$\begin{bmatrix} X_1 \\ X_2 \end{bmatrix} = \begin{bmatrix} X \\ \dot{X}/\omega_0 \end{bmatrix}, \qquad \begin{bmatrix} Y_1 \\ Y_2 \end{bmatrix} = \begin{bmatrix} Y \\ \dot{Y}/\omega_g \end{bmatrix} \tag{22}$$

and the new forcing function parameters

$$a_1 = \frac{\alpha_1}{\omega_0\omega_g}, \qquad a_3 = \frac{\alpha_3}{\omega_0\omega_g}, \qquad b_1 = \frac{\beta_1}{\omega_0}, \qquad b_3 = \frac{\beta_3\omega_g^2}{\omega_0} \tag{23}$$

Equations (18) and (19) can be rewritten in incremental form as

$$d\begin{bmatrix} X_1 \\ X_2 \end{bmatrix} = \begin{bmatrix} 0 & \omega_0 \\ -\omega_0 & -2\zeta_0\omega_0 \end{bmatrix}\begin{bmatrix} X_1 \\ X_2 \end{bmatrix} dt + \omega_g\begin{bmatrix} 0 \\ a_1 Y_1 + a_3 Y_1^3 + b_1 Y_2 + b_3 Y_2^3 \end{bmatrix} dt \tag{24}$$

$$d\begin{bmatrix} Y_1 \\ Y_2 \end{bmatrix} = \begin{bmatrix} 0 & \omega_g \\ -\omega_g & -2\zeta_g\omega_g \end{bmatrix}\begin{bmatrix} Y_1 \\ Y_2 \end{bmatrix} dt + \begin{bmatrix} 0 \\ 2\sqrt{\zeta_g\omega_g} \end{bmatrix} dW \tag{25}$$

Either of the angular frequencies ω_0 and ω_g can be used as a time scale, and thus eqns (24) and (25) only depend, in an essential way, on the frequency ratio $R = \omega_0/\omega_g$, the damping ratios ζ_0, ζ_g and on the forcing function coefficients a_1, a_3, b_1, b_3. In terms of these parameters eqn (14) for the moment $m(i, j; p, q)$ takes the form

$$
\begin{aligned}
(2\zeta_0 Rj &+ 2\zeta_g q)m(i, j; p, q) \\
&+ Rjm(i + 1, j - 1; p, q) - Rim(i - 1, j + 1; p, q) \\
&+ qm(i, j; p + 1, q - 1) - pm(i, j; p - 1, q + 1) \\
&= 2\zeta_g q(q - 1)m(i, j; p, q - 2) \\
&+ a_1 jm(i, j - 1; p + 1, q) + a_3 jm(i, j - 1; p + 3, q) \\
&+ b_1 jm(i, j - 1; p, q + 1) + b_3 jm(i, j - 1; p, q + 3) \qquad (26)
\end{aligned}
$$

For the moments of class $M(r, l)$ the sums $r = i + j$ and $l = p + q$ are fixed, and the left hand side of eqn (26) is in the form of a five point difference operator. Kotulski & Sobczyk[4] and Grigoriu & Ariaratnam[5] considerd the case in which $Y(t)$ is the Ornstein–Uhlenbeck process produced by a first order filter. In that case there is no index corresponding to q, and the equation systems are tri-diagonal.

Two forcing functions are considered, namely the power y^3 and a polynomial approximation to $y|y|$. The stochastic process $Y|Y|$ is representative of the drag force on a slender structural member, when $Y(t)$ is a normal process (Borgman[11]). When $Y|Y|$ is replaced by a polynomial, approximations are introduced, both with regard to the spectral density of the process and the moments of the probability density function. The covariance function of the process $Y|Y|$ is a function of the covariance function of the process Y. Replacing this function with its Taylor expansion corresponds to expanding $Y|Y|$ in terms of Hermite polynomials $He_j(\ )$, see for example Madsen et al.,[8] p. 325. The two-term expansion is

$$
\begin{aligned}
Y|Y| &\approx \sqrt{\frac{2}{\pi}}[2He_1(Y) + \tfrac{1}{3}He_3(Y)] \\
&= \frac{1}{3}\sqrt{\frac{2}{\pi}}(3Y + Y^3) \qquad (27)
\end{aligned}
$$

The variance of the polynomial approximation (27) is 0.99 times the variance of $Y|Y|$.

The normalized spectral density of the stochastic processes Y, Y^3,

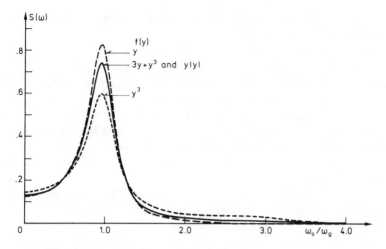

Fig. 1. Normalized spectral density of excitation processes.

$3Y + Y^3$ and $Y|Y|$ are shown in Fig. 1 for damping ratio $\zeta_g = 0 \cdot 2$ in the filter (19). It is seen that the spectral density of $Y|Y|$ and the polynomial approximation (27) are nearly indistinguishable. The damping ratio $\zeta_g = 0 \cdot 2$ was chosen with a view to the Pierson–Moskowitz spectrum. Matching of the corresponding covariance functions at a time separation of one half or a full mean period gives $\zeta = 0 \cdot 170$ and $\zeta = 0 \cdot 178$, respectively.

The deviation from normality is characterized by the higher moments, primarily via the kurtosis

$$\beta = \frac{E[X^4]}{E[X^2]^2} \tag{28}$$

or the coefficient of excess, $\gamma = \beta - 3$. The kurtosis of the excitation processes are given in Table 2. It is seen that although (27) is a good

Table 2
Kurtosis of excitation processes

Y	3
Y^3	46·2
$3Y + Y^3$	16·6
$Y|Y|$	11·7

second moment representation the fourth order moments show considerable difference.

The second and fourth order response moments have been calculated for the non-normal excitation processes Y^3 and $3Y + Y^3$ with filter damping ratio $\zeta_g = 0\cdot5$ and $\zeta_g = 0\cdot2$, representing a moderately wide band and a narrow band process, respectively. The system damping covered the values $\zeta_0 = 0\cdot01$, $0\cdot02$, $0\cdot10$, $0\cdot20$, and thereby the interval of most interest for the response of engineering structures. Results in the form of the coefficient of excess of the response, normalized with respect to the excess of the excitation process are shown in Figs 2–5.

The curves in Figs 2–5 show, as expected, that the coefficient of excess of the response approaches that of the excitation process for large values of the frequency ratio ω_0/ω_g. This corresponds to quasistatic response. The increase is most rapid for large system damping ζ_0. At the other extreme, $\omega_0/\omega_g \approx 0$, the coefficient of excess vanishes, suggesting normal response. With the exception of the very small system damping $\zeta_0 = 0\cdot01$ all the curves overshoot the quasistatic value. The amount of overshoot increases with increasing damping ζ_0.

The response curves in Figs 2 and 3 corresponding to the excitation Y^3 show very little dependence on the bandwidth of the process $Y(t)$ as expressed by ζ_g. On the other hand the excitation process $3Y + Y^3$ leads to an increase in the coefficient of excess of the response with

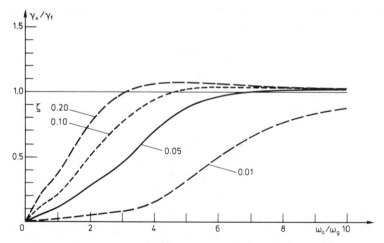

Fig. 2. Coefficient of excess for response to Y^3, damping ratio $\zeta_g = 0\cdot50$.

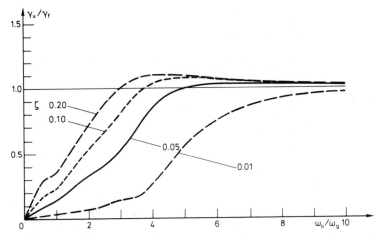

Fig. 3. Coefficient of excess for response to Y^3, damping ratio $\zeta_g = 0.20$.

decreasing bandwidth of $Y(t)$, as seen from Figs 4 and 5. It is notable that the coefficient of excess is considerably less than its quasistatic value at full resonance, i.e. for $\omega_0/\omega_g \approx 1$.

The results in Figs 2–5 can be compared directly with results by Lutes & Hu[2] for the case in which a non-normal white noise process is

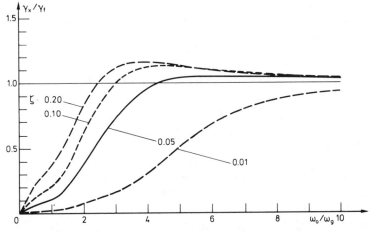

Fig. 4. Coefficient of excess for response to $3Y + Y^3$, damping ratio $\zeta_g = 0.50$.

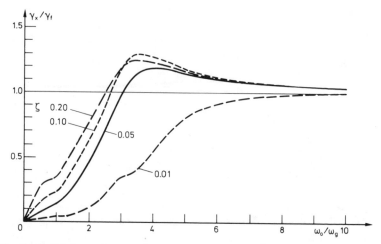

Fig. 5. Coefficient of excess for response to $3Y + Y^3$, damping ratio $\zeta_g = 0\cdot 20$.

passed through a second order filter and then used directly as input to a second order system without use of a non-linear function. The curves are qualitatively similar, but the non-normality generated by the functions Y^3 and $3Y + Y^3$ gives less overshoot and also requires a higher value of the frequency ratio ω_0/ω_g before the quasistatic limit value is reached. The excess of the response of a second order system to the second power of the Ornstein–Uhlenbeck process, considered by Kotulski & Sobczyk[4] and Grigoriu & Ariaratnam,[5] does not show any overshoot at all.

5 APPLICATION TO FATIGUE

The fatigue damage accumulated by a structure exposed to random loads may be severely influenced by deviations from normality. Often the damage accumulation is assumed to be governed by the stress amplitude raised to some power b, and the importance of deviations from normality then increases with increasing value of the exponent b. Typical values are $b \approx 3\text{--}4$. For quasistatic response to a drag force of the form $Y|Y|$ with a narrow band process $Y(t)$ Brouwers & Verbeek[10] found that the damage increases with a factor $2\cdot 4$ for $b = 3$ and $5\cdot 4$ for $b = 4$. This result is a special case of a process $X(t)$ that is a monotonic function of a standard normal process $U(t)$,

$$X(t) = h(U(t)) \tag{29}$$

The function $h(\ )$ performs a mapping of the normal process $U(t)$ whereby the extremes of $U(t)$ are mapped on the extremes of $X(t)$. Winterstein[12] proposed to characterize this mapping by the coefficient of skewness and the kurtosis β of $X(t)$ and found that the non-normality of the process $X(t)$ would increase the damage rate $E[D(t)]$ by the factor

$$\frac{E[D(t)]}{E[D(t) \mid X(t) \text{ normal}]} = \left[1 + \frac{b(b-1)}{24}(\beta - 3)\right] \tag{30}$$

This result is based on a narrow band assumption on $U(t)$ and moderate deviation from normality. Simulation results for processes of the type (29) by Lutes *et al.*[13] support the applicability of the formula (30) for kurtosis in the range $\beta = 3 \cdot 0 – 4 \cdot 5$.

In essence the formula (30) is based on the assumption that the function $h(U)$ can be replaced by a third degree polynomial, the coefficients of which are fitted via the moments of $X(t)$ up to the fourth order. For an odd function this is

$$X(t)/\sigma_x \simeq U(t) + \frac{\beta - 3}{24} He_3(U(t))$$

$$= U(t) + \frac{\beta - 3}{24}[U(t)^3 - 3U(t)] \tag{31}$$

In order to investigate the hypothesis (31) when applied to the response $X(t)$ of the second order system (18) calculation of the fourth and sixth order moments were carried out. The forcing function was

$$f(Y) = Y + \alpha Y^3 \tag{32}$$

where the parameter α was used to change the kurtosis of the excitation. Narrow band response was obtained by use of the parameters $\zeta_g = 0 \cdot 2$, $\zeta_0 = 0 \cdot 01$ and $\omega_0/\omega_g = 1 \cdot 0$. Figure 6 shows the sixth order moment calculated from the response by use of the moment equations (26) and estimated from the kurtosis by the polynomial approximation (31). The agreement is very good, supporting the usefulness of the formula (30) in this case. However, the magnitude of the kurtosis is somewhat limited for this particular type of loading—$\omega_0/\omega_g \approx 1$. It follows from combination of the information

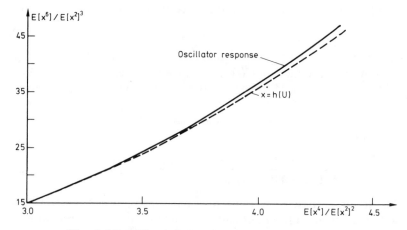

Fig. 6. Normalized sixth order moments of response.

on Y^3 from Fig. 3 and Table 2 that $3 \cdot 0 \leqslant \beta < 4 \cdot 5$. Incidentally this is also the interval covered by Lutes *et al.*[13] and Winterstein.[12]

6 CONCLUSIONS

A recursive algorithm for the calculation of moments of increasing order of the response of a linear system to excitation in the form of a polynomial of normal processes is established. Examples demonstrate that the non-normality of the response of a second order system, as measured by the coefficient of excess, increases with increasing system damping. There is a smaller influence from the specific form of the non-linear forcing function and the bandwidth of the excitation. For certain frequency ratios the coefficient of excess of the response exceeds that of the excitation except for very lightly damped systems. However, the magnitude of this excess is less than suggested by a previous model based on excitation by filtered non-normal white noise. A limited study indicates the usefulness of the coefficient of excess in estimating fatigue damage for moderately non-normal, narrow band response.

REFERENCES

1. Parzen, E., *Stochastic Processes*, Holden-Day, San Francisco, 1962.
2. Lutes, L. D. & Hu, S. J., Non-normal stochastic response of linear systems. *ASCE, Journal of Engineering Mechanics*, **112** (2) (1986) 127–41.
3. Grigoriu, M., White noise processes. *ASCE, Journal of Engineering Mechanics*, **113** (5) (1987) 757–65.
4. Kotulski, Z. & Sobczyk, K., Linear systems and normality. *Journal of Statistical Physics*, **24** (2) (1981) 359–73.
5. Grigoriu, M. & Ariaratnam, S. T., Stationary response of linear systems to non-gaussian excitations. *Reliability and Risk Analysis in Civil Engineering 2, ICASP5*, 1987, ed. N. C. Lind, Institute of Risk Research, University of Waterloo, 1987, pp. 718–24.
6. Wu, W. F. & Lin, Y. K., Cumulant-neglect closure for non-linear oscillators under random parametric and external excitations. *International Journal of Non-Linear Mechanics*, **19** (4) (1984) 349–62.
7. Gardiner, C. W., *Handbook of Stochastic Methods*, 2nd edn. Springer-Verlag, New York, 1985.
8. Madsen, H. O., Krenk, S. & Lind, N. C., *Methods of Structural Safety*, Prentice-Hall, Englewood Cliffs, N. J., 1986.
9. Spanos, P.-T. D., Spectral moments calculation of linear system output. *Journal of Applied Mechanics*, **50** (1983) 901–3.
10. Brouwers, J. J. H. & Verbeek, P. H. J., Expected fatigue damage and expected extreme response for Morison-type wave loading. *Applied Ocean Research*, **5** (1983) 129–33.
11. Borgman, L. E., Random hydrodynamic forces on objects. *Annals of Mathematical Statistics*, **38** (1967) 37–51.
12. Winterstein, S. R., Non-normal responses and fatigue damage. *ASCE, Journal of Engineering Mechanics*, **111** (10) (1985) 1291–5.
13. Lutes, L. D., Corazao, M., Hu, S. J. & Zimmerman, J., Stochastic fatigue damage accumulation. *ASCE, Journal of Structural Engineering*, **110** (11) (1984) 2585–601.

12

Nonstationary Response of Oscillators with Bilinear Hysteresis to Random Excitation

J. B. ROBERTS & A. H. SADEGHI

School of Engineering and Applied Sciences, University of Sussex, Brighton, UK

ABSTRACT

The method of stochastic averaging is applied to the case of an oscillator with bilinear hysteresis excited by wide band random excitation. Through a consideration of the energy lost per cycle, due to hysteresis, expressed as a fraction of the total energy in a cycle, the conditions under which the response amplitude process, a(t), can be modelled as a one-dimensional Markov process are examined in some detail. Suitable Fokker–Planck–Kolmogorov (FPK) diffusion equations for a(t) are proposed which allow one to predict the probabilistic response to both stationary and nonstationary random excitation. A simple numerical solution scheme, based on an explicit finite difference approximation of the governing diffusion equation, is given; it is shown that this is equivalent to a discrete, random walk analogue of the continuous process, a(t). Some typical numerical results, for both stationary and nonstationary response, are presented and compared with corresponding digital simulation estimates.

1 INTRODUCTION

In recent years the method of stochastic averaging[1,2] has proved to be a powerful method of analysing the response of nonlinear oscillators to random excitation. In situations where the excitation is wide band, and the damping is light, one can usually approximate the energy envelope

197

(or, alternatively, the amplitude envelope) of the oscillator response as a one-dimensional Markov process, governed by an appropriate Fokker–Planck–Kolmogorov (FPK) equation. The solution of this equation enables not only the distribution of the energy or amplitude envelope to be found but also the joint distribution of the response displacement and velocity. From this information reliability statistics, such as level crossing rates, can be calculated.

It has been shown by a number of workers[3–9] that, under suitable conditions, the stochastic averaging method is applicable to oscillators with various kinds of hysteretic restoring force. In particular, it has been demonstrated by the present author[3,4] that the standard stochastic averaging technique (which effectively neglects any 'stiffness type' nonlinearity) can yield good results when applied to the special case of an oscillator with bilinear hysteresis.

All previous work in this area has concentrated on estimating the stationary response of hysteretic oscillators to stationary random excitation. It is the purpose of the present paper to show that the earlier theory presented in Refs 2 and 3, for the bilinear oscillator, can be readily extended to the nonstationary case, through an appropriate modification of the governing FPK equation. Nonstationary solutions to this partial differential equation are found by using a generalisation of the random walk numerical scheme proposed earlier by one of the authors,[10] in the context of nonhysteretic oscillators. Comparisons between theoretical predictions of mean-square response and corresponding simulation estimates are presented, for both stationary and nonstationary cases.

2 THE BILINEAR OSCILLATOR

An oscillator with a bilinear hysteretic restoring force, excited by a random process, $n(t)$, is governed by an equation of motion of the following form:

$$m\ddot{x} + h(x, \dot{x}) + kF(x, t) = n(t) \tag{1}$$

Here m denotes the mass, k is a stiffness parameter, $h(x, \dot{x})$ represents nonlinear damping of an arbitrary form, and $F(x, t)$ has a bilinear characteristic, with a primary slope of unity and a secondary slope of α (tangent), with a yield amplitude x^*. $n(t)$ is assumed to be, in general, wide band and nonstationary, with a zero mean and a time-dependent power spectrum,[11] $S_n(\omega, t)$, where

$$S_n(\omega, t) = |A(t, \omega)|^2 \, \bar{S} \,|\,(\omega) \qquad (2)$$

$A(t, \omega)$ and $\bar{S}(\omega)$ are appropriate functions and the symbol $\|$ denotes the modulus of the complex function. In the special case where $n(t)$ is stationary,

$$S_n(\omega, t) = S_n(\omega) \qquad (3)$$

independent of t, where $S_n(\omega)$ is the usual power spectrum.

It is possible to nondimensionalise eqn (1) by introducing the nondimensional amplitude $y = x/x^*$ and the nondimensional time $\tau = \omega_0 t$ (where $\omega_0 = (k/m)^{1/2}$). Equation (1) can then be rewritten as

$$\ddot{y} + H(y, \dot{y}) + f(y, \tau) = N(\tau) \qquad (4)$$

where

$$H(y, \dot{y}) = h(y, \dot{y})/(mx^*\omega_0^2) \qquad (5)$$

$$f(y, \tau) = F(y, \tau)/x^* \qquad (6)$$

$$N(\tau) = n(t)/(mx^*\omega_0^2) \qquad (7)$$

The nondimensional bilinear characteristic $f(y, \tau)$ is sketched in Fig. 1. As noted by Suzuki & Minai,[12] this can be expressed in the following form

$$f(y, \tau) = \alpha y + (1 - \alpha)z(y, \tau) \qquad (8)$$

where $z(y, \tau)$ is the constitutive law for purely elasto-plastic hysteresis

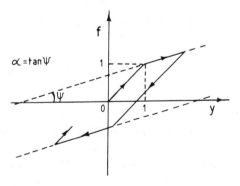

Fig. 1. The bilinear hysteretic characteristic.

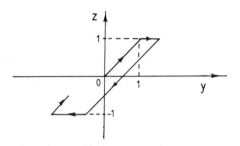

Fig. 2. The elasto-plastic characteristic.

(zero secondary slope) as sketched in Fig. 2. $z(y, \tau)$ is governed by the following differential relationship:

$$\dot{z} = \dot{y}[1 - U(\dot{y})U(y - 1) - U(-\dot{y})U(-y - 1)] \tag{9}$$

where $U(\cdot)$ denotes the unit step function, namely; $U(y) = 1$ for $y \geq 0$ and $U(y) = 0$ for $y < 0$.

On combining eqns (4) and (8), the governing equation of motion becomes

$$\ddot{y} + H(y, \dot{y}) + \alpha y + (1 - \alpha)z(y, \tau) = N(\tau) \tag{10}$$

3 MODELLING THE RESPONSE AS A MARKOV PROCESS

The stochastic averaging method is applicable when the response process, $y(\tau)$, is narrow band in character. In this situation the amplitude process $a(\tau)$, and the phase process, $\phi(\tau)$, defined by

$$y = a \cos[\omega(a)\tau + \phi]$$
$$\dot{y} = -a\omega(a) \sin[\omega(a)\tau + \phi] \tag{11}$$

are, by definition, slowly varying functions of time. Figure 3 shows the relationship between the envelope process, $a(\tau)$ and $y(\tau)$, in the narrow band response case. For a wide band excitation process, with a correlation time scale, τ_c, such that $\tau_c \ll 1/\omega(a)$, then the amplitude of a particular peak in $y(\tau)$ will depend only on the amplitude of the preceding peak. This follows from the fact that the joint process $[y(\tau), \dot{y}(\tau)]$ is approximately a two-dimensional Markov process when the excitation is wide band. Thus, if a peak in $y(\tau)$ occurs at $\tau = \tau_n$, say, and the next peak occurs at $\tau = \tau_{n+1}$, then $a(\tau_n) = y(\tau_n)$, $a(\tau_{n+1}) =$

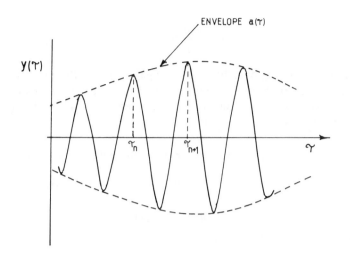

Fig. 3. Relationship between $a(\tau)$ and $y(\tau)$.

$y(\tau_{n+1})$, $\dot{y}(\tau_n) = \dot{y}(\tau_{n+1}) = 0$. For a known time difference $\Delta\tau = \tau_{n+1} - \tau_n$, it follows that $a(\tau_{n+1})$ depends only on $a(\tau_n)$. In the random case, $\Delta\tau$ are random, but have a small dispersion when the response is narrow band. Thus the discrete process $a(\tau_n)$ can be expected to approximate to a Markov process. This rough, heuristic argument is substantiated by the powerful limit theorem due to Stratonovich and Khasminskii (e.g. see Ref. 1), which shows that the continuous process $a(\tau)$ converges to a continuous Markov process, under suitable narrow band conditions.

An apparent complication in the application of the Stratonovich–Khasminskii limit theorem is that the frequency $\omega(a)$ in eqn (11) is amplitude dependent. However, as shown in Refs 3 and 4, this complication disappears when the conditions for the response to be narrow band are considered in depth.

For a narrow band response the energy dissipated per cycle due to damping must be a relatively small fraction of the energy associated with that cycle. The energy loss per cycle, due to hysteresis alone, is given by

$$E(a) = \oint f(y, \tau)\, dy \qquad (12)$$

or

$$E(a) = \oint f(y, \tau)\dot{y}\, d\tau \tag{13}$$

where integration is over one cycle. Assuming, for the moment, that narrow band conditions pertain, then $E(a)$ can be evaluated, approximately, by using eqn (11), and treating $a(\tau)$ and $\phi(\tau)$ as constants, over one cycle. Thus

$$E(a) = -a \oint_0^{2\pi} f(y, \tau) \sin \Phi\, d\Phi \tag{14}$$

or

$$E(a) = 2\pi a s(a) \tag{15}$$

where

$$s(a) = -\frac{1}{2\pi} \oint f(a \cos \Phi, \tau) \sin \Phi\, d\Phi \tag{16}$$

and

$$\Phi = \omega(a)\tau + \phi \tag{17}$$

For the bilinear oscillator, $s(a)$ has been evaluated by Caughey.[13] The result is

$$s(a) = \frac{2(1 - \alpha)(a - 1)}{\pi a} \qquad (a > 1)$$

$$= 0 \qquad\qquad (a < 1) \tag{18}$$

The energy lost per cycle, expressed as a fraction of the energy in the cycle $(=\tfrac{1}{2}\dot{y}^2 \max = \tfrac{1}{2}\omega^2(a)a^2)$ is thus given by

$$R(a) = \frac{E(a)}{\tfrac{1}{2}\omega^2(a)a^2} = \frac{8(1 - \alpha)(a - 1)}{a^2\omega^2(a)} \qquad (a > 1)$$

$$= 0 \qquad\qquad (a < 1) \tag{19}$$

Now $R(a)$ will, on average, be small, if at least one of the following conditions is satisfied:

1. The secondary slope α is close to unity, such that

$$\epsilon = (1 - \alpha)^{1/2} \tag{20}$$

is a small quantity. Then $R(a)$ will be of order ϵ^2 and, if the nonlinear damping term $h(y, \dot{y})$ is also of order ϵ^2 it follows (see Ref. 3) that $a(\tau)$ converges to a Markov process as $\epsilon \to 0$. The error associated with neglecting the variation of $\omega(a)$ with a, which is of order ϵ^2, is of higher order than the error inherent in the Markov approximation for $a(\tau)$. Thus one can set $\omega(a) = 1$, with an error consistent with the errors in the Markov approximation. The condition that ϵ in eqn (20) be small is independent of the level of the excitation.

2. The excitation level is sufficiently low for the rate at which yielding occurs to be relatively small. There are then only occasional, small excesses of unity by $a(\tau)$ and $R(a)$ is small, on average. In this situation, typical hysteresis loops (when they occur) are slim, as shown in Fig. 4, and the energy dissipated through hysteresis is small. For the special case of stationary response it is shown in Ref. 4 that, if the excitation is of order ϵ, the average rate of energy loss, due to hysteresis, is of order ϵ^2. Hence, if $h(y, \dot{y})$ is also small, and of order ϵ^2, $a(\tau)$ converges to a Markov process, according to the limit theorem. Since most of the response is in the elastic regime, the predominant frequency of the response, ω_{eq} say, is equal to unity, i.e.

$$\omega_{eq} = 1 \qquad (21)$$

Errors associated with neglecting the variation of $\omega(a)$ with a,

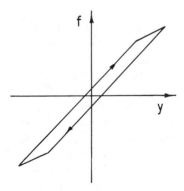

Fig. 4. Symmetrical hysteresis loop for low excitation levels.

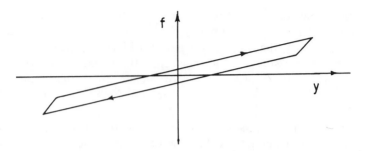

Fig. 5. Symmetrical hysteresis loop for high excitation levels.

in this case, are again of higher order than errors inherent in the Markov approximation. The condition that the excitation be weak is independent of the value of α.

3. The excitation level is so high that most of the oscillator response is in the plastic regime. From eqn (19) it follows that, as a becomes large, $R(a)$ is of order $1/a$ and therefore becomes small. Thus hysteresis loops are slim, as shown in Fig. 5, and the response is narrow band in nature. It follows that, as a becomes large, $a(\tau)$ approaches a Markov process, independently of the value of α. Since the oscillator response is mostly in plastic regime one can set $\omega(a)$ to a constant, ω_p, where ω_p is the natural frequency associated with the secondary slope, i.e.

$$\omega_p = \alpha^{1/2} \tag{22}$$

Once again, the error associated with treating $\omega(a)$ as a constant, in the Markov approximation of $a(\tau)$, is consistent with the inherent errors in that approximation.

When at least one of the above three conditions are satisfied the response is narrow band and the initial assumption that $a(\tau)$ and $\phi(\tau)$ are slowly varying (used to calculate $s(a)$) is justified. One can conclude that a Markov approximation for $a(\tau)$ is likely to be reasonably accurate except when (a) α is not close to unity *and* (b) the excitation is neither very weak, or very strong. When conditions (a) and (b) apply the response process is in fact wide band in character, as simulation studies verify, and a Markov approximation for the amplitude envelope is inappropriate.

3.1 FPK Equation for Estimating Stationary Response

It has been shown that $\omega(a)$ in eqn (11) can be replaced by unity, if the excitation is very weak and/or α is close to unity. If the excitation is very strong then $\omega(a)$ can be replaced by $\alpha^{1/2}$.

It is convenient, in practice, to replace $\omega(a)$ by a frequency, ω_{eq} say, which depends only on the level of excitation and varies *continuously* from unity, at lower excitation levels, to $\alpha^{1/2}$ at high excitation levels. One can then formulate a theory which gives results under all conditions, with good accuracy when the response is narrow band, and less accuracy when the response is wide band.

If one is interested only in the stationary response to stationary excitation it is natural to use equivalent linearisation to determine ω_{eq}. A method of computing this on the basis of the Krylov–Bogoliubov approximation is discussed in Ref. 3 (see also Ref. 13). Replacing eqn (11) by

$$\left. \begin{aligned} y &= a \cos \Phi \\ \dot{y} &= -a\omega_{eq} \sin \Phi \end{aligned} \right\} \tag{23}$$

where

$$\Phi = \omega_{eq}\tau + \phi \tag{24}$$

and transforming eqn (10) from y, $\dot{y}$ variables to a, ϕ variables, the following pair of first-order equations emerges:

$$\dot{a} = [H(a \cos \Phi, -a\omega_{eq}a \sin \Phi) + (\alpha - \omega_{eq}^2)a \cos \Phi + (1 - \alpha)z]$$
$$\times \frac{\sin \Phi}{\omega_{eq}} - \frac{N(\tau) \sin \Phi}{\omega_{eq}} \tag{25}$$

and

$$a\dot{\phi} = [H(a \cos \Phi, -a\omega_{eq}a \sin \Phi) + (\alpha - \omega_{eq}^2)a \cos \Phi + (1 - \alpha)z]$$
$$\times \frac{\cos \Phi}{\omega_{eq}} - \frac{N(\tau) \cos \Phi}{\omega_{eq}} \tag{26}$$

If the terms not involving $N(\tau)$ explicitly are now averaged over one cycle, eqns (25) and (26) simplify considerably, as follows:

$$\dot{a} = -\frac{[s(a) + g(a)]}{\omega_{eq}} - \frac{N(\tau) \sin \Phi}{\omega_{eq}} \tag{27}$$

$$a\dot{\phi} = -\frac{[l(a) + k(a)]}{\omega_{eq}} - \frac{N(\tau) \cos \Phi}{\omega_{eq}} \tag{28}$$

where

$$g(a) = -\frac{1}{2\pi} \oint_0^{2\pi} H(a \cos \Phi, -a\omega_{eq} \sin \Phi) \sin \Phi \, d\Phi \qquad (29)$$

$$l(a) = -\frac{1}{2\pi} \oint_0^{2\pi} [f(a \cos \Phi, \tau) - \omega_{eq}^2 a \cos \Phi] \cos \Phi \, d\Phi \qquad (30)$$

$$k(a) = -\frac{1}{2\pi} \oint_0^{2\pi} H(a \cos \Phi, -a\omega_{eq} \sin \Phi] \cos \Phi \, d\Phi \qquad (31)$$

and $s(a)$ is given by eqn (16) (see also eqn (18)).

As a final stage of the averaging operation, the Stratonovitch–Khasminskii limit theorem can be applied to eqns (27) and (28). This shows that, if $y(\tau)$ is narrow band, the joint process $[a(\tau), \phi(\tau)]$ converges weakly to a Markov process and enables one to write an FPK equation for this process. If $p(a, \phi; a_0, \phi_0 : \tau)$ denotes the appropriate transition density function then the FPK equation is as follows:

$$\frac{\partial p}{\partial \tau} = -\frac{\partial}{\partial a} \left[\left\{ -\frac{\Lambda(a)}{\omega_{eq}} + \frac{\pi S(\omega_{eq}, \tau)}{2a\omega_{eq}^2} \right\} p \right] + \frac{\Xi(a)}{a\omega_{eq}} \frac{\partial p}{\partial \phi}$$
$$+ \frac{\pi S(\omega_{eq}, \tau)}{2\omega_{eq}^2} \left[\frac{\partial^2 p}{\partial a^2} + \frac{1}{a^2} \frac{\partial^2 p}{\partial \phi^2} \right] \qquad (32)$$

where

$$\Lambda(a) = s(a) + g(a) \qquad (33)$$
$$\Xi(a) = l(a) + k(a) \qquad (34)$$

and $S(\omega, \tau)$ is the time-dependent power spectrum of $N(\tau)$ defined similarly to $S_n(\omega, t)$; see eqn (2).

An inspection of eqn (32) shows that the amplitude process $a(\tau)$ is *uncoupled* from the phase process, $\phi(\tau)$, i.e. $a(\tau)$ is a one-dimensional Markov process, with a transition density function $p(a; a_0 : \tau)$ governed by the following FPK equation:

$$\frac{\partial p}{\partial \tau} = -\frac{\partial}{\partial a} \left[\left\{ -\frac{\Lambda(a)}{\omega_{eq}} + \frac{\pi S(\omega_{eq}, \tau)}{2a\omega_{eq}^2} \right\} p \right] + \frac{\pi S(\omega_{eq}, \tau)}{2\omega_{eq}^2} \frac{\partial^2 p}{\partial a^2} \qquad (35)$$

A general solution of eqns (32) and (35) is difficult. However, if $a_0 = 0$ at $\tau = 0$ then $p(a; \phi : \tau) = p(a, \phi; 0, \phi_0, \tau)$, the marginal density function, is of separable form and the phase $\phi(\tau)$ is uniformly distributed, for all τ.[1] Thus

$$p(a; \phi : \tau) = \frac{1}{2\pi} p(a : \tau) \qquad (36)$$

where $p(a : \tau)$ is governed by eqn (35), with zero initial conditions. From $p(a; \phi : \tau)$ one can determine the joint marginal density function, $p(y, \dot{y} : \tau)$ of the process $[y, \dot{y}]$, using eqn (23) again. Thus

$$p(y, \dot{y} : \tau) = \frac{p(a; \phi : \tau)}{\omega_{eq} a} = \frac{p(a : \tau)}{2\pi \omega_{eq} a} \qquad (37)$$

From this result various statistics of the response can be calculated, including the standard deviation, σ_y, where

$$\sigma_y^2(\tau) = E(y^2(\tau)) \qquad (38)$$

Thus

$$\sigma_y^2(\tau) = \int_{-\infty}^{\infty} \int_{-\infty}^{\infty} y^2 p(y, \dot{y}) \, dy \, d\dot{y} \qquad (39)$$

Using eqn (37), this last expression can be written alternatively as

$$\sigma_y^2(\tau) = \frac{1}{2} \int_0^{\infty} p(a : \tau) a^2 \, da \qquad (40)$$

When the excitation is stationary then $S(\omega_{eq}, \tau) = S(\omega_{eq})$, independent of τ, and

$$p(a; \phi : \tau) \to w(a, \phi) = \frac{w(a)}{2\pi}, \text{ as } \tau \to \infty \qquad (41)$$

where $w(a)$ is the stationary density function for $a(\tau)$. Setting $\partial p / \partial \tau = 0$ in eqn (35), the following equation for $w(a)$ emerges:

$$0 = -\frac{\partial}{\partial a} \left[\left\{ -\frac{\Lambda(a)}{\omega_{eq}} + \frac{\pi S(\omega_{eq})}{2 a \omega_{eq}^2} \right\} w \right] + \frac{\pi S(\omega_{eq})}{2 \omega_{eq}^2} \frac{\partial^2 w}{\partial a^2} \qquad (42)$$

This equation is readily solved to give

$$w(a) = ca \, \exp \left\{ -\frac{2\omega_{eq}}{\pi S(\omega_{eq})} \int_0^a \Lambda(\xi) \, d\xi \right\} \qquad (43)$$

where c is a normalisation constant. In the special case where $h(y, \dot{y}) = 0$ (i.e. $\Lambda(\xi) = s(\xi)$) a simple explicit expression for $w(a)$ (and hence $w(y, \dot{y})$) can be found.[3]

3.2 FPK Equations for Estimating Nonstationary Response

The equations given in the preceding section are not entirely appropriate for the nonstationary case, since ω_{eq} is evaluated on the basis of a stationary response to a stationary input. This difficulty can be easily overcome by simply replacing ω_{eq} by $\omega(a)$, where $\omega(a)$ can be found by the Krylov–Bogoliubov equivalent linearisation method. This shows that (e.g. see Ref. 14),

$$\omega^2(a) = \frac{c(a)}{a} \tag{44}$$

where

$$c(a) = \frac{1}{\pi} \oint_0^{2\pi} f(a\cos\Phi,\,\tau)\cos\Phi\,\mathrm{d}\Phi \tag{45}$$

Then eqn (35) becomes

$$\frac{\partial p}{\partial \tau} = -\frac{\partial}{\partial a}\left[\left\{-\frac{\Lambda(a)}{\omega(a)} + \frac{\pi S[\omega(a),\,\tau]}{2a\omega^2(a)}\right\}p\right] + \frac{\pi S[\omega(a),\,\tau]}{2\omega^2(a)} \cdot \frac{\partial^2 p}{\partial a^2} \tag{46}$$

Once again it is noted that the error in this approximation is of the same order as the error inherent in the Markov approximation. In other words, the 'error difference' between eqns (35) and (44) is compatible with the basic accuracy of the Markov approximation of $a(\tau)$. The solution of (46), for the case $a_0 = 0$ at $\tau = 0$, can be used in conjunction with eqns (36) and (37) to compute statistics for the oscillator response.

An alternative to eqn (46), which again has an error consistent with the Markov approximation, is as follows:

$$\frac{\partial p}{\partial \tau} = -\frac{\partial}{\partial a}\left[\left\{-\frac{\Lambda(a)}{\omega(a)} + \frac{\pi S[\omega(a),\,\tau]}{2a\omega^2(a)}\right\}p\right] + \frac{1}{2}\frac{\partial^2}{\partial a^2}\left[\frac{\pi S[\omega(a),\,\tau]}{\omega^2(a)} \cdot p\right] \tag{47}$$

As will be shown shortly, this latter equation can be readily solved, using a physically based explicit finite difference scheme, and will be used here in preference to eqn (46).

3.3 Special Cases

In the following, attention will be restricted to the special case where $h(y,\dot{y}) = 0$. Then $\Lambda(a) = s(a)$ and, for stationary excitation, eqn (35) can be written as

$$\frac{\partial p}{\partial \tau^*} = -\frac{\partial}{\partial a}\left[\left\{-s^1(a) + \frac{2\beta}{\pi a}\right\}p\right] + \frac{2\beta}{\pi}\frac{\partial^2 p}{\partial a^2} \tag{48}$$

where

$$\tau^* = \frac{(1-\alpha)\tau}{\omega_{eq}} \qquad (49)$$

$$\beta = \frac{\pi^2 S(\omega_{eq})}{4\omega_{eq}(1-\alpha)} \qquad (50)$$

and

$$s^1(a) = \frac{s(a)}{(1-\alpha)} \qquad (51)$$

The stationary solution to eqn (48), $w(a)$, depends only on the parameter β.[3]

For nonstationary excitation, of the form of modulated white noise,

$$N(\tau) = \chi(\tau)w(\tau) \qquad (52)$$

where $w(\tau)$ is a stationary white noise, with spectrum $S(\omega) = S_0$, a constant, and $\chi(\tau)$ is a deterministic modulating function. Then (see Ref. 11)

$$S(\omega, \tau) = \chi^2(\tau)S_0 \qquad (53)$$

and, with $h(y, \dot{y}) = 0$, eqn (47) becomes

$$\frac{\partial p}{\partial \tau^1} = -\frac{\partial}{\partial a}[A(a, \tau^1)p] + \frac{1}{2}\frac{\partial^2}{\partial a^2}[B(a, \tau^1)p] \qquad (54)$$

where

$$A(a, \tau^1) = -\frac{s^1(a)}{\omega(a)} + \frac{2\beta^1\chi^2(\tau^1)}{\pi a\omega^2(a)} \qquad (55)$$

$$B(a, \tau^1) = \frac{4\beta^1\chi^2(\tau^1)}{\pi\omega^2(a)} \qquad (56)$$

$$\beta^1 = \frac{\pi^2 S_0}{4(1-\alpha)} \qquad (57)$$

and

$$\tau^1 = (1-\alpha)\tau \qquad (58)$$

The frequency $\omega(a)$ can be evaluated from eqn (44) where, as shown by Caughey,[13]

$$c(a) = \frac{a}{\pi} \left[(1 - \alpha)\theta^* + \alpha\pi - \frac{(1 - \alpha)}{2} \sin^2 \theta^* \right] \qquad a > 1 \qquad (59)$$

$$= a, \qquad\qquad\qquad\qquad\qquad\qquad a < 1 \qquad (60)$$

and

$$\cos \theta^* = 1 - \frac{2}{a} \qquad (61)$$

It follows from eqns (59) and (61) that, as expected, $\omega(a) \to 1$ as $a \to 0$ (actually here $\omega(a) = 1$ for $a < 1$) and $\omega(a) \to \alpha^{1/2}$ as $a \to \infty$.

4 NUMERICAL SOLUTION FOR THE NONSTATIONARY CASE

A diffusion equation of the general form given by eqn (54) can be solved by a simple, explicit finite difference scheme.

If the time τ^1 is discretised into times $\tau^1 j = j\, \delta\tau^1$, and the amplitude a is discretised into the values $a_k = k\, \delta a$, where $\delta\tau^1$ and δa are (small) constant increments, then a suitable finite difference approximation to eqn (54) is as follows:

$$\frac{p(k, j + 1) - p(k, j)}{\delta\tau^1}$$

$$= -\frac{1}{2\, \delta a} \{ p(k + 1, j)A(k + 1, j) - p(k - 1, j)A(k - 1, j) \}$$

$$+ \frac{1}{2\delta a^2} \{ p(k + 1, j)B(k + 1, j) - 2p(k, j)B(k, j)$$

$$+ p(k - 1, j)B(k - 1, j) \} \qquad (62)$$

Here $p(k, j)$ denotes $p(a, \tau^1 j)$, and similarly for $A(k, j)$ and $B(k, j)$.

After some rearrangement, eqn (62) can be expressed as follows:

$$p(k, j + 1) = p(k - 1, j)\rho_{k-1,j} + p(k, j)\nu_{k,j} + p(k + 1, j)\mu_{k+1,j} \qquad (63)$$

where

$$\rho_{k,j} = \frac{\lambda}{2} [B(k, j) + A(k, j)\, \delta a] \qquad (64)$$

$$v_{k,j} = 1 - \lambda B(k, j) \tag{65}$$

$$\mu_{k,j} = \frac{\lambda}{2}[B(k, j) - A(k, j)\,\delta a] \tag{66}$$

and

$$\lambda = \frac{\delta \tau^1}{\delta a^2} \tag{67}$$

Equation (63) gives a simple method of computing the values $p(k, j+1)$ at times τ^1_{j+1} from the values $p(k, j)$ at the preceding time, τ^1_j. Thus the solution can be marched forward in time, from some initial condition.

A slightly different interpretation of eqn (63) is obtained by multiplying throughout by δa. Then

$$P(k, j+1) = P(k-1, j)\rho_{k-1,j} + P(k, j)v_{k,j} + P(k+1, j)\mu_{k+1,j} \tag{68}$$

where

$$P(k, j) = p(k, j)\,\delta a \tag{69}$$

$P(k, j)$ can be interpreted as the (finite) probability of being in state (a_k, τ^1_j); thus the original, continuous probability 'mass' is discretised into a set of 'lumped masses', at the nodal points $a_k = k\,\delta a$. Clearly, for any j,

$$\sum_k P(k, j) = 1 \tag{70}$$

Thus eqn (68) represents an equation for diffusing finite probability 'masses' over the discrete space a_k, τ^1_j.

The discrete process obeying eqn (68) is, in fact, a *random walk* process. Thus, consider a random walk, $R(\tau^1_j)$, which can only assume the discrete amplitudes a_k. If the process is at state (a_k, τ^1_j) then assume that it can do one of the following:

(a)   move to state (a_{k+1}, τ^1_{j+1}), with probability $\rho_{k,j}$
(b)   move to state (a_k, τ^1_{j+1}), with probability $v_{k,j}$
(c)   move to state (a_{k-1}, τ^1_{j+1}), with probability $\mu_{k,j}$

This is represented graphically in Fig. 6(a). Since only these three possibilities are allowed it is evident that one must have

$$\rho_k + \mu_k + v_k = 1 \tag{71}$$

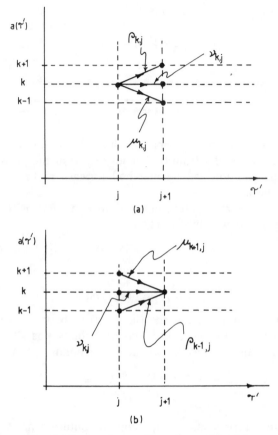

Fig. 6. The random walk analogue for $a(\tau)$.

It follows from the above that state (a_k, τ^1_{j+1}) can only be reached from (a_{k+1}, τ^1_j), (a_k, τ^1_j) or (a_{k-1}, τ^1_j), as illustrated in Fig. 6(b). Thus the probability, $P(k, j+1)$, of being in state a_k, τ^1_{j+1} is, in fact, given by eqn (68). Hence ρ_k, ν_k and μ_k in eqn (68) *can* be interpreted as transition probabilities and it is evident from eqns (64)–(66) that eqn (71) is indeed satisfied, at every j.

This random walk analogue of the continuous process $a(\tau^1)$ is a generalisation of that discussed earlier by one of the authors,[10] where $\nu_{k,j}$ was assumed to be zero, and $A(a, \tau^1)$ and $B(a, \tau^1)$ were independent of τ^1 (stationary excitation). It is evident from eqn (65) that, if $v_k = 0$,

$$\lambda = \frac{1}{B(k, j)} \qquad (72)$$

Since λ must be a constant, eqn (72) can only be satisfied if $B(k, j)$ is independent of both k and j. The inclusion of a nonzero transition probability, $v_{k,j}$, into the numerical scheme allows one to solve the more general type of diffusion equation, given by eqn (54), where $B(a, \tau^1)$ is not a constant.

It is noted that, in the special case of stationary excitation, the transition probabilities are independent of j and therefore need only be computed once, at the beginning of the numerical diffusion process. In the more general, nonstationary excitation case, they must be recalculated at every time step.

In implementing the random walk scheme it is important to choose $\delta\tau^1$ and δa (and hence λ) such that the transition probabilities remain positive, otherwise there is a strong tendency towards numerical instability. Care must also be exercised in the vicinity of the lower amplitude limit $a = 0$, since $A(a, \tau^1) \to \infty$ as $a \to 0$. This difficulty can be overcome by treating $a = 0$ as a reflecting barrier, and accordingly modifying the transition probabilities in the vicinity of this barrier.[10]

5 COMPARISON OF THEORY WITH SIMULATION

To give some idea of the accuracy of the proposed theoretical method of estimating the probabilistic response of bilinear oscillators to wide band random excitation, a few typical comparisons between theoretical predictions and corresponding digital simulation estimates will be given here. Only results for the standard deviation of the response, $\sigma_y(t)$, will be presented; it is noted, however, that, from the theoretical relationships given in Section 3, the joint distribution of the displacement and velocity, as a function of time, can be computed. As already mentioned, from this information, many other response statistics (e.g. level crossing rates) can be calculated.

5.1 Simulation Procedure
The excitation was assumed to be a modulated white noise. Thus $N(\tau)$ was of the form given by eqn (52), where $w(\tau)$ was a stationary white noise, with a uniform spectral height, S_0. Two particular forms of $\chi(\tau)$

were chosen:

(a) $\chi(\tau) = U(\tau)$ (the unit step function)
(b) $\chi(\tau) = \exp(-\rho\tau)$ $\quad \tau \geqslant 0$
$\qquad \quad = 0 \qquad \qquad \tau < 0$

where ρ is a constant. The first case corresponds to 'switched on' stationary excitation, and the resulting transient response eventually reaches stationary conditions.

Realisations of $w(\tau)$ were obtained, approximately, by computing functions which jumped from one, constant, level to another, at intervals $\Delta\tau$; the levels were obtained from a sequence of independent, Gaussian random numbers, suitably scaled. In the present investigation $\Delta\tau = 0.05$.

Corresponding sample functions of the response process, $y(\tau)$, at times $i\Delta\tau$ ($i = 1, 2, \ldots$) were obtained by numerically integrating eqns (9) and (10), using the fourth order Runge–Kutta algorithm.

To obtain estimates of the stationary standard deviation, σ_y, for a particular set of parameters, a single long realisation of $y(\tau)$ was generated and a running estimate of σ_y was calculated. Integration was continued until a reasonably stable estimate of σ_y was obtained (typically 500 000 samples generated). To obtain the nonstationary response, in terms of the variation of $\sigma_y(t)$ with time, for both cases (a) and (b), given above, 500 independent sample functions of the response were computed, over an appropriate time interval, and estimates of $\sigma_y(t)$, at $i\Delta\tau$, were computed by ensemble averaging.

5.2 Stationary Response

Figure 7 shows the variation of the stationary standard deviation, σ_y, plotted against the parameter β (computed using ω_{eq}, from Ref. 3), for various levels of secondary slope, α. The full lines show the theoretical predictions, computed according to eqns (54)–(56), these are compared with various relevant digital simulation estimates, taken from Ref. 3. As $\alpha \to 1$, the present theory converges to a result which is identical to that obtained by using a constant frequency, ω_{eq}, rather than $\omega(a)$, as in Ref. 3. This follows from the fact that, as $\alpha \to 1$, $\omega(a) \to \omega_{eq} \to 1$. It is seen that, as the arguments given earlier lead one to expect, the theory agrees well with the digital simulation estimates except when α is not close to unity *and* β is in the intermediate range (say 0·1–10).

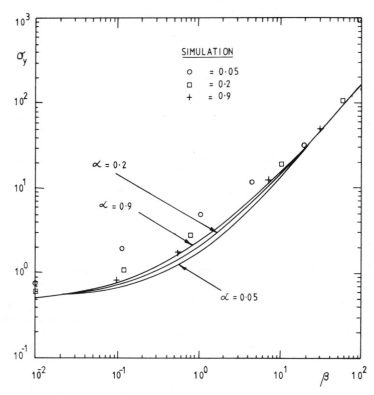

Fig. 7. Variation of the stationary standard deviation, σ_y, with β, for various values of α. Comparison with digital simulation estimates (from Ref. 3).

5.3 Nonstationary Response

Typical results for nonstationary response are given in Figs 8 and 9. Here the time-dependent standard deviation, $\sigma_y(\tau)$ is plotted against τ for two different values of ρ ($\rho = 0$ corresponding to the case of unit step modulation). Figure 8 gives results for the case of a fairly high level of excitation, such that the response is mainly in the plastic regime. It is seen that good agreement between theory and simulation is obtained, for both types of input modulation. For unit step modulation, $\sigma_y(\tau)$ approaches its stationary value as time elapses, whereas for exponential modulation the level of the response decays slowly, as τ increases.

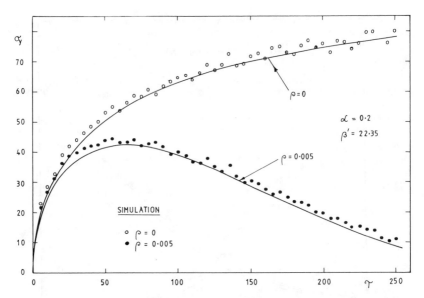

Fig. 8. Variation of the nonstationary standard deviation, $\sigma_y(\tau)$, with τ. $\alpha = 0.2$, $\beta^1 = 22.35$. $\rho = 0$ and $\rho = 0.005$. Comparison with digital simulation estimates.

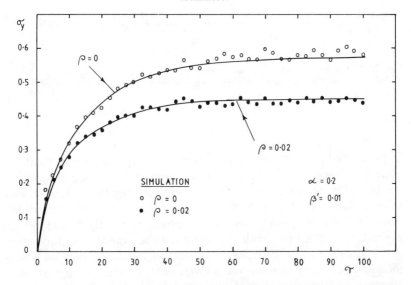

Fig. 9. Variation of the nonstationary standard deviation, $\sigma_y(\tau)$, with τ. $\alpha = 0.7$, $\beta^1 = 0.01$. $\rho = 0$ and $\rho = 0.02$. Comparison with digital simulation estimates.

Figure 9 gives a similar comparison, for the case of a fairly low level of excitation, such that the response is mainly in the elastic region. Again the present theory gives a good agreement with digital simulation estimates. It is interesting to observe that, in the case of the exponentially modulated input (here $\rho = 0 \cdot 02$), the standard deviation of the response does *not* decay to zero, as time elapses, as one might at first expect. The reason for this is that, as the input level decreases, due to $\chi(\tau)$ decreasing, the probability of the response reaching the plastic regime reduces effectively to zero. Thus the response stays in the elastic range and there is no energy dissipation due to hysteresis loops. Since other forms of damping are also absent there is no mechanism for energy extraction and thus the response level, as measured by σ_y, does not decay away to zero.

6 CONCLUSIONS

It has been demonstrated that the method of stochastic averaging is applicable to the problem of computing the probabilistic response of bilinear oscillators to wide band random excitation, of both stationary and nonstationary types. Appropriate diffusion equations for the amplitude of the response have been given, from which the time-dependent joint distribution of the displacement and velocity of the response can be estimated. Comparisons between theoretical predictions and digital simulation estimates, for the standard deviation of the response, indicate that the proposed theory is useful when the response is narrow band in nature.

REFERENCES

1. Roberts, J. B. & Spanos, P. D., Stochastic averaging: an approximate method of solving random vibration problems. *Int. J. Nonlinear Mechs.*, **21** (2) (1986) 111–34.
2. Lin, Y. K., Some observations on the stochastic averaging method. *Prob. Engng Mechs.*, **1** (1) (1986) 23–7.
3. Roberts, J. B., The response of an oscillator with bilinear hysteresis to stationary random excitation. *J. Appl. Mech., ASME,* **45** (4) (1978) 923–8.
4. Roberts, J. B., The yielding behaviour of a randomly excited elasto-plastic structure. *J. Sound Vib.,* **72** (1) (1980) 71–85.

5. Spanos, P. D., Hysteretic structural vibrations under random load. *J. Acoust. Soc. Amer.*, **65** (1979) 404–10.
6. Yar, M. & Hammond, J. K., Modelling and response of bilinear hysteretical systems. *J. Engng Mech., ASCE*, **113** (7) (1987) 1000–13.
7. Yar, M. & Hammond, J. K., Stochastic response of an exponentially hysteretic system through stochastic averaging. *Prob. Engng Mech.*, **2** (3) (1987) 147–55.
8. Roberts, J. B., Application of averaging methods to randomly excited hysteretic systems. *Proc. IUTAM Symposium on Nonlinear Stochastic Dynamic Engineering Systems*, Innsbruck, Austria, 1987, Springer-Verlag, Berlin, 1988, pp. 361–79.
9. Zhu, W. Q. & Lei, Y., Stochastic averaging of the energy envelope of bilinear hysteretic systems. *Proc. IUTAM Symposium on Nonlinear Stochastic Dynamic Engineering Systems*, Innsbruck, Austria, 1987, Springer-Verlag, Berlin, 1988, pp. 381–91.
10. Roberts, J. B., First-passage time for oscillators with nonlinear damping. *J. Appl. Mech., ASME*, **45** (1) (1978) 175–80.
11. Priestley, M. B., Power spectral analysis of nonstationary random processes. *J. Sound Vib.*, **6** (1967) 86–97.
12. Suzuki, Y. & Minai, R., Application of stochastic differential equations to seismic reliability analysis of hysteretic structures. Presented at the US-Japan Joint Seminar on Stochastic Approaches in Earthquake Engineering, Florida Atlantic University, Boca Raton, Florida, 1987.
13. Caughey, T. K., Random excitation of a system with bilinear hysteresis. *J. Appl. Mech., ASME*, **27** (1960) 649–52.
14. Iwan, W. D. & Spanos, P. D., Response envelope statistics for nonlinear oscillators with random excitation. *J. Appl. Mech., ASME* **45** (1978) 170–4.

13

Non-Gaussian Response of Systems under Dynamic Excitation

G. I. SCHUËLLER & C. G. BUCHER

Institute of Engineering Mechanics, University of Innsbruck, Austria

ABSTRACT

In many practical applications the excitation processes show non-Gaussian properties. In addition non-normal response properties result from non-linear systems, even if the input is of Gaussian nature. Hence various representations of stochastic variates, such as series expansions, maximum entropy and cumulant neglect closure are discussed. In addition the non-normal random processes are described by higher order correlations and spectra. Based on this the response of linear systems to non-normal loading is calculated. Furthermore, non-normal response properties of PDF's for non-linear systems under normal loading are analysed, where non-linearities of restoring force and damping are considered. The properties of non-normality are shown to affect the reliability estimates considerably. Finally, the evolution of the PDF's of non-linear systems under non-stationary loading is analysed.

1 INTRODUCTION

In linear random vibration analysis, generally Gaussian properties of the excitation processes are assumed. Although this assumption—which certainly simplifies the computational procedures considerably—is justified in many cases, the analyses of load histories of earthquake accelerations, wind pressures and water waves, quite frequently reveal considerable non-normal characteristics. This fact is reflected in the

219

respective properties of higher moments of the excitation process which directly cause non-Gaussian properties of the linear structural response process, and consequently affect the exceedance probabilities of thresholds, i.e. reliability estimates.

Non-Gaussian properties of the response process may also result from non-linear structures subjected to normal or non-normal excitation respectively, where in this context the present work concentrates on the first aspect. Non-linear effects may be due to non-linear damping and/or restoring force.

It is well known that there are a number of possibilities of representing mathematically the non-normal properties of a random process. With reference to their convenience in application, only three types, i.e. the Gram–Charlier series, the cumulant neglect-closure and the maximum entropy criteria are discussed here in some detail.

In the following some of the existing approaches for analysing non-Gaussian response are critically discussed as well as new procedures suggested.

2 MATHEMATICAL REPRESENTATION OF NON-NORMAL RANDOM VARIATES

2.1 General

A random variate can be characterized optimally in terms of a probability density function (PDF) as it contains all information. However, in many practical cases the information available is limited to a few statistical moments of low order (i.e. mean, variance, skewness, etc.). This is due to either a limited amount of data available (e.g. measurements of environmental loads) or the mathematical complexity of transforming known distributions (e.g. non-linear systems subjected to normal load process). In both cases, the available information on the statistical moments may be utilized to approximate the unknown PDF in closed form. Generally, such an approximation cannot be unique since a PDF is defined by infinitely many statistical moments. Widely used models are based on the assumption that in cases where only means and variances are known the approximated PDF should be normal. The representation of non-normal distributions, however, may be achieved either through series expansions or maximum entropy criteria.

2.2 Series Expansions

A computationally convenient way of approximating an unknown PDF is the Gram–Charlier (Type A) series expansion.[1] The unknown PDF $p(x)$ is expressed as

$$p(x) = \varphi(x)\left[1 + \sum_{k=3}^{n} \frac{C_k}{k!} H_k(x)\right] \tag{1}$$

in which $\varphi(x)$ defines the normal distribution and $H_k(x)$ are the kth Hermite polynomials. The coefficients C_k can be calculated from the statistical moments of the random variate x. The choice of Hermite polynomials is convenient in the sense that the moments up to nth order remain unchanged if additional terms are added in the series expansion. This means that normalization of the density is not affected by Hermite polynomials which, in turn, implies that the PDF may in some regions also attain negative values. This is easily shown by an example (Fig. 1). With respect to reliability considerations, i.e. probabilities of exceeding a given threshold value, the results may

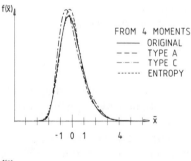

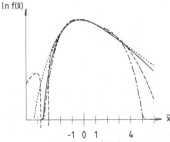

Fig. 1. Approximation of a lognormal PDF by utilizing Gram–Charlier Series Expansions A and C and maximum entropy method.

become inconclusive since the cumulative distribution functions may become larger than unity which in turn results in negative exceedance probabilities. This fact indicates that series expansions should be treated carefully for prediction of larger values, i.e. the tail region of the random variates under consideration (see e.g. Ref. 2)

A similar expansion using Hermite polynomials is known as (Gram–Charlier) series Type C. The unknown PDF is approximated by

$$p(x) = C_0 \exp\left[\sum_{k=1}^{n} C_k H_k(x)\right] \tag{2}$$

in which, again, the coefficients C_k are determined from the statistical moments. Obviously this PDF is always positive, but the approximation may behave rather badly in the tails (Fig. 1).

Another, more recent, approach utilizes the fact that for normal variates the cumulants of order >2 vanish. Thus it is reasonable to define non-normal variates in terms of higher order cumulants. This so-called cumulant-neglect closure method[3,4] assumes that cumulants higher than a certain order vanish. Mathematically, this means that the characteristic function, i.e. the Fourier transform $P(\omega)$ of the unknown density $p(x)$ is approximated by

$$P(\omega) = \exp\left[\sum_{k=1}^{n} \frac{K_k}{k!} (i\omega)^k\right] \tag{3}$$

in which the coefficients K_k are defined in terms of the cumulants. The density $p(x)$ may be obtained by inverse Fourier Transform (FT) of eqn (3). However, this is generally not feasible in closed form. Applying discrete Fourier Transform (FFT) yields in some regions negative values for $p(x)$. Hence cumulant-neglect closure, which in fact is comparable to the Gram–Charlier Type A series, also does not provide suitable means for estimating exceedance probabilities of high threshold values of a random variable.

2.3 Maximum Entropy Method (MEM)
A somewhat different approach to define an approximation of the PDF is to maximize the entropy of the random variate. This means that the random variable is left as 'uncertain' as possible.[5,6] The MEM leads to an exponential PDF of the form

$$p(x) = \exp\left(\sum_{k=0}^{n} C_k x^k\right) \tag{4}$$

in which the coefficients C_k are determined to satisfy the given statistical moments and where n is the number of known statistical moments. It is clearly seen that only positive values for $p(x)$ can result from eqn (4). However, if n is an odd number, $p(x)$ may become unbounded at either $x \to -\infty$ or $x \to +\infty$. This seems to be a rather severe restriction of the applicability of this method, but in many cases this difficulty may be circumvented if at least a rough estimate of the next higher moment of even order is known. From a computational point of view it should be noted that each additional term in eqn (4) changes all statistical moments so that computational procedures become rather involved. As mentioned above, truncation of the series in eqn (4) at $n = 2$ (i.e. only mean and variance are known) renders a normal density. In some cases this method may give far better results than the series expansions (see Fig. 1)

3 STATISTICAL DESCRIPTION OF NON-NORMAL RANDOM PROCESSES

3.1 General
It is well known that a random process is described completely by all possible joint probability densities at different times (which are of course uncountably many for a continuous time process). Even for cases where the assumptions of stationarity and ergodicity are appropriate, it is usually impossible to determine the required joint densities from experimental data. On the other hand, it is comparatively easy to estimate moment functions (mean, autocorrelation, third order correlation, etc.) up to some order with sufficient accuracy. In the stationary case these moment functions may be represented in the frequency domain yielding higher order spectra. This spectral approach is especially useful for linear transformations of the non-normal random process (comparable to the Power Spectral Method in linear stochastic dynamics).

3.2 Time Domain Representation
Representations for mean and autocorrelation are sufficiently known, hence they need not be reiterated here. Higher order correlations have similar representations, so, for example, the third order correlations R_{xxx} which is defined by

$$R_{xxx}(t_1, t_2, t_3) = E[x(t_1)x(t_2)x(t_3)] \tag{5}$$

in which $E[\cdot]$ denotes mathematical expectation and $x(\cdot)$ is a zero mean process.

Assuming stationarity, the number of time variables is reduced by one, i.e. only the time differences are important. In this case the third order correlation becomes

$$R_{xxx}(\tau_1, \tau_2) = E[x(t)x(t + \tau_1)x(t + \tau_2)] \qquad (6)$$

Analogously the fourth order correlation becomes

$$R_{xxxx}(\tau_1, \tau_2, \tau_3) = E[x(t)x(t + \tau_1)x(t + \tau_2)x(t + \tau_3)] \qquad (7)$$

If, additionally, ergodicity is assumed, these correlations may be expressed in terms of time averages rather than ensemble averages, i.e. by integrating powers of $x(t)$. For zero time lags, the correlation functions yield the one time (central) statistical moments of the random process x

$$\mu_{3,x} = R_{xxx}(0, 0) = E[x(t)^3] \qquad (8)$$

$$\mu_{4,x} = R_{xxxx}(0, 0, 0) = E[x(t)^4] \qquad (9)$$

These moments provide some measure of the deviation of $x(t)$ from a Gaussian process, although they do not contain complete information. It is of some importance to note that the third order correlation is not entirely symmetric in the (τ_1, τ_2)-domain. Although

$$R_{xxx}(\tau_1, \tau_2) = R_{xxx}(\tau_2, \tau_1) \qquad (10)$$

which means symmetry with respect to the axis $\tau_1 = \tau_2$, this in general does not hold for the coordinate axes nor for the axis $\tau_1 = -\tau_2$, i.e.

$$R_{xxx}(\tau_1, \tau_2) \neq R_{xxx}(-\tau_1, \tau_2) \qquad (11a)$$

$$R_{xxx}(\tau_1, \tau_2) \neq R_{xxx}(\tau_1, -\tau_2) \qquad (11b)$$

$$R_{xxx}(\tau_1, \tau_2) \neq R_{xxx}(-\tau_1, -\tau_2) \qquad (11c)$$

3.3 Frequency Domain Representation

In complete analogy to the well known Wiener–Khintchine relations between autocorrelation (second order correlation) and power spectrum, multiple Fourier transforms may be applied to higher order correlation functions. This yields higher order spectra which may be interpreted as distribution of higher order moments in a multidimensional frequency domain. Corresponding to eqns (6) and (7), the Bi-spectrum $S_{xxx}(\omega_1, \omega_2)$ and the Tri-spectrum $S_{xxxx}(\omega_1, \omega_2, \omega_3)$ may be defined by means of

$$S_{xxx}(\omega_1, \omega_2) = \frac{1}{4\pi^2} \int_{-\infty}^{\infty} \int_{-\infty}^{\infty} R_{xxx}(\tau_1, \tau_2) e^{-i\omega_1\tau_1} e^{-i\omega_2\tau_2} \, d\tau_1 \, d\tau_2 \quad (12)$$

$$S_{xxxx}(\omega_1, \omega_2, \omega_3) = \frac{1}{8\pi^3} \int_{-\infty}^{\infty} \int_{-\infty}^{\infty} \int_{-\infty}^{\infty} R_{xxxx}(\tau_1, \tau_2, \tau_3)$$
$$e^{-i\omega_1\tau_1} e^{-i\omega_2\tau_2} e^{-i\omega_3\tau_3} \, d\tau_1 \, d\tau_2 \, d\tau_3 \quad (13)$$

In view of the properties (11a)–(11c) it becomes clear that in general the higher order spectra are complex rather than real quantities. The inverse relation to eqn (12) yields

$$R_{xxx}(\tau_1, \tau_2) = \int_{-\infty}^{\infty} \int_{-\infty}^{\infty} S_{xxx}(\omega_1, \omega_2) e^{i\omega_2\tau_2} e^{i\omega_1\tau_1} \, d\omega_1 \, d\omega_2 \quad (14)$$

and for $\tau_1 = \tau_2 = 0$ the third moment follows from

$$\mu_{x,3} = R_{xxx}(0, 0) = \int_{-\infty}^{\infty} \int_{-\infty}^{\infty} S_{xxx}(\omega_1, \omega_2) \, d\omega_1 \, d\omega_2 \quad (15)$$

Since $\mu_{x,3}$ is a real quantity, from eqn (15) it follows that the integral on the imaginary part of $S_{xxx}(\omega_1, \omega_2)$ vanishes.

For the interpretation of these higher order spectra it might be useful to note that the Bi-spectrum vanishes identically for a Gaussian process $x(t)$, since the third order correlation function $R_{xxx}(\tau_1, \tau_2)$ vanishes identically.

3.4 Examples

The statistical analysis of wind pressure data[7] shows that the density function is skewed, i.e. the third central moment is not zero. The third order correlation function is plotted in Fig. 2 with the corresponding Bi-spectrum.

It is seen that the Bi-spectrum exhibits a sharp peak at low frequencies. An analytical expression for the third order correlation and the Bi-spectrum of wind pressure p may be derived assuming that the wind velocity v is a stationary Gaussian process with mean wind speed $\bar{v}$ and variance σ_v^2. Since p may be expressed as

$$p = cv^2 \quad (16)$$

in which c is a constant, the higher order correlation functions of v may be expressed in terms of the autocorrelation function R_{vv} of v.

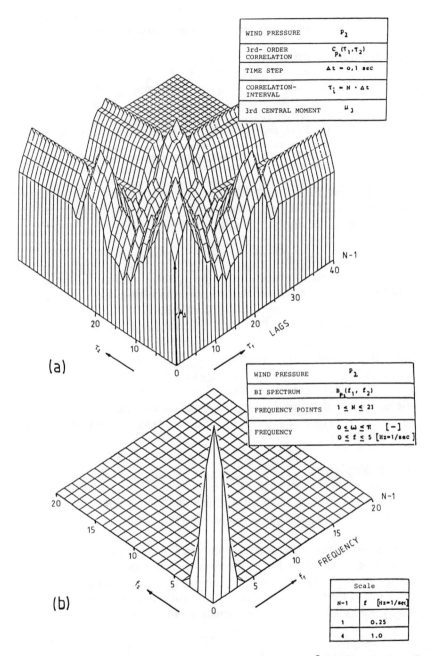

Fig. 2. (a) Third order correlation of wind pressure data.[7] (b) Bi-spectrum of wind pressure data.[7]

This results in

$$
\begin{aligned}
R_{ppp}(\tau_1, \tau_2) = 8c^6 \bar{v}^2 [& R_{vv}(\tau_1)R_{vv}(\tau_2) \\
& + R_{vv}(\tau_1)R_{vv}(\tau_2 - \tau_1) + R_{vv}(\tau_2 - \tau_1)R_{vv}(\tau_2 - \tau_1)] \\
& + 8c^6 R_{vv}(\tau_1)R_{vv}(\tau_2)R_{vv}(\tau_2 - \tau_1)
\end{aligned} \tag{17}
$$

In the above equation the last term on the right hand side is of minor importance since, in general $\sigma_v^2 \ll \bar{v}^2$.

Neglecting this term, the Bi-spectrum S_{ppp} of the wind pressure becomes

$$
\begin{aligned}
S_{ppp}(\omega_1, \omega_2) = 8c^6 \bar{v}^2 [& S_{vv}(\omega_1)S_{vv}(\omega_2) \\
& + S_{vv}(\omega_1)S_{vv}(\omega_1 + \omega_2) + S_{vv}(\omega_2)S_{vv}(\omega_1 + \omega_2)]
\end{aligned} \tag{18}
$$

Since experimentally measured wind spectra exhibit peaks at low frequencies, it is obvious that the same property holds for Bi-spectra of wind pressure data if eqn (18) is considered.

An analytical example for a Tri-spectrum may be found by investigating the fourth order correlation of a Gaussian process. Since for jointly Gaussian variables x_1, x_2, x_3, x_4

$$
\begin{aligned}
E[x_1 x_2 x_3 x_4] = E[x_1 x_2]E[x_3 x_4] \\
+ E[x_1 x_3]E[x_2 x_4] + E[x_1 x_4]E[x_2 x_3]
\end{aligned} \tag{19}
$$

it is easily shown that for a Gaussian process with autocorrelation function $R(\tau)$

$$
\begin{aligned}
R_{xxx}(\tau_1, \tau_2, \tau_3) = R(\tau_1)R(\tau_3 - \tau_2) + R(\tau_2)R(\tau_3 - \tau_1) \\
+ R(\tau_3)R(\tau_2 - \tau_1)
\end{aligned} \tag{20}
$$

Applying Fourier transformation to eqn (17) it follows that

$$
\begin{aligned}
S_{xxxx}(\omega_1, \omega_2, \omega_3) = S(\omega_1)S(\omega_2)\,\delta(\omega_2 + \omega_3) \\
+ S(\omega_2)S(\omega_3)\,\delta(\omega_3 + \omega_1) \\
+ S(\omega_3)S(\omega_1)\,\delta(\omega_1 + \omega_2)
\end{aligned} \tag{21}
$$

in which $S(\cdot)$ denotes the power spectral density of $x(t)$. Hence, in this case, the Tri-spectrum is a real quantity with Dirac-Delta singularities along the planes $\omega_1 = -\omega_2$, $\omega_2 = -\omega_3$, $\omega_3 = -\omega_1$ which coincide at the origin. From this result it may be expected that slightly non-Gaussian processes have similar Tri-spectra.

4 RESPONSE OF LINEAR SYSTEMS TO NON-NORMAL LOADING

4.1 General Representation

Since for linear systems the principle of superposition holds, statistical properties of the response may be expressed in terms of the respective properties of the input and the impulse response function $h(t)$ of the system. Let y be the response of a linear system with $h(t)$ to the input x. Then the correlation functions of the response may be written in integral form

$$E[y(t_1)y(t_2)y(t_3)] = \int_{-\infty}^{t_3} \int_{-\infty}^{t_2} \int_{-\infty}^{t_1} h(t_1 - \tau_1)h(t_2 - \tau_2)h(t_3 - \tau_3)$$
$$\times E[x(\tau_1)x(\tau_2)x(\tau_3)] \, d\tau_1 \, d\tau_2 \, d\tau_3 \quad (22)$$

An analogous relation may be established for the fourth order correlation. If eqn (22) is transformed into the frequency domain the higher order spectra of the response become[8]

$$S_{yyy}(\omega_1, \omega_2) = H(\omega_1)H(\omega_2)\bar{H}(\omega_1 + \omega_2)S_{xxx}(\omega_1, \omega_2) \quad (23)$$
$$S_{yyyy}(\omega_1, \omega_2, \omega_3) = H(\omega_1)H(\omega_2)H(\omega_3)$$
$$\times \bar{H}(\omega_1 + \omega_2 + \omega_3)S_{xxxx}(\omega_1, \omega_2, \omega_3) \quad (24)$$

in which $H(\cdot)$ denotes the complex transfer function of the system and $\bar{H}(\cdot)$ its complex conjugate. In view of eqns (8) and (9) it may be seen that the statistical moments of the response process follow from integrating eqns (23) and (24) over the entire frequency domain.

4.2 'White Noise' Approximations

Due to the numerical efforts required to carry out integrations over multidimensional domains in order to obtain higher order moments from a spectral approach it is desirable to obtain a closed form approximation. In analogy to the white noise approximation of the power spectral density for input processes with short correlation time, it may be assumed that higher order spectra are constant within the entire frequency domain.

As the wind pressure example (Sect. 3.4) shows, the Bi-spectrum can be expressed in terms of the power spectral density. Using equivalent white noise (in terms of the variance of the response), eqn (18) results in a Bi-spectrum which is actually constant in the entire (ω_1, ω_2)-domain.

The results for the higher statistical moments of the response based on

$$S_{xxx}(\omega_1, \omega_2) = S_{02} = \text{const.} \tag{25}$$

$$S_{xxxx}(\omega_1, \omega_2, \omega_3) = S_{03} = \text{const.} \tag{26}$$

are given below. If the equation of motion reads

$$\ddot{y} + 2D\omega_0\dot{y} + \omega_0^2 y = x \tag{27}$$

(ω_0 is natural frequency, D is damping ratio) the higher statistical moments of the response are

$$\mu_{y,3} = \frac{\pi^2 S_{02}}{\omega_0^4} \frac{2}{1 + 8D^2} \tag{28}$$

$$\mu_{y,4} = \frac{\pi^3 S_{03}}{\omega_0^5} \frac{3}{8D} \frac{1}{1 + 3D^2} \tag{29}$$

It is to be noted that eqns (28) and (29) were not obtained through integrating eqns (23) and (24) but by using moment equations from a state vector approach analogous to the Lyapunov equation for the second moments.[9] By means of this method only a set of linear equations has to be solved by hand. An integration approach (e.g. method of residues) is untractable for higher statistical moments.

It should be mentioned that the assumptions underlying eqns (25) and (26) must be checked for each specific problem, since they need not necessarily be consistent with the physical reality as the Gaussian example shows.

4.3 Example
The following analytical example shows the procedure to find the fourth central moment of the response from the frequency domain approach. Using eqn (24) and the Tri-spectrum S_{xxxx} defined by eqn (21), the Tri-spectrum S_{yyyy} of the response of a linear system is obtained. Integration yields

$$\mu_{y,4} = \int\!\!\!\int\!\!\!\int_{-\infty}^{\infty} [S(\omega_1)S(\omega_2)\,\delta(\omega_2 + \omega_3) + S(\omega_2)S(\omega_3)\,\delta(\omega_3 + \omega_1)$$

$$+ S(\omega_3)S(\omega_1)\,\delta(\omega_1 + \omega_2)]H(\omega_1)H(\omega_2)H(\omega_3)$$

$$\times \bar{H}(\omega_1 + \omega_2 + \omega_3)\,d\omega_1\,d\omega_2\,d\omega_3 = I_1 + I_2 + I_3 \tag{30}$$

The transformation

$$\omega_1 + \omega_2 = u \qquad \omega_2 + \omega_3 = v \qquad \omega_3 + \omega_1 = w \qquad (31)$$

reduces the first summand of eqn (30) to

$$I_1 = \frac{1}{2} \int\limits_{-\infty}^{\infty}\!\!\int S\left(\frac{u}{2}+\frac{w}{2}\right) S\left(\frac{u}{2}-\frac{w}{2}\right) H\left(\frac{u}{2}+\frac{w}{2}\right)$$

$$\times H\left(\frac{u}{2}-\frac{w}{2}\right) H\left(-\frac{u}{2}+\frac{w}{2}\right) \bar{H}\left(\frac{u}{2}+\frac{w}{2}\right) du\ dw \qquad (32)$$

Another transformation

$$\frac{u}{2} - \frac{w}{2} = z, \qquad \frac{u}{2} + \frac{w}{2} = t \qquad (33)$$

yields

$$I_1 = \int\limits_{-\infty}^{\infty}\!\!\int S(z)H(z)\bar{H}(z)S(t)H(t)H(-t)\ dz\ dt = (\sigma_y^2)^2 \qquad (34)$$

Since the remaining two summands in eqn (30) are obtained by interchanging the indices, the final result is

$$\mu_{y,4} = 3[\sigma_y^2]^2 \qquad (35)$$

which is an inherent property of a Gaussian process. This simple example shows that Tri-spectra with Delta singularities have a physical meaning.

5 RESPONSE OF NON-LINEAR SYSTEMS TO NORMAL LOADING

5.1 General

The only powerful tool available for obtaining closed form results of the response of non-linear systems is the Markov vector approach.[9] This requires the excitation to be Gaussian white noise (although non-white processes can be modeled by passing white noise through a filter, e.g. a Kanai–Tajimi filter). In that case the joint density of the response vector components including filter state variables are governed by the Fokker–Planck equation. However, it is impossible to find closed form solutions for this equation in general cases, especially for

MDOF systems. Even in the SDOF case only few stationary solutions are known (e.g. Refs 9–12). This means that for more general cases approximate methods, such as equivalent linearization, e.g. Ref. 13 or Fokker–Planck based moment equations utilizing approximate densities (see Section 2.2) have to be used considering the equation of motion (27). In many cases the sources of non-linearities are the stiffness or damping terms respectively. Hence these two types of non-linearities will be discussed in some detail.

5.2 Non-linear Restoring Force
If the equation of motion may be written in the form

$$\ddot{y} + 2D\omega_0\dot{y} + \omega_0^2 f(y) = x \tag{36}$$

in which x is Gaussian white noise and $f(y)$ stabilizes the system, i.e.

$$\lim_{y \to \pm\infty} f(y) = \geqslant 0 \tag{37}$$

then the stationary joint density of the response vector components y, $\dot{y}$ is obtained by integration, i.e.

$$p(y, \dot{y}) = C \exp\left(-\frac{\dot{y}^2}{2\sigma_{\dot{y}}^2}\right) \exp\left(-B \int f(y)\, dy\right) \tag{38}$$

in which $\sigma_{\dot{y}}^2$ and B follow from the system parameters and the intensity of the excitation. C is a normalizing constant. For the special case of the Duffing oscillator

$$f(y) = y + \varepsilon y^3 \tag{39}$$

the resulting PDF of the displacement, $p_Y(y)$, is plotted in Fig. 3 along

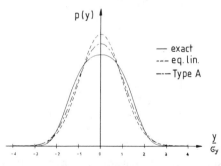

Fig. 3. PDF of displacement of Duffing oscillator using various approaches.

with the results obtained by equivalent linearization and Gram–Charlier series expansions. Note that for this specific example the maximum entropy method using the first four statistical moments yields the exact results.

5.3 Non-linear Damping
If the non-linearity in the damping term may be expressed in the form

$$\ddot{y} + h(E)\dot{y} + g(y) = x \qquad (40)$$

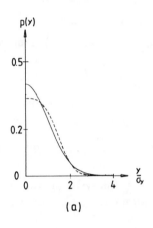

(a)

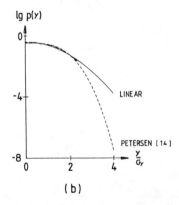

(b)

Fig. 4. PDF of displacement of oscillator with non-linear damping (a) linear scale; (b) log scale.

in which E is the total energy

$$E = \tfrac{1}{2}\dot{y}^2 + \int g(y)\,\mathrm{d}y \tag{41}$$

the joint density of the response vector components is given by

$$p(y,\dot{y}) = C\exp\!\left(-B\int_0^E h(\bar{E})\,\mathrm{d}(\bar{E})\right) \tag{42}$$

in which B follows from the intensity of the excitation and C is a normalizing constant. Generalizations of this case may be found in Refs 11 and 12.

The special case

$$g(y) = \omega_0^2 y \tag{43a}$$

$$h(E) = C \cdot E \tag{43b}$$

represents a damping model for structural steel as suggested by Petersen.[14] The resulting PDF of the displacement is plotted in Fig. 4 along with a normal distribution having the same variance. The tail behavior is clearly seen to be quite different.

6 SOME SPECIAL ASPECTS

6.1 Excursion Probability

For reliability estimates it is most essential to obtain the probability by which the response process exceeds a certain (critical) threshold level within a given time interval. Under the restriction of white noise excitation this excursion probability is governed by the Kolmogorov backward equation.[9] However, closed form solutions for this problem do not exist, although there are some numerical (FEM) results (see Ref. 15) available. Generally the problem is reduced to excursions of a suitably defined envelope which, for small damping, approximately forms a Markov process.[16] If furthermore the envelope is discretized in time (at the expected points of extreme values of the response itself) and the statistical dependence of subsequent extremes is accounted for, a close approximation of the exact exceedance probability is obtained.[17] This approach uses the concept of a 'two-step-memory' process for modeling the statistical dependence between successive non-excursion events[18] which is somewhat superior to Markov or

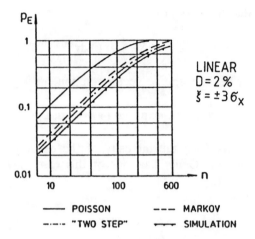

Fig. 5. Exceedance probability for a linear system using various approaches.

Poisson assumptions.[19] Figure 5 compares the approximate results for the exceedance probability of a linear system obtained by different approaches to simulation results. Similar results are also obtained for non-linearly damped systems.[17,18] In Fig. 6 the excursion probabilities of a non-linearly damped system (eqns (43)) and the equivalent linear system (identical variance of response) are compared. The difference in results is shown to be significant.

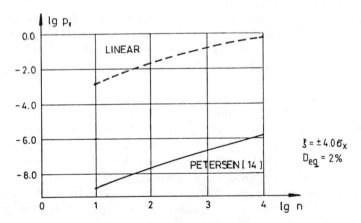

Fig. 6. Excursion probability for linearly and non-linearly damped system with identical variances of the response.

6.2 Non-stationary Processes

It is a well known fact that for linear systems subjected to Gaussian input the type of distribution is always Gaussian, i.e. it does not change in the non-stationary state. This, however, is not true for non-stationary states of non-linear systems. An approximate closed

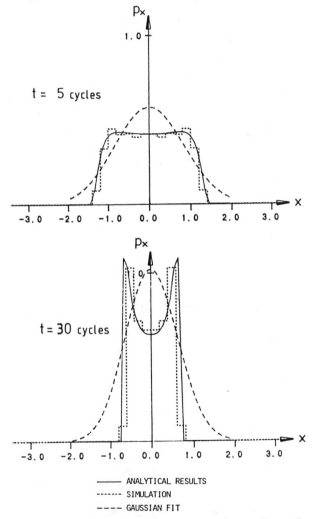

Fig. 7. PDF of displacement of non-linear oscillator (free vibration with stationary initial conditions).

form solution for the system described by eqns (40) and (43) obtained by stochastic averaging shows that in the state of free vibration with stationary initial conditions the shape of the PDF of the response changes considerably, i.e. the PDF becomes bimodal and is restricted in its range. Both phenomena are not possible for linear systems. Figure 7 shows the PDF of the displacement during the decay phase.[20] Figure 7 compares the analytical with the simulation results (2000

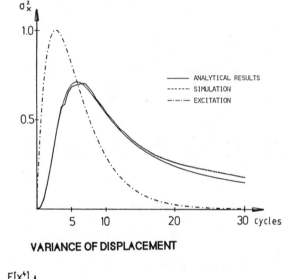

VARIANCE OF DISPLACEMENT

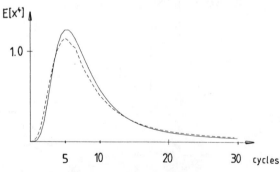

FOURTH MOMENT OF DISPLACEMENT

Fig. 8. Non-stationary second and fourth moment of displacement of non-linearly damped system.

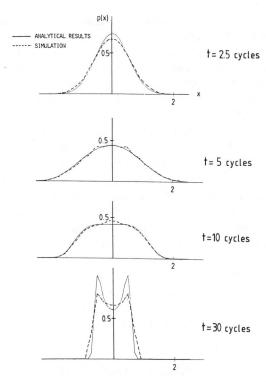

Fig. 9. Non-stationary PDF of response of non-linearly damped system.

samples) and a Gaussian fit with the same variance, i.e. an equivalent linear result.

For general non-stationary cases, however, a closed form solution is not available. It is shown in Ref. 21 that moment closure using the stationary solution (which is either known or more easily obtained) yields good approximations for the statistical moments of the response even for highly non-stationary cases. Using those moments for a maximum entropy fit may give satisfactory results for non-stationary PDF's of non-linearly damped systems (see Figs 8 and 9).

7 CONCLUSIONS

It has been shown that non-normal properties of the response processes of linear as well as non-linear systems significantly influence the respective reliability estimates. Higher order correlations and spectra are utilized to represent these processes in time and frequency ranges respectively when analysing linear structures. To simplify the computational effort in carrying out integrations over multidimensional domains, the 'white noise' approximation proves to be very useful. Yet, its consistency with physical reality has to be verified for each specific problem.

Gaussian white noise is assumed to analyse nonlinear systems for obtaining closed form solutions, i.e. the Markov vector approach is applied. To exemplify the effect of the non-linear restoring force, a Duffing oscillator is utilized. The non-normal properties of the PDF of the response are shown by applying various approaches such as Gram–Charlier series expansion and equivalent linearization. The non-linear damping effect was also analysed by the Fokker–Planck equation. Again, the tail behavior of PDF of the response is shown to be quite different from Gaussian.

Finally, the change of characteristics of the probability density functions evolving in time for non-linear systems under non-stationary excitation is discussed. It is interesting to note, that in such cases the PDF may become bimodal and restricted in its range.

ACKNOWLEDGEMENT

This research is supported by the Austrian Research Council (FWF) under contract No. S30-03 which is gratefully acknowledged by the authors.

REFERENCES

1. Charlier, C. V. L., A new form of the frequency function. *Meddelande fran Astronomiska Observatorium*, Ser. II 51 (1928), 1–28.
2. Shinozuka, M. & Nishimura, A., On general representation of a density function. Technical Report No. 14, Institute for the Study of Fatigue and Reliability, Columbia University, 1965.
3. Zeman, J. L., Zur Lösung nichtlinearer stochastischer Probleme der Mechanik. *Acta Mechanica*, **14** (1972) 157–69.
4. Wu, W. F. & Lin, Y. K., Cumulant neglect closure for non-linear oscillators under random parametric and external excitation. *International Journal of Non-Linear Mechanics*, **19** (1984) 349–62.
5. Papoulis, A., *Probability, Random Variables, and Stochastic Processes*, 2nd Edition. McGraw-Hill, Singapore, 1984.

6. Dowson, D. C. & Wragg, A., Maximum-entropy distributions having prescribed first and second moments. *IEEE Trans. on Information Theory* (Sept., 1973) 689–93.
7. Melzer, H.-J., Tragwerksschwingungen unter Zufallslast mit nichtgaußischer Wahrscheinlichkeitsverteilung, Dissertation, Technical University Munich, 1981.
8. Melzer, H.-J. & Schuëller, G. I., On the reliability of flexible structures under non-normal loading processes. *Proc. IUTAM-Symp. on Random Vibrations and Reliability,* ed. K. Hennig. Akademie Verlag, Berlin, 1983, pp. 73–83.
9. Lin, Y. K., *Probabilistic Theory of Structural Dynamics.* McGraw-Hill, New York, 1967; Reprint Robert E. Krieger, Huntington, 1976.
10. Roberts, J. B., Stationary response of oscillators with non-linear damping to random excitation. *J. Sound Vibration,* **50** (1977) 145–56.
11. Caughey, T. K. & Fai, M. A., The exact steady state solution of a class of non-linear stochastic systems. *Int. J. Non-Linear Mech.,* **17**(3), (1982) 137–42.
12. Dimentberg, M. F., An exact solution to a certain non-linear random vibration problem. *Int. J. Non-Linear Mech.,* **17**(4), (1982) 231–6.
13. Baber, T. T. & Wen, Y. K., Stochastic equivalent linearization for hysteretic degrading, multistorey structures. Structural Research Series No. 471, University of Illinois at Urbana-Champaign, Urbana, 1979.
14. Petersen, Ch., Aerodynamische und seismische Einflüsse auf die Schwingungen insbesondere schlanker Bauwerke. *Fortschrittber. d. VDI Zeitschr.,* Reihe 11, *Schwingungstechnik-Lärmbekämpfung,* Nr. 11, (1971), 326 pp.
15. Bergman, L. A. & Spencer, B. F., Jr., Solution of the first passage problem for simple linear and nonlinear oscillators by the finite element method. T. & A. M. Report No. 461, University of Illinois at Urbana-Champaign, Urbana, 1983.
16. Stratonovich, R. L., *Topics in the Theory of Random Noise.* Gordon and Breach, New York, Vol. 1, 1963, Vol. 2, 1967.
17. Schuëller, G. I. & Bucher, C. G., Nonlinear damping and its effect on the reliability estimates of structures. *Random Vibration—Status and Recent Developments,* ed. I. Elishakoff and R. H. Lyon. Elsevier, Amsterdam, 1986, pp. 277–86.
18. Bucher, C. G., Zuverlässigkeit von mechanischen Systemen mit nichtlinearen Dämpfungseigenschaften. Dissertation, Universität Innsbruck, 1986.
19. Yang, J. N. & Shinozuka, M., On the first excursion probability in stationary narrow band random process. *J. Appl. Mech.,* **38**(4), (1971) 1017–22.
20. Bucher, C. G., Response of nonlinearly damped systems to nonstationary excitation. *Proc. 2nd Int. Workshop on Stochastic Methods in Struct. Mechanics,* Pavia, Italy, eds F. Casciati & L. Faravelli, S. E. A. G., Pavia, Italy, 1986, pp. 305–15.
21. Bucher, C. G., Instationäre Reaktion nichtlinear gedämpfter mechanischer Systeme auf stochastische Erregung. *ZAMM,* **67** (1987), T65-T67.

14

Stochastic Finite Element Analysis: an Introduction

Masanobu Shinozuka*

Department of Civil Engineering and Engineering Mechanics, Columbia University, New York, USA

&

Fumio Yamazaki

Ohsaki Research Institute, Shimizu Corporation, Tokyo, Japan

ABSTRACT

With the aid of the finite element method, the present paper deals with the problem of structural response variability resulting from the spatial variability of material properties of structures, when they are subjected to static loads of a deterministic nature. Several forms of spatial variability of Young's modulus and Poisson's ratio are considered; they are assumed to be two-dimensional Gaussian or non-Gaussian stochastic fields. The finite element discretization is performed in such a way that the size of each element is sufficiently small. A Neumann expansion method is developed and used in deriving the finite element solution of such stochastic systems within the framework of the Monte Carlo method. Then the results from the Neumann expansion method are compared with those from the first- and second-order perturbation approximation methods and direct Monte Carlo simulation method with respect to accuracy, convergence and computational efficiency.

1 INTRODUCTION

It is only in the last several years that the finite element methods have begun to be used for solving the problems of system stochasticity. In most of these solutions, the first-order perturbation approximation has been utilized. Cambou[1] appears to be the first to apply a first-order

---

* Present address: Department of Civil Engineering and Operations Research, Princeton University, Princeton, New Jersey 08544, USA.

perturbation approximation in seeking finite element solutions of linear static problems involving loading and system stochasticity. Similar techniques were also utilized by Baecher & Ingra[2] for geotechnical problems, and by Handa & Andersson[3] for static problems of beam and frame structures. Hisada & Nakagiri further introduced a second-order perturbation approximation, first for static problems (Hisada & Nakagiri[4]), and then for eigenvalue and other problems of system stochasticity (Nakagiri & Hisada[5]). With respect to the eigenvalue problems of stochastic systems, variabilities of eigenvalues and eigenvectors were studied earlier by Fox & Kapoor,[6] Collins & Thomson,[7] Hoshiya & Shah,[8] Hasselman & Hart[9] and Shinozuka & Astill.[10] In these papers, the first-order perturbation approximation was also employed.

However, the perturbation method has rarely been successfully applied to dynamic response of stochastic systems. Liu *et al.*[11,12] recently showed some examples along this line. Also, Hisada & Nakagiri[13] provided time history responses of stochastic systems, in which only the damping coefficient is treated to be uncertain, however. This lack of research on dynamic response variability of stochastic systems reflects the fact that the perturbation method cannot very well deal with the response variability when the natural frequency of the system changes, and indeed this is pointed out in a state-of-the-art paper by Vanmarcke *et al.*[14] Considering all these, Monte Carlo simulation methods can provide a useful alternative for the problems to which the perturbation method does not apply very well. First-order perturbation solutions appear to be reasonably accurate when the material property variation is of small degree. However, the perturbation method usually suffers from questions on their accuracy, convergence and computational efficiency. These questions become more crucial as higher-order solutions are sought, as the degree of the material property variability becomes more pronounced and when dynamic and/or nonlinear problems must be considered.

The finite element discretization is also an issue when we utilize the finite element analysis involving stochastic fields. Whether or not the size of each element (of appropriate dimensionality) is sufficiently small must be confirmed from the spatial variability of material properties as well as stress and strain gradients point of view. In this regard, Shinozuka & Deodatis[15] discussed the proper element size of statically determinate bars when Young's modulus is spatially stochastic using a concept of the 'scale of correlation' recently studied by Harada & Shinozuka.[16] Similar concepts are also found in the work by

Tatarski,[17] Monin & Yaglom,[18] Lumley[19] and Vanmarcke[20] to represent the scale of turbulence or the scale of fluctuation.

In the present paper, the primary focus is placed on the stochastic finite element method involving static linear elastic problems of structural mechanics. In doing so, perturbation techniques, and Neumann expansion and direct Monte Carlo simulation methods are used as major tools for analytical and numerical solutions. While the problems of geometrical variability can also be treated in a similar fashion, these are not specifically mentioned in the present paper.

2 FINITE ELEMENT FORMULATION

The standard finite element analysis method (see for example, Zienkiewicz;[21] Bathe[22]) for linear static problems is briefly described here in order to facilitate the introduction of that which follows. A rectangular element with four nodes as shown in Fig. 1 is used in this study primarily for its simplicity, although the methodology to be presented is not restricted to this particular element.

Each node has two in-plane displacement components and the element displacement vector can be constructed for each element out of these nodal displacements at the four corner nodes in the following form:

$$\mathbf{u}_e = [u_{x1}u_{y1}u_{x2}u_{y2}u_{x3}u_{y3}u_{x4}u_{y4}]^T \tag{1}$$

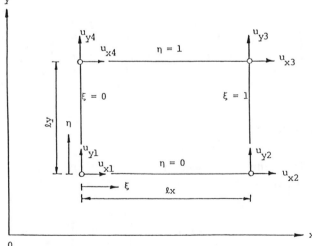

Fig. 1. Rectangular element with four nodes.

The nondimensional coordinates whose origin is taken at the lower left corner node (x_1, y_1) are introduced as

$$\xi = \frac{x - x_1}{l_x} \quad \text{and} \quad \eta = \frac{y - y_1}{l_y} \tag{2}$$

in which l_x and l_y represent the lengths of the element in the x- and y-directions, respectively. The shape function which satisfies the assumption of linearly varying boundary displacements may be taken as

$$\mathbf{N} = \begin{bmatrix} N_1 & 0 & N_2 & 0 & N_3 & 0 & N_4 & 0 \\ 0 & N_1 & 0 & N_2 & 0 & N_3 & 0 & N_4 \end{bmatrix} \tag{3}$$

with

$$N_1 = (1 - \zeta)(1 - \eta); \quad N_2 = \xi\eta; \quad N_3 = (1 - \xi)\eta; \quad N_4 = \xi(1 - \eta) \tag{4}$$

and the displacement $\mathbf{u}^e = [u_x u_y]^T$ within element e is represented by

$$\mathbf{u}^e = \mathbf{N}\mathbf{u}_e \tag{5}$$

By differentiating eqn (5), the strain $\boldsymbol{\varepsilon}_e = [\varepsilon_{xx}\varepsilon_{yy}\gamma_{xy}]^T$ within element e can be obtained as

$$\boldsymbol{\varepsilon}_e = \mathbf{B}_e\mathbf{u}_e \tag{6}$$

where

$$\mathbf{B}_e = \begin{bmatrix} \dfrac{-1+\eta}{l_x} & 0 & \dfrac{1-\eta}{l_x} & 0 & \dfrac{\eta}{l_x} & 0 & \dfrac{-\eta}{l_x} & 0 \\[2mm] 0 & \dfrac{-1+\xi}{l_y} & 0 & \dfrac{-\xi}{l_y} & 0 & \dfrac{\xi}{l_y} & 0 & \dfrac{1-\xi}{l_y} \\[2mm] \dfrac{-1+\xi}{l_y} & \dfrac{-1+\eta}{l_x} & \dfrac{-\xi}{l_y} & \dfrac{1-\eta}{l_x} & \dfrac{\xi}{l_y} & \dfrac{\eta}{l_x} & \dfrac{1-\xi}{l_y} & \dfrac{-\eta}{l_x} \end{bmatrix} \tag{7}$$

The stress $\boldsymbol{\sigma}_e = [\sigma_{xx}\sigma_{yy}\tau_{xy}]^T$ within element e is given by the stress–strain relationship:

$$\boldsymbol{\sigma}_e = \mathbf{D}_e\boldsymbol{\varepsilon}_e \tag{8}$$

in which $\mathbf{D}_e$ is the elasticity matrix and in the case of plane stress problems, $\mathbf{D}_e$ is known as:

$$\mathbf{D}_e = \frac{E_e}{1 - v_e^2} \begin{bmatrix} 1 & v_e & 0 \\ & 1 & 0 \\ & & \dfrac{1 - v_e}{2} \\ \text{sym.} & & \end{bmatrix} \tag{9}$$

in which E_e is Young's modulus and ν_e is Poisson's ratio of element e. The element stiffness matrix $\mathbf{k}_e$ can be obtained by applying the virtual work principle over the volume of the element as

$$\mathbf{k}_e = \int_{\text{vol.}} \mathbf{B}_e^T \mathbf{D}_e \mathbf{B}_e \; dV = t l_x l_y \int_0^1 \int_0^1 \mathbf{B}_e^T \mathbf{D}_e \mathbf{B}_e \; d\xi \, d\eta \qquad (10)$$

in which t is the thickness of the element and taken as unity in the present study.

The two-fold integration in eqn (10) can be performed in closed form with the aid of the symbolic manipulation program SMP (Inference Corporation[23]). For ease of writing, $\mathbf{D}_e$ is expressed in the following general form for two-dimensional elastic problems:

$$\mathbf{D}_e = \begin{bmatrix} a & b & 0 \\ & a & 0 \\ \text{sym.} & & c \end{bmatrix} \qquad (11)$$

in which a, b and c are constants containing Young's modulus and Poisson's ratio. For plane stress problems, an explicit expression for a, b and c can be readily obtained by comparing eqn (9) with eqn (11). Substituting eqns (7) and (11) into eqn (10) and performing the integration, the element stiffness matrix is obtained as

$$\mathbf{k}_e = t \begin{bmatrix}
\frac{av}{3}+\frac{cu}{3} & \frac{b}{4}+\frac{c}{4} & -\frac{av}{3}+\frac{cu}{6} & \frac{b}{4}-\frac{c}{4} & -\frac{av}{6}-\frac{cu}{6} & \frac{b}{4}-\frac{c}{4} & \frac{av}{6}-\frac{cu}{4} & -\frac{b}{4}+\frac{c}{4} \\[2mm]
& \frac{au}{3}+\frac{cv}{3} & -\frac{b}{4}+\frac{c}{4} & \frac{au}{6}-\frac{cv}{3} & -\frac{b}{4}-\frac{c}{4} & -\frac{au}{6}-\frac{cv}{6} & \frac{b}{4}-\frac{c}{4} & -\frac{au}{3}+\frac{cv}{6} \\[2mm]
& & \frac{av}{3}+\frac{cu}{3} & -\frac{b}{4}-\frac{c}{4} & \frac{av}{6}-\frac{cu}{3} & \frac{b}{4}-\frac{c}{4} & -\frac{av}{6}-\frac{cu}{6} & \frac{b}{4}+\frac{c}{4} \\[2mm]
& & & \frac{au}{3}+\frac{cv}{3} & -\frac{b}{4}+\frac{c}{4} & -\frac{au}{3}+\frac{cv}{6} & \frac{b}{4}+\frac{c}{4} & -\frac{au}{6}-\frac{cv}{6} \\[2mm]
& & & & \frac{av}{3}+\frac{cu}{3} & \frac{b}{4}+\frac{c}{4} & -\frac{av}{3}+\frac{cu}{6} & \frac{b}{4}-\frac{c}{4} \\[2mm]
& & & & & \frac{au}{3}+\frac{cv}{3} & -\frac{b}{4}+\frac{c}{4} & \frac{av}{6}-\frac{cu}{3} \\[2mm]
& & & & & & \frac{av}{3}+\frac{cu}{3} & -\frac{b}{4}-\frac{c}{4} \\[2mm]
\text{sym.} & & & & & & & \frac{au}{3}+\frac{cv}{3}
\end{bmatrix}$$

$$(12)$$

in which $u = l_x/l_y$ and $v = l_y/l_x$.

The global stiffness matrix $\mathbf{K}$ is obtained by assembling the element stiffness matrices over the entire region as

$$\mathbf{K} = \sum_{e=1}^{n} \mathbf{k}_e \tag{13}$$

in which n is the total number of finite elements and the summation implies the addition of the appropriate element stiffness matrices at appropriate locations within the global stiffness matrix.

The equilibrium equation for static problems is then written as

$$\mathbf{K}\mathbf{U} = \mathbf{F} \tag{14}$$

in which $\mathbf{U}$ is the displacement vector and $\mathbf{F}$ is the force vector. In order to apply the boundary condition of the system, eqn (14) should be rearranged and divided into two parts with respect to the displacement as

$$\begin{bmatrix} \mathbf{K}_{aa} & \mathbf{K}_{ab} \\ \mathbf{K}_{ba} & \mathbf{K}_{bb} \end{bmatrix} \begin{Bmatrix} \mathbf{U}_a \\ \mathbf{U}_b \end{Bmatrix} = \begin{Bmatrix} \mathbf{F}_a \\ \mathbf{F}_b \end{Bmatrix} \tag{15}$$

where $\mathbf{U}_a$ represents the unknown displacement, $\mathbf{U}_b$ the known displacement, $\mathbf{F}_a$ the known external force and $\mathbf{F}_b$ the unknown reaction force. The stiffness matrix is also divided correspondingly. Equation (15) may be divided into two equations as

$$\mathbf{K}_{aa}\mathbf{U}_a = \mathbf{F}_a - \mathbf{K}_{ab}\mathbf{U}_b \ . \tag{16}$$

$$\mathbf{F}_b = \mathbf{K}_{ba}\mathbf{U}_a + \mathbf{K}_{bb}\mathbf{U}_b \tag{17}$$

Under fixed displacement boundary conditions, $\mathbf{U}_b$ is a zero vector. Hence, eqn (16) is reduced to the following simpler form suppressing the subscripts.

$$\mathbf{K}\mathbf{U} = \mathbf{F} \tag{18}$$

The unknown displacement $\mathbf{U}$ is obtained by solving this reduced equilibrium equation.

3 NEUMANN EXPANSION METHOD

The equilibrium equation (18) formulated by the finite element method can obviously be solved by taking the inverse of the stiffness matrix as

$$\mathbf{U} = \mathbf{K}^{-1}\mathbf{F} \tag{19}$$

However, it is well-known that matrix inversions require a large amount of CPU time. Also, $\mathbf{K}^{-1}$ is no longer banded, although $\mathbf{K}$ is usually narrowly banded. Hence, the multiplication on the right-hand side of eqn (19) cannot be performed efficiently if the number of degrees of freedom is large.

An alternative to directly solving eqn (18) is to first take the Cholesky decomposition of $\mathbf{K}$ and to obtain the lower triangular matrix $\mathbf{L}$ as

$$\mathbf{L}\mathbf{L}^T = \mathbf{K} \tag{20}$$

and then to solve the following equations with respect to the unknowns $\mathbf{X}$ and $\mathbf{U}$ in turn.

$$\mathbf{L}\mathbf{X} = \mathbf{F} \tag{21}$$

$$\mathbf{L}^T\mathbf{U} = \mathbf{X} \tag{22}$$

It is important to note that, since the lower triangular matrix $\mathbf{L}$ preserves the same bandwidth as $\mathbf{K}$, $\mathbf{X}$ and $\mathbf{U}$ can be solved from eqns (21) and (22) efficiently. Under the assumption that $\mathbf{K}$ contains parameters which are subjected to spatial variabilities, the solution method in which eqns (20)–(22) are applied sample by sample is referred to as 'the direct Monte Carlo simulation'.

On the other hand, in 'the Neumann expansion method', $\mathbf{K}$ is decomposed into two matrices:

$$\mathbf{K} = \mathbf{K}_0 + \Delta\mathbf{K} \tag{23}$$

where $\mathbf{K}_0$ represents the stiffness matrix in which the spatially variable parameters are replaced by their representative values (mean values or medians in this study) and $\Delta\mathbf{K}$ consists of components representing the 'deviatoric parts' of the corresponding components in $\mathbf{K}$: $\Delta\mathbf{K} = \mathbf{K} - \mathbf{K}_0$. The solution $\mathbf{U}_0$ which corresponds to $\mathbf{K}_0$ can be obtained as

$$\mathbf{K}_0\mathbf{U}_0 = \mathbf{F} \quad \text{or} \quad \mathbf{U}_0 = \mathbf{K}_0^{-1}\mathbf{F} \tag{24}$$

The Neumann expansion of $\mathbf{K}^{-1}$ takes the following form:

$$\mathbf{K}^{-1} = (\mathbf{K}_0 + \Delta\mathbf{K})^{-1} = (\mathbf{I} - \mathbf{P} + \mathbf{P}^2 - \mathbf{P}^3 + \ldots)\mathbf{K}_0^{-1} \tag{25}$$

with

$$\mathbf{P} = \mathbf{K}_0^{-1}\Delta\mathbf{K} \tag{26}$$

Introducing eqn (25) into eqn (19) and using eqn (24), the solution vector $\mathbf{U}$ is represented by the following series as

$$\begin{aligned} \mathbf{U} &= \mathbf{U}_0 - \mathbf{P}\mathbf{U}_0 + \mathbf{P}^2\mathbf{U}_0 - \mathbf{P}^3\mathbf{U}_0 + \ldots \\ &= \mathbf{U}_0 - \mathbf{U}_1 + \mathbf{U}_2 - \mathbf{U}_3 + \ldots \end{aligned} \tag{27}$$

This series solution is equivalent to the following recursive equation:

$$\mathbf{K}_0 \mathbf{U}_i = \Delta \mathbf{K} \mathbf{U}_{i-1} \qquad (i = 1, 2, \ldots) \tag{28}$$

Thus, once the Cholesky decomposition of $\mathbf{K}_0$ is performed as

$$\mathbf{L}_0 \mathbf{L}_0^T = \mathbf{K}_0 \tag{29}$$

and once $\mathbf{U}_0$ is obtained using the algorithm represented by eqns (21) and (22), then the same algorithm can be used to obtain $\mathbf{U}_i$ iteratively with the aid of eqn (28). The expansion series in eqn (27) may be terminated after a few terms if convergence of the series is confirmed by the following criterion:

$$\frac{\|\mathbf{U}_i\|_2}{\left\| \sum_{k=0}^{i} (-1)^k \mathbf{U}_k \right\|_2} \leq \delta_{\text{err}} \tag{30}$$

where δ_{err} is the allowable error for convergence ($\delta_{\text{err}} = 0 \cdot 01$ is used in the present study), and $\|\cdot\|_2$ is the vector norm (length) defined by

$$\|\mathbf{U}\|_2 = \sqrt{\mathbf{U}^T \mathbf{U}} \tag{31}$$

It is pointed out that if the number of degrees of freedom is large, calculating the vector norm may take nontrivial CPU time. In such a case, a less stringent but more efficient convergence criterion using only the largest component of the displacement vector (instead of the norm) may be used.

The most outstanding feature of this approach in the case of Monte Carlo simulation for the spatial variation of material properties is that matrix factorization is required only once for all samples, and the rest of the computational process can fully utilize the banded characteristics of $\Delta \mathbf{K}$ and $\mathbf{L}_0$. Therefore, the computational time and costs can usually be reduced considerably.

It is known that the Neumann expansion shown in eqn (25) converges if the absolute values of all the eigenvalues of the product $\mathbf{P} = \mathbf{K}_0^{-1} \Delta \mathbf{K}$ are less than 1 and it can be proved as follows. The eigenvalues and eigenvectors of $\mathbf{P}$ are obtained as real numbers by solving the following eigenequation:

$$\mathbf{P}\boldsymbol{\Phi}_P = \boldsymbol{\Phi}_P\boldsymbol{\Lambda}_P \tag{32}$$

in which $\boldsymbol{\Phi}_P = [\boldsymbol{\Phi}_1\boldsymbol{\Phi}_2 \ldots \boldsymbol{\Phi}_{ND}]^T$ is the modal matrix of dimensions $ND \times ND$ (ND indicates the number of degrees of freedom) whose column vectors are orthogonal to each other and have Euclidean lengths of 1 as below.

$$\boldsymbol{\Phi}_P^T\boldsymbol{\Phi}_P = \mathbf{I} \tag{33}$$

In eqn (33), $\mathbf{I}$ is the identity matrix and $\boldsymbol{\Lambda}_P$ the diagonal eigenvalue matrix as shown below:

$$\boldsymbol{\Lambda}_P = \begin{bmatrix} \lambda_1 & 0 & \ldots & 0 \\ 0 & \lambda_2 & \ldots & 0 \\ & & \ldots & \\ 0 & 0 & \ldots & \lambda_{ND} \end{bmatrix} \tag{34}$$

where the λs represent the real eigenvalues that satisfy the following conditions.

$$|\lambda_1| \geqslant |\lambda_2| \geqslant \quad \ldots \ldots \quad \geqslant |\lambda_{ND}| \tag{35}$$

Although $\mathbf{P}$ is not necessarily symmetric, these eigenvalues are proved to be real by way of the similarity of eqn (32) with an eigenequation $\mathbf{M}\boldsymbol{\Phi} = \mathbf{K}\boldsymbol{\Phi}\boldsymbol{\Lambda}$ for vibration systems. Using eqns (32) and (33), $\mathbf{P}$ is written as

$$\mathbf{P} = \boldsymbol{\Phi}_P\boldsymbol{\Lambda}_P\boldsymbol{\Phi}_P^T \tag{36}$$

Thus, $\mathbf{P}^l$ is obtained as

$$\mathbf{P}^l = \boldsymbol{\Phi}_P\boldsymbol{\Lambda}_P^l\boldsymbol{\Phi}_P^T \tag{37}$$

with

$$\boldsymbol{\Lambda}_P^l = \begin{bmatrix} \lambda_1^l & 0 & \ldots & 0 \\ 0 & \lambda_2^l & \ldots & 0 \\ & & \ldots & \\ 0 & 0 & \ldots & \lambda_{ND}^l \end{bmatrix} \tag{38}$$

Equations (37) and (38) indicate that the Neumann expansion in eqn (27) converges if the absolute values of all the eigenvalues of $\mathbf{P}$ are less than 1.

However, this convergence criterion can be easily met irrespective of how large each component of the deviation matrix $\Delta\mathbf{K}$ is in comparison with the corresponding component of $\mathbf{K}$. This can be done by choosing, for each sample, a reference matrix $\mathbf{K}_0^*$ for expansion in such a way that

$$\mathbf{K} = \mathbf{K}_0^* + \Delta\mathbf{K}^* \tag{39}$$

and

$$\mathbf{K}_0^* = m\mathbf{K}_0 \tag{40}$$

where m is a scalar which is chosen to satisfy the convergence criterion of the sample. Then, eqns (27) and (28) must be modified into the following form:

$$\mathbf{U} = \mathbf{U}_0^* - \mathbf{U}_1^* + \mathbf{U}_2^* - \mathbf{U}_3^* + \ldots \tag{41}$$

$$\mathbf{K}_0^*\mathbf{U}_i^* = \Delta\mathbf{K}^*\mathbf{U}_{i-1}^* \qquad (i = 1, 2, \ldots) \tag{42}$$

However, this change in the reference matrix from $\mathbf{K}_0$ to $\mathbf{K}_0^*$ induces practically no additional computational effort to derive the solution because of the following relationships:

$$\mathbf{U}_0^* = \frac{1}{m}\mathbf{U}_0 \tag{43}$$

$$\mathbf{K}_0\mathbf{U}_i^* = \frac{1}{m}\Delta\mathbf{K}^*\mathbf{U}_{i-1}^* = \left(\frac{1}{m}\mathbf{K} - \mathbf{K}_0\right)\mathbf{U}_{i-1}^* \tag{44}$$

Thus, by replacing $\mathbf{U}_0$ and $\Delta\mathbf{K}$ with $\mathbf{U}_0/m$ and $(\mathbf{K}/m - \mathbf{K}_0)$ respectively, the same decomposition algorithm involving $\mathbf{L}_0$ as utilized in eqns (28) and (29) can be applied to eqn (42) for iterative solutions.

The possible range of m is determined by eigenvalue analysis. Equation (32) can be written for each eigenvalue as

$$[\mathbf{K}_0^{-1}\Delta\mathbf{K}]\mathbf{\Phi}_k = \lambda_k\mathbf{\Phi}_k \qquad (k = 1, 2, \ldots, ND) \tag{45}$$

If the λs are such that $|\lambda_k| \geqslant 1$, the reference matrix of expansion must be changed from $\mathbf{K}_0$ to $\mathbf{K}_0^*$ in order to make the expansion convergent. Then, the deviatoric part of $\mathbf{K}$ is written as

$$\Delta\mathbf{K}^* = \mathbf{K} - \mathbf{K}_0^* = (1 - m)\mathbf{K}_0 + \Delta\mathbf{K} \tag{46}$$

The eigenvalue and eigenvector of the product of $(\mathbf{K}_0^*)^{-1}\Delta\mathbf{K}^*$ are also represented by an equation similar to eqn (45) as

$$[(\mathbf{K}_0^*)^{-1}\Delta\mathbf{K}^*]\mathbf{\Phi}_{k'}^* = \lambda_{k'}^*\mathbf{\Phi}_{k'}^* \qquad (k' = 1, 2, \ldots, ND) \tag{47}$$

Introducing eqns (40) and (46) into eqn (47), we obtain

$$\left[\frac{1}{m}\mathbf{K}_0^{-1}\Delta\mathbf{K} + \frac{1-m}{m}\mathbf{I}\right]\mathbf{\Phi}_{k'}^* = \lambda_{k'}^*\mathbf{\Phi}_{k'}^* \tag{48}$$

Rearranging eqn (48),

$$[\mathbf{K}_0^{-1}\Delta\mathbf{K}]\mathbf{\Phi}_{k'}^* = (m\lambda_{k'}^* - 1 + m)\mathbf{\Phi}_{k'}^* \tag{49}$$

By comparing eqns (45) and (49), the following relationships are obtained.

$$\mathbf{\Phi}_{k'}^* = \mathbf{\Phi}_k \tag{50}$$

and

$$\lambda_{k'}^* = \frac{\lambda_k + 1 - m}{m} \tag{51}$$

If m is chosen in such a way that it satisfies the inequality $|\lambda_{k'}^*| < 1$ for all the k's, we obtain

$$|m| > |\lambda_k + 1 - m| \qquad (k = 1, 2, \ldots, ND) \tag{52}$$

The area of m which satisfies eqn (52) in the $\lambda_k - m$ plane is shown in Fig. 2. Because of the positive definiteness of $\mathbf{K}$, all the λ_ks in eqn (45) are larger than -1. If the largest positive eigenvalue is known as $\mathrm{Max}[\lambda_k]$, we can select a range of m which satisfies eqn (52) in the

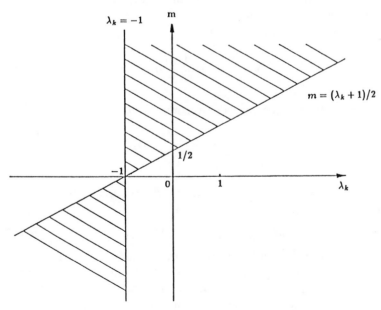

Fig. 2. Area of m satisfying $|m| > |\lambda_k + 1 - m|$.

λ_k–m plane as

$$m > \frac{\text{Max}[\lambda_k] + 1}{2} \tag{53}$$

Then, such an m satisfies eqn (52) for all the λ_ks. Therefore, the existence of m which makes Neumann expansion possible is guaranteed, irrespective of how large the deviation $\Delta\mathbf{K}$ is.

The practical determination of m without using eigenvalue analysis is also possible. For example, in the present study, the following m is used in the numerical example dealing with the variation of Young's modulus.

$$m = \frac{E_{\min} + E_{\max}}{2E_0} \tag{54}$$

in which E_0, $E_{\min}$ and $E_{\max}$ are respectively the median, minimum and maximum values among a sample of n Young's moduli generated for n finite elements. It can be shown that m as defined by eqn (54) satisfies eqn (53).

Either using the direct Monte Carlo method or the Neumann expansion method, sample displacement vectors $\mathbf{U}^{(i)}$ $(i = 1, 2, \ldots, N_s)$ are obtained with N_s denoting the sample size. The expected value $E[\mathbf{U}]$ of $\mathbf{U}$ is estimated by the estimator:

$$E[\mathbf{U}] \approx \bar{\mathbf{U}} = \frac{1}{N_s} \sum_{i=1}^{N_s} \mathbf{U}^{(i)} \tag{55}$$

The variance vector of $\mathbf{U}$ is also estimated by

$$\text{Var}[\mathbf{U}] \approx \mathbf{V}_U = \frac{1}{N_s} \sum_{i=1}^{N_s} \text{diag}[(\mathbf{U}^{(i)} - \bar{\mathbf{U}})(\mathbf{U}^{(i)} - \bar{\mathbf{U}})^T] \tag{56}$$

in which diag $[\mathbf{A}]$ denotes a vector whose components consist of the diagonal members of the square matrix $\mathbf{A}$. We may also define the estimator of the standard deviation of $\mathbf{U}$ as

$$\boldsymbol{\sigma}_X = [\sqrt{V_{U_1}} \sqrt{V_{U_2}} \ldots \sqrt{V_{U_{ND}}}]^T \tag{57}$$

with V_{U_j} $(j = 1, 2, \ldots, ND)$ indicating the jth term of the vector $\mathbf{V}_U$.

Once a sample displacement vector is obtained, corresponding sample strain and stress of each element are obtained by utilizing eqns (6) and (8). The statistical characteristics of strains and stresses can also be evaluated in a manner similar to that for the displacement.

4 PERTURBATION METHOD

4.1 Perturbation Formulation for Stochastic Systems

The stochastic finite element method for static problems based on perturbation techniques has been developed by several researchers such as Cambou,[1] Baecher & Ingra,[2] Handa & Andersson,[3] Hisada & Nakagiri[4,13] as mentioned earlier. All these techniques are basically the same, although Hisada and Nakagiri's approach is well-organized and covers a wide range of structural engineering problems. In order to demonstrate the difference between the Neumann expansion method and perturbation methods, a perturbation method involving terms up to the second-order approximation is described below, primarily following Hisada and Nakagiri's notation.

The equation of equilibrium for static problems is represented by

$$\mathbf{K}\mathbf{U} = \mathbf{F} \tag{58}$$

and the stiffness matrix $\mathbf{K}$ involves the nondimensional random variables α_i $(i = 1, 2, \ldots, N)$ which represent the spatial variation of the material properties. $\mathbf{K}$ can be expanded in the following form with the assumption that each α_i is small $(\alpha_i \ll 1)$ and has zero mean.

$$\mathbf{K} = \mathbf{K}^0 + \sum_{i=1}^{N} \mathbf{K}_i^{\mathrm{I}} \alpha_i + \tfrac{1}{2} \sum_{i=1}^{N} \sum_{j=1}^{N} \mathbf{K}_{ij}^{\mathrm{II}} \alpha_i \alpha_j + \ldots \tag{59}$$

in which $\mathbf{K}^0$ is the stiffness matrix evaluated at $\boldsymbol{\alpha} = \mathbf{0}$, $\mathbf{K}_i^{\mathrm{I}}$ is the first partial derivative of $\mathbf{K}$ with respect to α_i and evaluated at $\boldsymbol{\alpha} = \mathbf{0}$ and $\mathbf{K}_{ij}^{\mathrm{II}}$ is the second partial derivative of $\mathbf{K}$ with respect to α_i and α_j and evaluated at $\boldsymbol{\alpha} = \mathbf{0}$ as follows:

$$\mathbf{K}_i^{\mathrm{I}} = \frac{\partial \mathbf{K}}{\partial \alpha_i}\bigg|_{\boldsymbol{\alpha}=0}; \qquad \mathbf{K}_{ij}^{\mathrm{II}} = \frac{\partial^2 \mathbf{K}}{\partial \alpha_i \, \partial \alpha_j}\bigg|_{\boldsymbol{\alpha}=0} \tag{60}$$

in which $\boldsymbol{\alpha} = [\alpha_1 \alpha_2 \ldots \alpha_N]^T$.

Similarly, the external force vector $\mathbf{F}$ may also involve the random variables α_i. Although that is not the case at this time, we derive a general perturbation formulation including such a case. Then, $\mathbf{F}$ can be expanded in the series form:

$$\mathbf{F} = \mathbf{F}^0 + \sum_{i=1}^{N} \mathbf{F}_i^{\mathrm{I}} \alpha_i + \tfrac{1}{2} \sum_{i=1}^{N} \sum_{j=1}^{N} \mathbf{F}_{ij}^{\mathrm{II}} \alpha_i \alpha_j + \ldots \tag{61}$$

with partial derivative vectors:

$$\mathbf{F}_i^{\mathrm{I}} = \frac{\partial \mathbf{F}}{\partial \alpha_i}\bigg|_{\alpha=0} \; ; \qquad \mathbf{F}_{ij}^{\mathrm{II}} = \frac{\partial^2 \mathbf{F}}{\partial \alpha_i \, \partial \alpha_j}\bigg|_{\alpha=0} \tag{62}$$

These partial derivatives of $\mathbf{F}$ become zero if $\mathbf{F}$ is deterministic.

The unknown displacement vector $\mathbf{U}$ is also expanded in a similar form:

$$\mathbf{U} = \mathbf{U}^0 + \sum_{i=1}^{N} \mathbf{U}_i^{\mathrm{I}}\alpha_i + \tfrac{1}{2}\sum_{i=1}^{N}\sum_{j=1}^{N} \mathbf{U}_{ij}^{\mathrm{II}}\alpha_i\alpha_j + \dots . \tag{63}$$

in which the coefficient vectors $\mathbf{U}^0$, $\mathbf{U}_i^{\mathrm{I}}$ and $\mathbf{U}_{ij}^{\mathrm{II}}$ can be represented by the following set of recursive equations:

$$\mathbf{U}^0 = (\mathbf{K}^0)^{-1}\mathbf{F}^0 \tag{64}$$

$$\mathbf{U}_i^{\mathrm{I}} = (\mathbf{K}^0)^{-1}(\mathbf{F}_i^{\mathrm{I}} - \mathbf{K}_i^{\mathrm{I}}\mathbf{U}^0) \tag{65}$$

$$\mathbf{U}_{ij}^{\mathrm{II}} = (\mathbf{K}^0)^{-1}(\mathbf{F}_{ij}^{\mathrm{II}} - \mathbf{K}_i^{\mathrm{I}}\mathbf{U}_j^{\mathrm{I}} - \mathbf{K}_j^{\mathrm{I}}\mathbf{U}_i^{\mathrm{I}} - \mathbf{K}_{ij}^{\mathrm{II}}\mathbf{U}^0) \tag{66}$$

Equation (65) is obtained by differentiating eqn (58) with respect to α_i as

$$\frac{\partial \mathbf{K}}{\partial \alpha_i}\mathbf{U} + \mathbf{K}\frac{\partial \mathbf{K}}{\partial a_i} = \frac{\partial \mathbf{F}}{\partial \alpha_i} \tag{67}$$

and setting $\alpha = \mathbf{0}$. Equation (66) is also obtained by differentiating eqn (67) with respect to α_j as

$$\frac{\partial^2 \mathbf{K}}{\partial \alpha_i \, \partial \alpha_j}\mathbf{U} + \frac{\partial \mathbf{K}}{\partial \alpha_i}\frac{\partial \mathbf{U}}{\partial \alpha_j} + \frac{\partial \mathbf{K}}{\partial \alpha_j}\frac{\partial \mathbf{U}}{\partial \alpha_i} + \mathbf{K}\frac{\partial^2 \mathbf{U}}{\partial \alpha_i \, \partial \alpha_j} = \frac{\partial^2 \mathbf{F}}{\partial \alpha_i \, \partial \alpha_j} \tag{68}$$

and setting $\alpha = \mathbf{0}$. Higher-order coefficient vectors can also be obtained in a similar manner if necessary.

The strain and stress of the eth element are calculated in terms of $\mathbf{u}_e$, a part of the solution vector $\mathbf{U}$ related to the eth element. $\mathbf{u}_e$ is also represented in the following series form by taking the appropriate part of eqn (63).

$$\mathbf{u}_e = \mathbf{u}_e^0 + \sum_{i=1}^{N} \mathbf{u}_{ei}^{\mathrm{I}}\alpha_i + \tfrac{1}{2}\sum_{i=1}^{N}\sum_{j=1}^{N} \mathbf{u}_{eij}^{\mathrm{II}}\alpha_i\alpha_j + \dots \tag{69}$$

in which $\mathbf{u}_e^0$, $\mathbf{u}_{ei}^{\mathrm{I}}$ and $\mathbf{u}_{eij}^{\mathrm{II}}$ are appropriate sub-vectors of $\mathbf{U}^0$, $\mathbf{U}_i^{\mathrm{I}}$ and $\mathbf{U}_{ij}^{\mathrm{II}}$, respectively.

Utilizing the standard finite element technique, the strain–displacement relationship is described as

$$\boldsymbol{\varepsilon}_e = \mathbf{B}_e\mathbf{u}_e \tag{70}$$

where $\mathbf{B}_e$ is given by eqn (7) and depends on the shape functions and geometric conditions of the element. It is important to note that $\mathbf{B}_e$ does not involve randomness of the material properties. The strain vector of the eth element, $\boldsymbol{\varepsilon}_e = [\varepsilon_{xx}\varepsilon_{yy}\gamma_{xy}]^T$, is then obtained by introducing eqn (69) into eqn (70) as

$$\boldsymbol{\varepsilon}_e = \boldsymbol{\varepsilon}_e^0 + \sum_{i=1}^{N} \boldsymbol{\varepsilon}_{ei}^I \alpha_i + \tfrac{1}{2}\sum_{i=1}^{N}\sum_{j=1}^{N} \boldsymbol{\varepsilon}_{eij}^{II}\alpha_i\alpha_j + \ldots \tag{71}$$

with

$$\boldsymbol{\varepsilon}_e^0 = \mathbf{B}_e\mathbf{u}_e^0; \quad \boldsymbol{\varepsilon}_{ei}^I = \mathbf{B}_e\mathbf{u}_{ei}^I; \quad \boldsymbol{\varepsilon}_{eij}^{II} = \mathbf{B}_e\mathbf{u}_{eij}^{II} \tag{72}$$

The stress vector of the eth element, $\boldsymbol{\sigma}_e = [\sigma_{xx}\sigma_{yy}\tau_{xy}]^T$, is represented by the stress–strain relationship:

$$\boldsymbol{\sigma}_e = \mathbf{D}_e\boldsymbol{\varepsilon}_e \tag{73}$$

in which $\mathbf{D}_e$ is the elasticity matrix of the eth element and can be expanded into the following series:

$$\mathbf{D}_e = \mathbf{D}_e^0 + \sum_{i=1}^{N} \mathbf{D}_{ei}^I \alpha_i + \tfrac{1}{2}\sum_{i=1}^{N}\sum_{j=1}^{N} \mathbf{D}_{eij}^{II}\alpha_i\alpha_j + \ldots \tag{74}$$

where $\mathbf{D}_e^0$ is the elasticity matrix evaluated at $\boldsymbol{\alpha} = \mathbf{0}$, $\mathbf{D}_{ei}^I$ is the first partial derivative of $\mathbf{D}_e$ with respect to α_i and evaluated at $\boldsymbol{\alpha} = \mathbf{0}$, and $\mathbf{D}_{eij}^{II}$ is the second partial derivative of $\mathbf{D}_e$ with respect to α_i and α_j and evaluated at $\boldsymbol{\alpha} = \mathbf{0}$:

$$\mathbf{D}_{ei}^I = \frac{\partial \mathbf{D}_e}{\partial \alpha_i}\bigg|_{\boldsymbol{\alpha}=\mathbf{0}}; \qquad \mathbf{D}_{eij}^{II} = \frac{\partial^2 \mathbf{D}_e}{\partial \alpha_i \, \partial \alpha_j}\bigg|_{\boldsymbol{\alpha}=\mathbf{0}} \tag{75}$$

Introducing eqns (71) and (74) into eqn (73), $\boldsymbol{\sigma}_e$ is obtained as

$$\boldsymbol{\sigma}_e = \left(\mathbf{D}_e^0 + \sum_{i=1}^{N} \mathbf{D}_{ei}^I \alpha_i + \tfrac{1}{2}\sum_{i=1}^{N}\sum_{j=1}^{N} \mathbf{D}_{eij}^{II}\alpha_i\alpha_j + \ldots \right)\left(\boldsymbol{\varepsilon}_e^0 + \sum_{i=1}^{N} \boldsymbol{\varepsilon}_{ei}^I \alpha_i \right.$$

$$\left. + \tfrac{1}{2}\sum_{i=1}^{N}\sum_{j=1}^{N} \boldsymbol{\varepsilon}_{eij}^{II}\alpha_i\alpha_j + \ldots \right)$$

$$= \boldsymbol{\sigma}_e^0 + \sum_{i=1}^{N} \boldsymbol{\sigma}_{ei}^I \alpha_i + \tfrac{1}{2}\sum_{i=1}^{N}\sum_{j=1}^{N} \boldsymbol{\sigma}_{eij}^{II}\alpha_i\alpha_j + \ldots \tag{76}$$

with the relationships:

$$\boldsymbol{\sigma}_e^0 = \mathbf{D}_e^0\boldsymbol{\varepsilon}_e^0 \tag{77}$$

$$\boldsymbol{\sigma}_{ei}^I = \mathbf{D}_e^0\boldsymbol{\varepsilon}_{ei}^I + \mathbf{D}_{ei}^I\boldsymbol{\varepsilon}_e^0 \tag{78}$$

$$\boldsymbol{\sigma}_{eij}^{II} = \mathbf{D}_e^0\boldsymbol{\varepsilon}_{eij}^{II} + 2\mathbf{D}_{ei}^I\boldsymbol{\varepsilon}_{ej}^I + \mathbf{D}_{eij}^{II}\boldsymbol{\varepsilon}_e^0 \tag{79}$$

If there is only one material property such as Young's modulus that forms a stochastic field, eqns (71) and (74) take much simpler forms:

$$\mathbf{\varepsilon}_e = \mathbf{\varepsilon}_e^0 + \mathbf{\varepsilon}_{ee}^{\mathrm{I}}\alpha_e + \tfrac{1}{2}\mathbf{\varepsilon}_{eee}^{\mathrm{II}}\alpha_e^2 + \ldots \tag{80}$$

$$\mathbf{D}_e = \mathbf{D}_e^0 + \mathbf{D}_{ee}^{\mathrm{I}}\alpha_e + \tfrac{1}{2}\mathbf{D}_{eee}^{\mathrm{II}}\alpha_e^2 + \ldots \tag{81}$$

Then, $\mathbf{\sigma}_e$ is also expressed in a simpler form:

$$\mathbf{\sigma}_e = \mathbf{\sigma}_e^0 + \mathbf{\sigma}_{ee}^{\mathrm{I}}\alpha_e + \tfrac{1}{2}\mathbf{\sigma}_{eee}^{\mathrm{II}}\alpha_e^2 + \ldots \tag{82}$$

with

$$\mathbf{\sigma}_{ee}^{\mathrm{I}} = \mathbf{D}_e^0\mathbf{\varepsilon}_{ee}^{\mathrm{I}} + \mathbf{D}_{ee}^{\mathrm{I}}\mathbf{\varepsilon}_e^0 \tag{83}$$

$$\mathbf{\sigma}_{eee}^{\mathrm{II}} = \mathbf{D}_e^0\mathbf{\varepsilon}_{eee}^{\mathrm{II}} + 2\mathbf{D}_{ee}^{\mathrm{I}}\mathbf{\varepsilon}_{ee}^{\mathrm{I}} + \mathbf{D}_{eee}^{\mathrm{II}}\mathbf{\varepsilon}_e^0 \tag{84}$$

The first-order approximation for the displacement is obtained by truncating the right-hand side of eqn (63) after the second term as

$$\mathbf{U} = \mathbf{U}^0 + \sum_{i=1}^{N} \mathbf{U}_i^{\mathrm{I}}\alpha_i \tag{85}$$

with expected value

$$E^{\mathrm{I}}[\mathbf{U}] = \mathbf{U}^0 \tag{86}$$

and covariance matrix

$$\mathrm{Cov}^{\mathrm{I}}[\mathbf{U}, \mathbf{U}] = E[(\mathbf{U} - E^{\mathrm{I}}[\mathbf{U}])(\mathbf{U} - E^{\mathrm{I}}[\mathbf{U}])^T] = \sum_{i=1}^{N}\sum_{j=1}^{N} \mathbf{U}_i^{\mathrm{I}}(\mathbf{U}_j^{\mathrm{I}})^T E[\alpha_i\alpha_j] \tag{87}$$

or the variance vector which is the diagonal component of $\mathrm{Cov}^{\mathrm{I}}[\mathbf{U}, \mathbf{U}]$

$$\mathrm{Var}^{\mathrm{I}}[\mathbf{U}] = \sum_{i=1}^{N}\sum_{j=1}^{N} \mathrm{diag}[\mathbf{U}_i^{\mathrm{I}}(\mathbf{U}_j^{\mathrm{I}})^T]E[\alpha_i\alpha_j] \tag{88}$$

where $E[\alpha_i\alpha_j]$ is determined from the autocorrelation function of the underlying stochastic field of α.

The second-order approximation for the displacement is obtained by truncating the right-hand side of eqn (63) after the third term as

$$\mathbf{U} = \mathbf{U}^0 + \sum_{i=1}^{N} \mathbf{U}_i^{\mathrm{I}}\alpha_i + \tfrac{1}{2}\sum_{i=1}^{N}\sum_{j=1}^{N} \mathbf{U}_{ij}^{\mathrm{II}}\alpha_i\alpha_j \tag{89}$$

with expected value

$$E^{\mathrm{II}}[\mathbf{U}] = E^{\mathrm{I}}[\mathbf{U}] + \tfrac{1}{2}\sum_{i=1}^{N}\sum_{j=1}^{N} \mathbf{U}_{ij}^{\mathrm{II}}E[\alpha_i\alpha_j] \tag{90}$$

and covariance matrix

$$\text{Cov}^{\text{II}}[\mathbf{U}, \mathbf{U}] = \text{Cov}^{\text{I}}[\mathbf{U}, \mathbf{U}]$$
$$+ \tfrac{1}{4} \sum_{i=1}^{N} \sum_{j=1}^{N} \sum_{k=1}^{N} \sum_{l=1}^{N} \mathbf{U}_{ij}^{\text{II}} (\mathbf{U}_{kl}^{\text{II}})^{T} (E[\alpha_i \alpha_l] E[\alpha_j \alpha_k]$$
$$+ E[\alpha_i \alpha_k] E[\alpha_j \alpha_l]) \tag{91}$$

or the variance vector

$$\text{Var}^{\text{II}}[\mathbf{U}] = \text{Var}^{\text{I}}[\mathbf{U}]$$
$$+ \tfrac{1}{4} \sum_{i=1}^{N} \sum_{j=1}^{N} \sum_{k=1}^{N} \sum_{l=1}^{N} \text{diag}[\mathbf{U}_{ij}^{\text{II}} (\mathbf{U}_{kl}^{\text{II}})^{T}] (E[\alpha_i \alpha_l] E[\alpha_j \alpha_k]$$
$$+ E[\alpha_i \alpha_k] E[\alpha_j \alpha_l]) \tag{92}$$

In the process of obtaining $\text{Cov}^{\text{II}}[\mathbf{U}, \mathbf{U}]$ or $\text{Var}^{\text{II}}[\mathbf{U}]$, the following well-known relationships for Gaussian random variables are used:

$$E[\alpha_i \alpha_j \alpha_k] = 0 \tag{93}$$
$$E[\alpha_i \alpha_j \alpha_k \alpha_l] = E[\alpha_i \alpha_j] E[\alpha_k \alpha_l] + E[\alpha_i \alpha_l] E[\alpha_j \alpha_k] + E[\alpha_i \alpha_k] E[\alpha_j \alpha_l] \tag{94}$$

If the αs are not Gaussian, one must evaluate these higher-order moments for the second-order approximation in accordance with their joint probability distribution. However, such an evaluation of higher-order moments may be practically impossible. Hence, the second-order approximation is usually restricted to stochastic fields following Gaussian distribution.

The equations for evaluating the first- and second-order moments of the displacement are equally applied to obtain the first- and second-order moments of the strains and stresses just by replacing $\mathbf{U}^0$, $\mathbf{U}_i^{\text{I}}$, $\mathbf{U}_{ij}^{\text{II}}$ with $\boldsymbol{\varepsilon}_e^0$, $\boldsymbol{\varepsilon}_{ei}^{\text{I}}$, $\boldsymbol{\varepsilon}_{eij}^{\text{II}}$ or with $\boldsymbol{\sigma}_e^0$, $\boldsymbol{\sigma}_{ei}^{\text{I}}$, $\boldsymbol{\sigma}_{eij}^{\text{II}}$ because of the identical analytical forms that eqns (63), (71) and (76) exhibit.

4.2 Partial Derivatives for Perturbation Formulation

The perturbation method described above requires that the coefficient matrices (partial derivatives with respect to random variables) of the expansion such as $\mathbf{D}_e^0$, $\mathbf{D}_{ee}^{\text{I}}$, $\mathbf{D}_{eee}^{\text{II}}$ in eqn (81) and $\mathbf{K}^0$, $\mathbf{K}_i^{\text{I}}$, $\mathbf{K}_{ij}^{\text{II}}$ in eqn (59) be evaluated. Obviously, these matrices depend on which material properties are spatially varying and in what way. In this study, the following four cases involving the homogeneous spatial variation of Young's modulus or Poisson's ratio are assumed.

(1) Variation of E

$$E_e = E_0(1 + \alpha_e); \qquad \nu_e = \nu_0 \qquad (95)$$

(2) Variation of $1/E$

$$\frac{1}{E_e} = \frac{1}{E_0}(1 + \alpha_e); \qquad \nu_e = \nu_0 \qquad (96)$$

(3) Variation of $\ln E$

$$E_e = E_0 \exp(\alpha_e); \qquad \nu_e = \nu_0 \qquad (97)$$

(4) Variation of ν

$$E_e = E_0; \qquad \nu_e = \nu_0(1 + \alpha_e) \qquad (98)$$

where E_0 and ν_0 are the medians of Young's modulus and Poisson's ratio, respectively, the α's are Gaussian random variables with zero mean, each associated with a corresponding finite element and consistent with the assumed stochastic field characteristics, and the suffix e indicates that α_e is associated with the eth element.

Under plane stress conditions, the elasticity matrix of the eth element is known as

$$\mathbf{D}_e = \frac{E_e}{1 - \nu_e^2} \begin{bmatrix} 1 & \nu_e & 0 \\ & 1 & 0 \\ & & \dfrac{1 - \nu_e}{2} \\ \text{sym.} & & \end{bmatrix} \qquad (99)$$

Substituting each set of the relationships represented by eqns (95)–(98) into eqn (99) and setting $\alpha_e = 0$, $\mathbf{D}_e^0$ is obtained as

$$\mathbf{D}_e^0 = \frac{E_0}{1 - \nu_0^2} \begin{bmatrix} 1 & \nu_0 & 0 \\ & 1 & 0 \\ & & \dfrac{1 - \nu_0}{2} \\ \text{sym.} & & \end{bmatrix} \qquad (100)$$

In the case of variation of E, $\mathbf{D}_{ee}^{\mathrm{I}}$ and $\mathbf{D}_{eee}^{\mathrm{II}}$ are easily obtained by substituting eqn (95) into eqn (99) and differentiating it with respect to α_e and setting $\alpha_e = 0$:

$$\mathbf{D}_{ee}^{\mathrm{I}} = \mathbf{D}_e^0; \qquad \mathbf{D}_{eee}^{\mathrm{II}} = 0 \qquad (101)$$

In the case of variation of $1/E$, $\mathbf{D}^{\mathrm{I}}_{ee}$ and $\mathbf{D}^{\mathrm{II}}_{eee}$ take the following forms:

$$\mathbf{D}^{\mathrm{I}}_{ee} = -\mathbf{D}^{0}_{e}; \qquad \mathbf{D}^{\mathrm{II}}_{eee} = 2\mathbf{D}^{0}_{e} \qquad (102)$$

Equation (102) can be easily obtained by using the following expression:

$$E_e = \frac{E_0}{1 + \alpha_e} = E_0(1 - \alpha_e + \alpha_e^2 - \ldots) \qquad (103)$$

In the case of variation of $\ln E$, $\mathbf{D}^{\mathrm{I}}_{ee}$ and $\mathbf{D}^{\mathrm{II}}_{eee}$ are obtained as

$$\mathbf{D}^{\mathrm{I}}_{ee} = \mathbf{D}^{0}_{e}; \qquad \mathbf{D}^{\mathrm{II}}_{eee} = \mathbf{D}^{0}_{e} \qquad (104)$$

In the case of variation of v, substituting eqn (98) into eqn (99) and differentiating with respect to α_e with the aid of the SMP program and setting $\alpha_e = 0$,

$$\mathbf{D}^{\mathrm{I}}_{ee} = \frac{E_0 v_0}{(1 - v_0^2)^2} \begin{bmatrix} 2v_0 & 1 + v_0^2 & 0 \\ & 2v_0 & 0 \\ \text{sym.} & & \dfrac{-1 + 2v_0 - v_0^2}{2} \end{bmatrix} \qquad (105)$$

$$\mathbf{D}^{\mathrm{II}}_{eee} = \frac{E_0 v_0^2}{(1 - v_0^2)^3} \begin{bmatrix} 2 + 6v_0^2 & 6v_0 + 2v_0^3 & 0 \\ & 2 + 6v_0^2 & 0 \\ \text{sym.} & & 2 - 4v_0 - v_0^2 + 3v_0^3 \end{bmatrix} \qquad (106)$$

Introducing eqn (81) into eqn (10), the element stiffness matrix of the eth element $\mathbf{k}_e$ is also represented by the following series when we have only one spatially variable parameter:

$$\mathbf{k}_e = \mathbf{k}^0_e + \mathbf{k}^{\mathrm{I}}_{ee}\alpha_e + \tfrac{1}{2}\mathbf{k}^{\mathrm{II}}_{eee}\alpha_e^2 + \ldots \qquad (107)$$

where

$$\mathbf{k}^0_e = \mathbf{k}_e \bigg|_{\alpha_e=0} = t l_x l_y \int_0^1 \int_0^1 \mathbf{B}_e^T \mathbf{D}^0_e \mathbf{B}_e \, \mathrm{d}\xi \, \mathrm{d}\eta \qquad (108)$$

$$\mathbf{k}^{\mathrm{I}}_{ee} = \frac{\partial \mathbf{k}_e}{\partial \alpha_e} \bigg|_{\alpha_e=0} = t l_x l_y \int_0^1 \int_0^1 \mathbf{B}_e^T \mathbf{D}^{\mathrm{I}}_{ee} \mathbf{B}_e \, \mathrm{d}\xi \, \mathrm{d}\eta \qquad (109)$$

$$\mathbf{k}^{\mathrm{II}}_{eee} = \frac{\partial^2 \mathbf{k}_e}{\partial \alpha_e^2} \bigg|_{\alpha_e=0} = t l_x l_y \int_0^1 \int_0^1 \mathbf{B}_e^T \mathbf{D}^{\mathrm{II}}_{eee} \mathbf{B}_e \, \mathrm{d}\xi \, \mathrm{d}\eta \qquad (110)$$

Equations (108)–(110) can be written in the following generalized form:

$$\mathbf{k}_e^* = tl_xl_y \int_0^1 \int_0^1 \mathbf{B}_e^T \mathbf{D}_e^* \mathbf{B}_e \, \mathrm{d}\xi \, \mathrm{d}\eta \tag{111}$$

and

$$\mathbf{D}_e^* = \begin{bmatrix} a & b & 0 \\ & a & 0 \\ \mathrm{sym.} & & c \end{bmatrix} \tag{112}$$

in which a, b and c are constants consisting of Young's modulus and Poisson's ratio. The result of the two-fold integration is already shown in eqn (12) with the aid of the SMP program.

The coefficient matrices $\mathbf{K}^0$, $\mathbf{K}_i^{\mathrm{I}}$, $\mathbf{K}_{ij}^{\mathrm{II}}$ of the global stiffness are obtained by assembling $\mathbf{k}_e^0$, $\mathbf{k}_{ei}^{\mathrm{I}}$, $\mathbf{k}_{eij}^{\mathrm{II}}$ over the entire region in the same manner as in eqn (13):

$$\mathbf{K}^0 = \sum_{e=1}^n \mathbf{k}_e^0 = \sum_{e=1}^n \mathbf{k}_e \Big|_{\alpha=0} \tag{113}$$

$$\mathbf{K}_i^{\mathrm{I}} = \sum_{e=1}^n \mathbf{k}_{ei}^{\mathrm{I}} = \sum_{e=1}^n \frac{\partial \mathbf{k}_e}{\partial \alpha_i} \Big|_{\alpha=0} \tag{114}$$

$$\mathbf{K}_{ij}^{\mathrm{II}} = \sum_{e=1}^n \mathbf{k}_{eij}^{\mathrm{II}} = \sum_{e=1}^n \frac{\partial^2 \mathbf{k}_e}{\partial \alpha_i \, \partial \alpha_j} \Big|_{\alpha=0} \tag{115}$$

with n = number of elements. If we consider the spatial variability of one material property parameter, the summations in eqns (114) and (115) have meaning only when $e = i$ or $e = i = j$ because of the following relationships:

$$\frac{\partial \mathbf{k}_e}{\partial \alpha_i} = 0 \quad \text{if } e \neq i; \qquad \frac{\partial^2 \mathbf{k}_e}{\partial \alpha_i \, \partial \alpha_j} = 0 \quad \text{if } e \neq i \text{ or } e \neq j \tag{116}$$

The methodology described here can be applied to other linear elastic problems, such as those involving plate bending, plane strain, axisymmetry or three-dimensional space, by replacing the elasticity matrix $\mathbf{D}_e$ and the displacement–strain matrix $\mathbf{B}_e$ with their appropriate alternatives.

5 NUMERICAL EXAMPLE

5.1 Finite Element Model and Assumptions

A computer program was developed by which we could estimate the response variability due to material property variations. This program estimates the expected value and standard deviation of the response by means of Monte Carlo techniques. In doing so, two different Monte Carlo methods are used for the solution of the equation of equilibrium: the direct method uses eqns (20), (21) and (22) for each sample, while the Neumann expansion method uses an iterative scheme either in the form of eqn (28) or in that of eqn (42). A number of sample global stiffness matrices are constructed on the basis of the sample stochastic fields generated by means of the Cholesky decomposition algorithm (see Yamazaki *et al.*[24]). Then, the direct Monte Carlo solution (direct MCS) is obtained each time by employing one of the sample global stiffness matrices and taking statistics on the resulting response quantities. On the other hand, the simulation solution based on the Neumann expansion, referred to as the Neumann expansion Monte Carlo simulation or, for brevity, expansion Monte Carlo simulation (expansion MCS), uses eqn (28) or (42) for Monte Carlo purposes. Another program based on the perturbation method, which involves first- and second-order perturbation approximations, was also developed. Numerical examples are presented here in order to examine the accuracy and efficiency of these methods.

The finite element model shown in Fig. 3 is adopted as an example. The particular structural geometry and loading conditions are chosen so that the response variability can be clearly highlighted by the material property variability. The model consists of one hundred (100) plane-stress, square finite elements with 121 nodes. Nodal displacements in the y-direction are constrained along the lower edge and nodal displacements in both directions are constrained at the left lower corner nodes. Under these boundary conditions, the number of degrees of freedom (DOF) of this structural system is 230. A uniformly distributed load is applied along the upper edge.

The five cases of variability of Young's modulus or Poisson's ratio are assumed with medians $E_0 = 1 \cdot 0$ and $v_0 = 0 \cdot 3$. In the present study, all the parameters are represented without units so that any units can be specified so long as they are used consistently. The stochastic field $\alpha(\mathbf{x})$ representing the deviatoric component of the material property is assumed to be isotropic with two-dimensional autocorrelation given by

$$R_{\alpha\alpha}(\xi) = \sigma_\alpha^2 \exp\left[-\left(\frac{|\xi|}{d}\right)^2\right] \qquad (117)$$

in which σ_α is the standard deviation and d is the correlation distance of the stochastic field, and $\xi = [\xi_x \xi_y]^T$ is the separation vector between

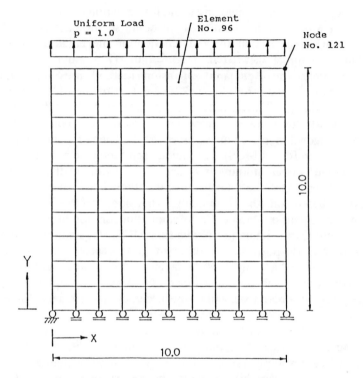

Fig. 3. Finite element model with 100 elements.

two points $\mathbf{x}$ and $\mathbf{x} + \boldsymbol{\xi}$. This stochastic field is discretized in accordance with the finite element mesh indicated in Fig. 3 and $\alpha(\mathbf{x})$ in element e is represented by the value α_e of $\alpha(\mathbf{x})$ at the centroid $\mathbf{x}_e$ of the element: $\alpha_e = \alpha(\mathbf{x}_e)$ where the assumption that the variation of α within each element is small enough is used. The proper mesh division depends primarily on the values of the standard deviation and correlation distance because of the particular structural geometry and loading conditions considered. Three values of σ_α, i.e. 0·1, 0·2 and 0·3, and $d = 2·0$ are used throughout the numerical example.

Our experience indicates that the element size in Fig. 3 is considered to be small enough in view of the fact that $d = 2·0$ in this case; the element size is 1/2 of the scale of correlation which has been studied by Harada and Shinozuka.[16] The question of 'in what sense is it small enough' is another matter to be studied as was done by Shinozuka &

Deodatis.[15] As mentioned previously, however, the major concern of the present study is to examine the accuracy of the various analytical and numerical methods of variability prediction. Thus, once the stochastic field is discretized in the manner indicated above, the accuracy problem can be dealt with as an issue separate from that of the mesh size.

The results of the numerical example will be shown at locations where significant output values are expected, i.e. node 121 for displacement U_y and element 96 for strains ε_{xx} and ε_{yy}.

5.2 Example for Variation of E

First, Young's modulus is assumed to be Gaussian as indicated below.

$$E_e = E_0(1 + \alpha_e) \tag{118}$$

where E_e represents the discretized version of $E(\mathbf{x})$ in element e, and α_e is defined above. In this case, the standard deviation of α_e represents the coefficient of variation of E_e. The assumption of Gaussian distribution implies the possibility of generating negative values of Young's modulus. In order to avoid this difficulty, the values of random variable α_e in the case of Monte Carlo simulation are confined to the range:

$$-1 + \varepsilon \leqslant \alpha_e \leqslant 1 - \varepsilon \tag{119}$$

in which ε is a positive constant. Introducing such a positive constant also removes mathematical complications that would arise if α_e indeed becomes zero. In this numerical example, we use $\varepsilon = 0.05$ which is in the range of ε values where the response statistics are not very sensitive to ε. Inequality on the right-hand side is also introduced to maintain the symmetry of its probability density function. The effect of truncation is smaller if the standard deviation of α_e is smaller. In an analytical approach such as the perturbation method, these negative values are implicitly included if a Gaussian distribution is assumed as in the case of eqns (95) and (96). It is noted that, when Monte Carlo simulation techniques are used to generate a discretized stochastic field, the generated sample field is discarded if at least one of α_es is out of the specified range as shown in eqn (119). This rule is also observed for the other two cases of material parameter variation studied in Sections 5.3 and 5.5.

Samples of α_e simulated by the Cholesky decomposition algorithm are depicted in Fig. 4. Using these samples, convergence of the

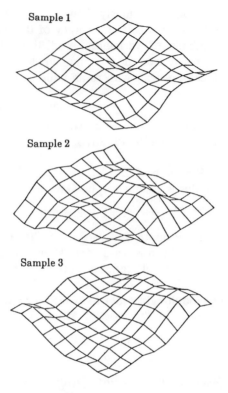

Fig. 4. Simulated sample variations of Young's modulus.

response values obtained by the Neumann expansion method with respect to the order of expansion is examined and typical results are shown in Figs 5 and 6. These figures indicate that the rate of convergence of the sample responses differs from sample to sample, and that, as higher-order expansion is used the results approach more closely those obtained from the direct method (use of eqn (19) sample by sample and hence considered to represent the expansion order of infinity).

Convergence of the standard deviation in the Neumann expansion simulation method with respect to the order of expansion is also examined. Using 100 samples, the results are plotted in Figs 7 and 8 along with those by the direct Monte Carlo simulation and perturbation method. The rate of convergence of the standard deviations is

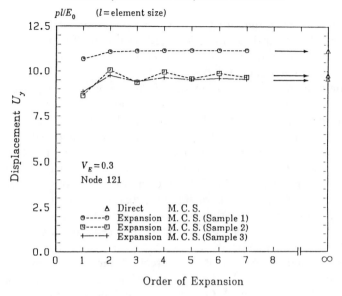

Fig. 5. Convergence of displacement U_y (variation of E).

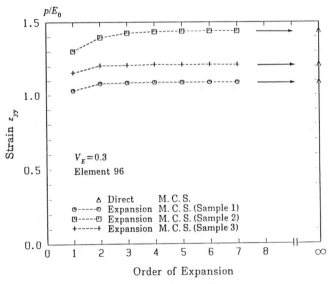

Fig. 6. Convergence of strain ε_{yy} (variation of E).

seen to be slower than that of the sample responses, although the results of the expansion simulation method eventually approach those from the direct Monte Carlo simulation method. This is partly due to the fact that each standard deviation plotted in Figs 7 and 8 results from only the deviatoric component of the sample response, while the sample responses shown in Figs 5 and 6 consist of mean values plus deviatoric components. It is also noted that the first- and second-order perturbation results are close to those of the first- and second-order Neumann expansion method, respectively. This implies that the accuracy of the perturbation method is comparable to the Neumann expansion simulation method of the same order although no explicit convergence criterion exists for the perturbation method.

In the case of Monte Carlo simulation, we must examine whether or not the sample size is large enough to represent the exact solution. The statistical fluctuation of the expected values and standard deviations evaluated by means of the direct Monte Carlo simulation is shown in Figs 9 and 10 as a function of the sample size (triangles connected by dashed lines). Observing these figures, especially Fig. 10, we decided to use a sample size equal to 100 for following reasons: (1)

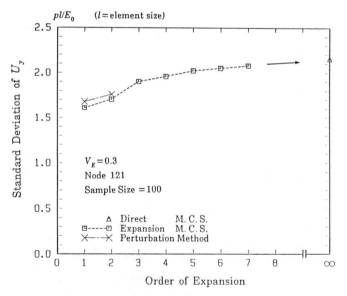

Fig. 7. Convergence of standard deviation of U_y (variation of E).

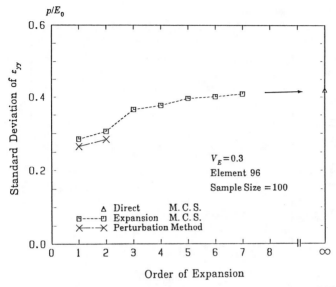

Fig. 8. Convergence of standard deviation of ε_{yy} (variation of E).

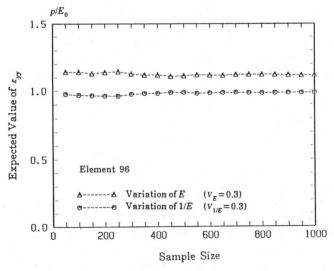

Fig. 9. Expected value of ε_{yy} by direct MCS as a function of sample size.

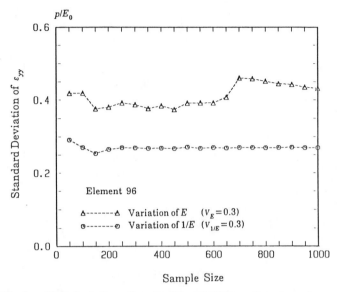

Fig. 10. Standard deviation of ε_{yy} by direct MCS as a function of sample size.

while the fluctuation is observed after 100 samples, it falls within a tolerable range (between 0·38 and 0·46); (2) Fig. 10 represents the worst case corresponding to a coefficient of variation V_E of Young's modulus equal to 0·3 (this value of V_E applied to Gaussian distribution before truncation); for smaller values of the coefficient of variation, statistical stability is much better; and most importantly, (3) with a sample size of 100, we can well demonstrate the difference in trend in which the standard deviation increases as V_E also increases. The expected value in Fig. 9 estimated by the direct Monte Carlo simulation method indicates that satisfactory stability can be obtained with even a smaller sample size and the fluctuation is much smaller.

In passing, it is mentioned that Fig. 10 also plots the standard deviations of ε_{yy} when $1/E$ rather than E is Gaussian (circles connected by dashed lines) and that in this case statistical stability is achieved much more rapidly and smoothly than in the case when E is Gaussian. Actually, when E is Gaussian with coefficient of variation V_E as large as 0·3, the statistical distribution of ε_{yy} becomes highly skewed as indicated by Fig. 11 (dashed line). This is due to the fact that the element stiffness can approach a very small value with a relatively

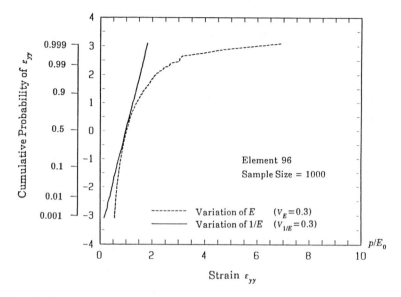

Fig. 11. Statistical distribution of ε_{yy} by direct MCS plotted on Gaussian probability paper.

large probability since V_E is as large as 0·3. On the other hand, the corresponding distribution function in the case of $1/E$ being Gaussian is close to Gaussian, as seen in Fig. 11 (solid line).

The direct and expansion MCS analyses and first-order and second-order perturbation analyses are performed for three values of V_E, 0·1, 0·2 and 0·3. Second-order perturbation analysis can be performed for only one element at a time because of the enormous CPU time and huge memory space the analysis requires.

The results of the computation by these four methods (direct MCS, expansion MCS, first- and second-order perturbation) are depicted in Figs 12–15 as a function of V_E. It is noted that in the range of small values of V_E, the results of all these methods are very close.

In Fig. 14, for example, a moderate increase is observed for the expected value of ε_{yy} as V_E increases. This observation applies to all the four methods used, except for the first-order perturbation method in which case the expected value remains constant regardless of the value of V_E. In Fig. 15, for example, the standard deviation of ε_{yy} is shown to increase linearly when estimated by the first-order perturbation

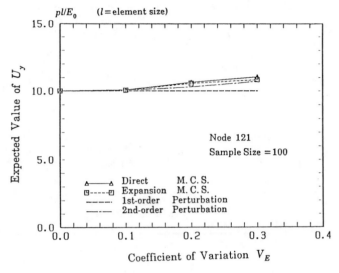

Fig. 12. Comparison of expected value of U_y (variation of E).

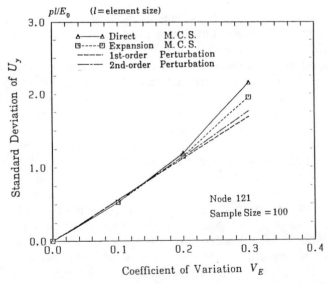

Fig. 13. Comparison of standard deviation of U_y (variation of E).

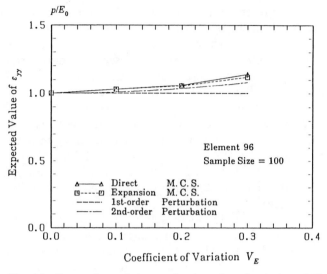

Fig. 14. Comparison of expected value of ε_{yy} (variation of E).

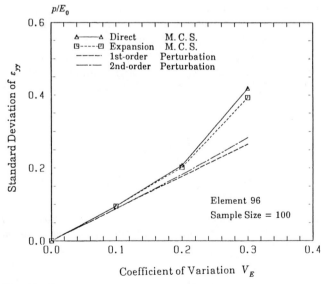

Fig. 15. Comparison of standard deviation of ε_{yy} (variation of E).

method. This represents a considerable underestimation of the standard deviation, which can be only slightly improved by the implementation of a second-order perturbation approximation as also observed in Fig. 15. This result obtained from the second-order perturbation approximation is disappointing, particularly because it requires an enormous amount of computational time as well as memory space. In contrast to the results provided by the perturbation methods, the standard deviation estimated by the Monte Carlo methods is shown in Fig. 15 to have an accelerated rate of increase as V_E increases. The estimated standard deviation by the expansion MCS method is close to that of the direct MCS method. Obviously, these two results can be made more close if we use a higher-order Neumann expansion at the expense of additional computer time.

It is observed in Fig. 15, for example, that perturbation methods, whether first- or second-order, underestimate the response variability for large values of V_E for which such a response variability analysis is particularly significant. It is important to note that this result is consistent with earlier results as demonstrated, for example, by Shinozuka & Astill,[10] Shinozuka & Wen[25] and Vaicaitis *et al.*[26]

Referring to the CPU time, Table 1 compares the CPU time on a VAX 11/750 in estimating the expected value and standard deviation for all the response quantities (displacement vectors as well as strain and stress tensors). The direct MCS method requires more time than the expansion MCS method does. How much more, however, depends on many factors including the order of expansion, variability of input (V_E), desired level of accuracy (eqn (30)), number of degrees of freedom (DOF) of the system, and bandwidth of the stiffness matrix.

Table 1
Comparison of CPU time (230 DOF model) (h : min : s)

Monte Carlo simulation (sample size = 100)		Perturbation method	
Direct method	*Neumann expansion method*	*First-order approximation*	*Second-order approximation*
00 : 15 : 05	00 : 09 : 01 (1st) 00 : 10 : 10 (2nd) 00 : 11 : 19 (3rd)	00 : 03 : 58	04 : 46 : 09 (for one element)

In the present example, the use of eqn (30) requires third-order expansion for the expansion MCS method in the case where $V_E = 0.3$ on the average. On the basis of counting the number of units of add–multiply operations involved in the algorithm mentioned earlier, it can be shown that, if the order of Neumann expansion is fixed and not too large, the expansion method will be more advantageous in terms of CPU time, as the number of DOF increases. For example, the ratio of CPU time required for the expansion MCS method to that for the direct MCS method is of the order of 1/50, when the sample size = 100, the order of expansion = 3, the number of DOF = 10 000 and the bandwidth = 1000 ~ 2500.

Table 1 also indicates that the first-order perturbation method requires the least amount of CPU time, as expected. However, second-order perturbation approximation, even for only one element, requires by far the greatest amount of CPU time, suggesting that this approximation method is quite impractical.

5.3 Example for Variation of 1/E

As the second model for the spatial variability of Young's modulus, the inverse of Young's modulus is assumed to be Gaussian as indicated below.

$$\frac{1}{E_e} = \frac{1}{E_0}(1 + \alpha_e) \tag{120}$$

The physical meaning of this assumption can be given as follows: if we try to measure the Young's modulus of a member of a structure, we will measure the displacements of the member under known load conditions. In such a situation, the measurable quantity is $1/E$, not E, thus assuming some distribution for $1/E$ is also reasonable. Such a model has been adopted by Vanmarcke and Grigoriu,[27] although the response variability of the particular structure they dealt with in that reference can be evaluated analytically, without recourse to the finite element or finite difference methods.

The response variabilities arising from these two different assumptions for spatial distributions of E are close if the standard deviation of α_e, σ_α, is not large. For assumed range of α_e such that $-1 + \varepsilon \leqslant \alpha_e \leqslant 1 - \varepsilon$ (eqn (119)), E_e is distributed between 0^+ and $2E_0^-$ in the case of variation of E. On the other hand, E_e is distributed between $0.5E_0^+$ and a very large value (E_0/ε) in the case of variation of $1/E$.

In the case of variation of $1/E$, the coefficient of variation $V_{1/E}$ of

$1/E$ is also σ_α. Assuming $|\alpha_e| < 1$, the expected value μ_E and standard deviation σ_E of E itself are derived by employing the expansion form of E_e (eqn (103)) as

$$\mu_E = E_0(1 + \sigma_\alpha^2 + 3\sigma_\alpha^4 + \ldots) \tag{121}$$

$$\sigma_E^2 = E_0^2(\sigma_\alpha^2 + 8\sigma_\alpha^4 + \ldots) \tag{122}$$

When obtaining the above equations, the following characteristics of Gaussian random variables are utilized.

$$E[\alpha_e^k] = 0 \ (k: \text{odd integer}); \ E[\alpha_e^2] = \sigma_\alpha^2; \ E[\alpha_e^4] = 3\sigma_\alpha^4 \tag{123}$$

Assuming $\sigma_\alpha = V_{1/E} = 0 \cdot 1, \ 0 \cdot 2, \ 0 \cdot 3, \ \mu_E$ is obtained as $1 \cdot 010, \ 1 \cdot 045,$ $1 \cdot 114E_0$ and σ_E is obtained as $0 \cdot 104, \ 0 \cdot 230, \ 0 \cdot 393E_0$, respectively (these values of $V_{1/E}$, μ_E and σ_E apply to the Gaussian distribution before truncation). Hence, the values of μ_E and σ_E are larger than those in the case of variation of E although the same variability of α_e is given for both cases.

The expected value and standard deviation of the strain under the assumption of variation of $1/E$ are depicted in Figs 16 and 17. It is indicated that both the standard deviation and expected value may be represented by a linear relationship with respect to the coefficient of variation $V_{1/E}$. These results can be estimated from the fact that the

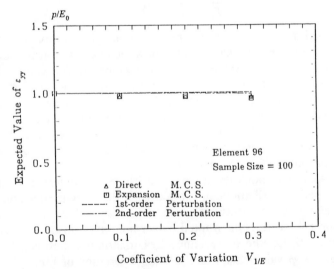

Fig. 16. Comparison of expected value of ε_{yy} (variation of $1/E$).

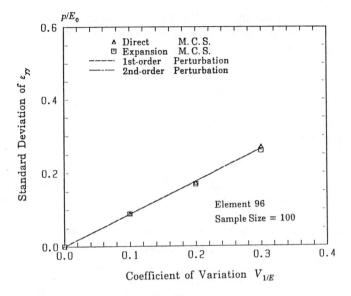

Fig. 17. Comparison of standard deviation of ε_{yy} (variation of $1/E$).

statistical strain distribution in Fig. 11 appears almost Gaussian, i.e. straight lines on Gaussian probability paper.

Such a linear relationship may be explained as follows: the displacement of static problems is obtained by $U = K^{-1}F$ and K^{-1} is highly correlated with the inverse of the elasticity matrix, $C_e = D_e^{-1}$, because a part of K (element stiffness matrix k_e) consists of the elasticity matrix D_e. In the case of a plane stress problem, C_e is known as

$$C_e = \frac{1}{E_e} \begin{bmatrix} 1 & -v_e & 0 \\ & 1 & 0 \\ \text{sym.} & & 2(1+v_e) \end{bmatrix} \tag{124}$$

Since C_e has a linear relationship with $1/E$, K^{-1} is also related linearly with $1/E$ at least in approximation. Once U has such a linear relationship with $1/E$, the same also applies to the strain by way of eqn (6).

In the case where the response distribution is close to the input distribution, the first-order perturbation method seems to yield a very good approximation of the exact solution, currently represented by the direct Monte Carlo simulation. It is suggested, however, that in order to verify the accuracy of the first-order perturbation analysis, a Monte Carlo simulation or analysis based on second-order approximation be performed at least once for each problem.

5.4 Example for Variation of ln E

As the third model for the spatial variability of Young's modulus, the natural logarithm of Young's modulus, ln E, is assumed to be Gaussian as below.

$$E_e = E_0 \exp(\alpha_e) \quad \text{or} \quad \ln E_e = \ln E_0 + \alpha_e \qquad (125)$$

where α_e is a Gaussian random variable which represents the variation of ln E at the centroid of element e with zero mean and standard deviation ζ_E. Then, the expected value λ_E of ln E is $\lambda_E = \ln E_0$ and the standard deviation of ln E is also ζ_E. There is an advantage that eqn (125) never generates negative values of E. Thus, a log-normal distribution is often considered to be a convenient idealization. Truncation of α_e in Monte Carlo simulation as represented by eqn (119) is not necessary and the perturbation method does not involve negative E if a log-normal distribution is assumed for E.

The expected value μ_E and standard deviation σ_E of E itself are derived as

$$\mu_E = \exp(\lambda_E + \tfrac{1}{2}\zeta_E^2) = E_0 \exp(\tfrac{1}{2}\zeta_E^2) \qquad (126)$$

$$\sigma_E^2 = \mu_E^2[\exp(\zeta_E^2) - 1] \qquad (127)$$

If we assume the variation of α_e in the same way as in previous sections ($\zeta_E = 0 \cdot 1$, $0 \cdot 2$, $0 \cdot 3$) and use the relationships of eqns (126) and (127), μ_E is obtained as $1 \cdot 005$, $1 \cdot 020$, $1 \cdot 046 E_0$ and σ_E is obtained as $0 \cdot 101$, $0 \cdot 206$, $0 \cdot 321 E_0$, respectively. For these assumed variations of E, numerical examples are demonstrated in Figs 18–21.

The variation of the standard deviations with respect to the sample size in the direct Monte Carlo simulation shown in Fig. 18 is very smooth and convergence is obtained when the sample size reaches 200. The strain distribution plotted in Fig. 19 indicates that it is also close to being log-normal. Hence, the standard deviations and expected values of the responses plotted as a function of the logarithmic standard deviation ζ_E can be approximated by a linear relationship, and the result of the first-order perturbation method gives a good agreement with those of Monte Carlo simulation methods as shown in Figs 20 and 21.

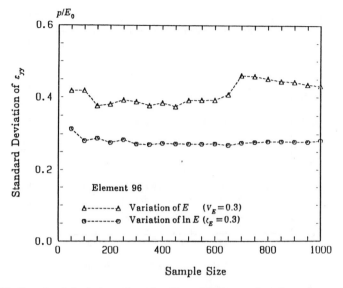

Fig. 18. Standard deviation of ε_{yy} by direct MCS as a function of sample size (variation of $\ln E$).

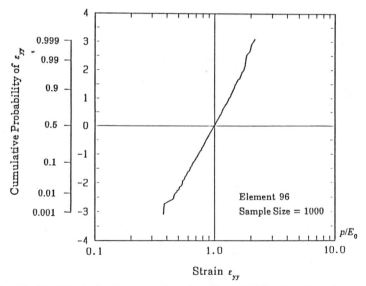

Fig. 19. Statistical distribution of ε_{yy} by direct MCS plotted on log-normal probability paper (variation of $\ln E$).

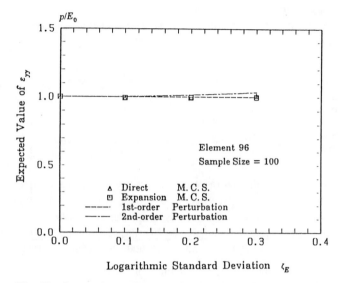

Fig. 20. Comparison of expected value of ε_{yy} (variation of $\ln E$).

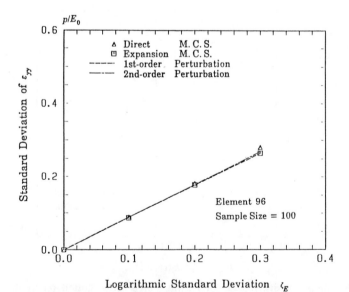

Fig. 21. Comparison of standard deviation of ε_{yy} (variation of $\ln E$).

5.5   Example for Variation of ν

The effects of spatial variation of Poisson's ratio are also investigated. The Poisson's ratio of element e is assumed to be

$$\nu_e = \nu_0(1 + \alpha_e) \tag{128}$$

where α_e is a Gaussian random variable with zero mean and standard deviation σ_α. Physically, Poisson's ratio should be positive and less than 0·5. In order to satisfy this condition, the range of α_e is limited in the case of Monte Carlo simulation as

$$-\tfrac{2}{3} < \alpha_e < \tfrac{2}{3} \tag{129}$$

Two values of the standard deviation of α_e (which is the coefficient of variation V_ν of ν), $\sigma_\alpha = 0\cdot1$ and $0\cdot2$, are used for numerical example. The maximum value of the standard deviation is smaller in this case than former cases because the possible range of α_e is narrower due to the condition specified in eqn (129). Young's modulus is assumed to be constant in this case.

Numerical examples are given in Figs 22–25 in the same form as in

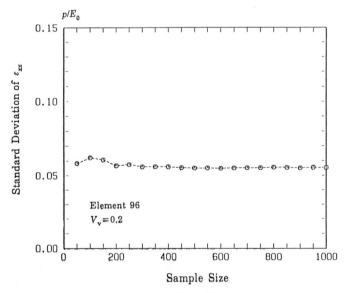

Fig. 22. Standard deviation of ε_{xx} by direct MCS as a function of sample size (variation of ν).

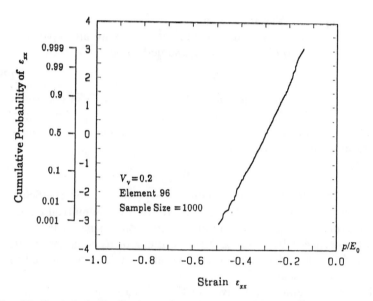

Fig. 23. Statistical distribution of ε_{xx} by direct MCS plotted on Gaussian probability paper (variation of v).

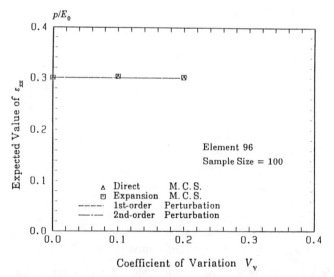

Fig. 24. Comparison of expected value of ε_{xx} (variation of v).

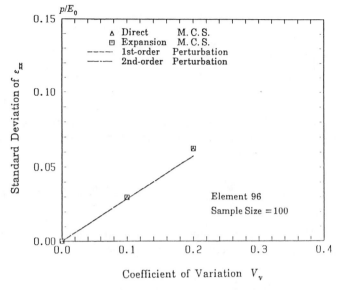

Fig. 25. Comparison of standard deviation of ε_{xx} (variation of v).

the other cases. Since the variation of output values is observed primarily in the transverse direction of the applied load in this case, only the response quantities predominant in the x-direction are shown.

The strain distribution shown in Fig. 23 can be approximated by a Gaussian distribution. This fact may result from the same reason as that in the case of variation of $1/E$, i.e. the inverse matrix of the elasticity matrix can be represented by a linear function of v_e as shown in eqn (124). Hence, the first-order perturbation method again produces a good approximation as verified by Monte Carlo simulation.

5.6 Example for NonGaussian Variation of E

As a final example of material property variability, Young's modulus is assumed to follow a beta distribution and it is represented by

$$E_e = E_0(1 + \alpha_e) \qquad (130)$$

where α_e is a random variable which represents the variation of E at the centroid of element e with zero mean and following the beta distribution.

A beta (β) distribution has the following probability density function:

$$f_A(\alpha) = \begin{cases} \dfrac{1}{B(q,r)} \dfrac{(\alpha - \alpha_{\min})^{q-1}(\alpha_{\max} - \alpha)^{r-1}}{(\alpha_{\max} - \alpha_{\min})^{q+r-1}} & (\alpha_{\min} \leq \alpha \leq \alpha_{\max}) \\ 0 & \text{(otherwise)} \end{cases}$$

(131)

in which $\alpha_{\min}$ and $\alpha_{\max}$ are the lower and upper bounds of the distribution, q and r are parameters of the distribution, and $B(q, r)$ is the beta function defined by

$$B(q, r) = \int_0^1 x^{q-1}(1-x)^{r-1}\,\mathrm{d}x = \frac{\Gamma(q)\Gamma(r)}{\Gamma(q+r)}$$

(132)

where $\Gamma(\cdot)$ is the gamma function. The probability distribution function corresponding to eqn (131) is obtained as

$$F_A(\alpha) = \frac{B_u(q, r)}{B(q, r)} \qquad (\alpha_{\min} \leq \alpha \leq \alpha_{\max})$$

(133)

in which $B_u(q, r)$ is the incomplete beta function which is defined by

$$B_u(q, r) = \int_0^u x^{q-1}(1-x)^{r-1}\,\mathrm{d}x$$

(134)

with u given by

$$u = \frac{(\alpha - \alpha_{\min})}{(\alpha_{\max} - \alpha_{\min})}$$

(135)

The mean and variance of the beta distribution are known as

$$\mu_\alpha = \alpha_{\min} + \frac{q}{q+r}(\alpha_{\max} - \alpha_{\min})$$

(136)

and

$$\sigma_\alpha^2 = \frac{qr(\alpha_{\max} - \alpha_{\min})^2}{(q+r)^2(q+r+1)}$$

(137)

There are four parameters, q, r, $\alpha_{\min}$ and $\alpha_{\max}$, in eqn (131). However, assuming that μ_α and σ_α^2 are specified and utilizing eqns (136) and (137), $\alpha_{\min}$ and $\alpha_{\max}$ can be written as

$$\alpha_{\min} = \mu_\alpha - \sigma_\alpha\sqrt{\frac{q(q+r+1)}{r}}; \qquad \alpha_{\max} = \mu_\alpha + \sigma_\alpha\sqrt{\frac{r(q+r+1)}{q}} \quad (138)$$

Since the beta distribution is bounded by the minimum value α_{min} and maximum value α_{max} and Young's modulus must be positive, we assume that $\alpha_{min} = -1$ and $\alpha_{max} = 1$ for the largest variability of Young's modulus $V_E = 0.3$ (the standard deviation σ_α of α is also 0·3). Thus, in this case, the truncation of the distribution is not necessary. Note, however, that not all the moments exists in this case. Assuming the symmetry of the probability density function ($q = r$), the parameters q and r for the beta distribution are obtained as $q = r = 5.056$. For the cases of smaller variability of Young's modulus ($V_E = 0.1$ and 0·2), the same values of q and r, $q = r = 5.056$, are also used. Then the range of the beta distribution becomes narrower by way of eqn (138), $\alpha_{min} = -0.333$ and $\alpha_{max} = 0.333$ for $V_E = 0.1$, and $\alpha_{min} = -0.667$ and $\alpha_{max} = 0.667$ for $V_E = 0.2$. The probability density and distribution functions for these three cases are depicted in Figs 26 and 27.

The probability density and distribution functions of the beta distribution for $V_E = 0.3$ are compared with those of the Gaussian distribution in Figs 28 and 29. These two distributions look similar. However, a difference is observed as α approaches to -1 where the beta distribution has almost zero density while the Gaussian distribution possesses nonzero density (this is the reason why the truncation is introduced for the Gaussian distribution).

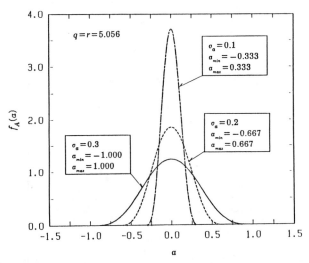

Fig. 26. Probability density functions of beta distribution.

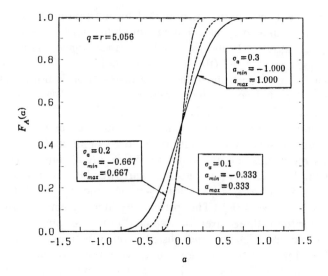

Fig. 27. Probability distribution functions of beta distribution.

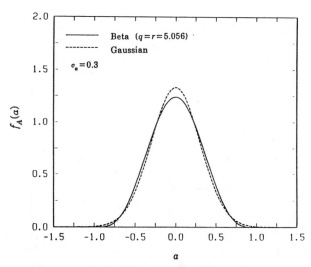

Fig. 28. Comparison of probability density function between beta and Gaussian distributions.

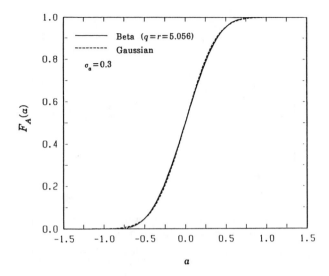

Fig. 29. Comparison of probability distribution function between beta and Gaussian distributions.

With the aid of the nonGaussian stochastic field simulation technique developed by Yamazaki & Shinozuka,[28] the sample fields of Young's modulus are generated and they are used for the direct and Neumann expansion Monte Carlo simulation. All the other conditions of the calculation are same as those for the previous examples. The results of the simulation analyses are shown in Figs 30–33. The fluctuation of the standard deviation of the strain ε_{yy} for the beta distribution is similar to that for the Gaussian distribution. Note that Gaussian sample fields are also generated by the spectral representation technique developed by Shinozuka[29] in Fig. 30. The statistical distribution of the strain plotted in Fig. 31 indicates that the strain obtained under the assumption of beta distribution for E is very close to that evaluated under the Gaussian assumption except for the range where the strain is very large. This exception results from the difference in their probability densities near $\alpha_e = -1$ as mentioned above.

The expected value and standard deviation when E follows the beta distribution are found to be close to those when E is Gaussian. In these figures, the first-order perturbation results are identical with

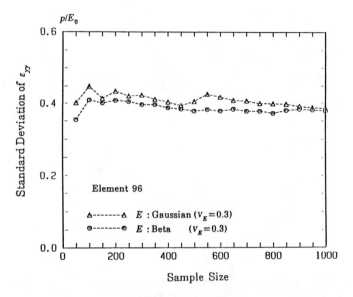

Fig. 30. Standard deviation of ε_{yy} by direct MCS as a function of sample size (E: beta distribution).

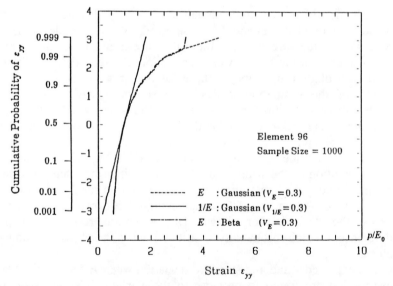

Fig. 31. Statistical distribution of ε_{yy} by direct MCS plotted on Gaussian probability paper (E: beta distribution).

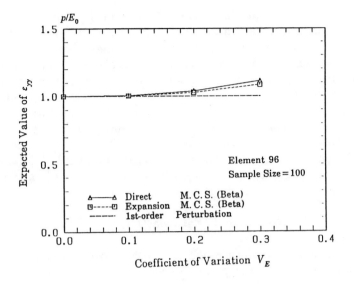

Fig. 32. Comparison of expected value of ε_{yy} (E: beta distribution).

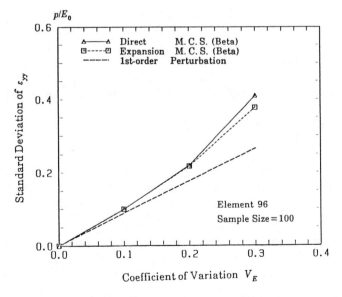

Fig. 33. Comparison of standard deviation of ε_{yy} (E: beta distribution).

those obtained for the case when E is Gaussian; the first-order perturbation approximation is not affected by the form of its distribution function as long as the stochastic fields have identical mean values and autocorrelation functions. However, second-order perturbation approximation cannot be constructed without further information about the third- and fourth-order moments of E, and it is not shown here. In this example where Young's modulus is assumed to follow a nonGaussian (beta) distribution, the first-order perturbation approximation underestimates the response variability when V_E becomes large. The result by the Neumann expansion method, however, exhibits good accuracy as verified by the direct Monte Carlo method.

6 CONCLUSION

The Neumann expansion method is used for evaluating the effect of spatially varying material properties within the framework of linear static structural analysis, and in conjunction with Monte Carlo simulation (MCS) methods. The advantage of this method is summarized as follows:

1. The expansion MCS method requires much smaller CPU time than does the direct MCS method. This reduction in CPU time results from the stiffness matrix factorization to be performed only once for the entire sample of structures.
2. The convergence of the series solution is guaranteed for each sample structure and can be confirmed numerically and graphically.
3. Existing deterministic finite element analysis programs can be readily utilized.

A first- and second-order approximation analysis program has also been developed based on the perturbation technique. Numerical examples have indicated the following features of the perturbation technique as compared with the simulation methods.

1. The first-order perturbation analysis shows a good agreement with the MCS result in the range of small variability of elastic moduli and requires a much smaller amount of CPU time than does the expansion MCS.
2. The first-order perturbation analysis does not agree with the

simulation result in the range of large variability of Young's modulus if Young's modulus is truncated Gaussian or follows a nonGaussian distribution, specifically a beta distribution.

3. The second-order perturbation analysis can improve the results of the first-order analysis only a little. In spite of this, the second-order analysis requires such an enormous amount of CPU time (much more than does the direct MCS solution) that it is indeed impractical.

4. The first-order perturbation analysis may be quite useful depending on the type of stochastic fields the material property in question subscribes to; for example, it provides a very good result when $1/E$, $\ln E$ and v are Gaussian. Nevertheless, it is strongly recommended that the results be verified by MCS techniques.

ACKNOWLEDGEMENT

This work was partially supported by the Ohsaki Research Institute, Shimizu Corporation, Tokyo, Japan and also partially supported by the National Center for Earthquake Engineering Research under NCEER Contract Number 86-3033 (NSF Master Contract Number ECE-86-07591).

REFERENCES

1. Cambou, B., Application of first-order uncertainty analysis in the finite element method in linear elasticity. *Proceedings of the 2nd International Conference on Application of Statistics and Probability in Soil and Structural Engineering,* Aachen, West Germany, Deutsche Gesellschaft für Grd-und Grundbau ev, Essen, FRG, 1975, pp. 67–87.
2. Baecher, G. B. & Ingra, T. S., Stochastic FEM in settlement predictions. *Journal of the Geotechnical Engineering Division, ASCE,* **107** (GT4) (1981) 449–63.
3. Handa, K. & Andersson, K., Application of finite element methods in the statistical analysis of structures. *Proceedings of the 3rd International Conference on Structural Safety and Reliability,* Elsevier, Amsterdam, 1981, pp. 409–17.
4. Hisada, T. & Nakagiri, S., Stochastic finite element method developed for structural safety and reliability. *Proceedings of the 3rd International Conference on Structural Safety and Reliability,* Elsevier, Amsterdam, 1981, pp. 395–408.

5. Nakagiri, S. & Hisada, T., *Introduction to Stochastic Finite Element Method*, Baifu-kan, Tokyo, 1985 (in Japanese).
6. Fox, R. L. & Kapoor, M. P., Rates of change of eigenvalues and eigenvectors. *AIAA Journal*, **6**(12) (1968) 2426–9.
7. Collins, J. D. & Thomson, W. T., The eigenvalue problem for structural systems with statistical properties. *AIAA Journal*, **7**(4) (1969) 642–8.
8. Hoshiya, M. & Shah, H. C., Free vibration of stochastic beam-column *Journal of the Engineering Mechanics Division, ASCE*, **97** (EM4) (1971) 1239–55.
9. Hasselman, T. K. & Hart, G. C., Modal analysis of random structural systems. *Journal of the Engineering Mechanics Division, ASCE*, **98** (EM3) (1972) 561–79.
10. Shinozuka, M. & Astill, C. J., Random eigenvalue problems in structural mechanics. *AIAA Journal*, **10**(4) (1972) 456–62.
11. Liu, W. K., Belytschko, T. & Mani, A., Probabilistic finite elements for nonlinear structural dynamics. *Computer Methods in Applied Mechanics and Engineering*, **56** (1986) 61–81.
12. Liu, W. K., Belytschko, T. & Mani, A., Random field finite elements. *International Journal for Numerical Methods in Engineering*, **23** (1986) 1831–45.
13. Hisada, T. & Nakagiri, S., Role of stochastic finite element method in structural safety and reliability. *Proceedings of the 4th International Conference on Structural Safety and Reliability*, Kobe, Japan, IASSAR, New York, 1985, pp. I 385–94.
14. Vanmarcke, E., Shinozuka, M., Nakagiri, S., Schuëller, G. I. & Grigoriu, M., Random fields and stochastic finite element. *Structural Safety*, **3** Elsevier, Amsterdam (1986) 143–66.
15. Shinozuka, M. & Deodatis, G., Response variability of stochastic finite element systems. *Stochastic Mechanics*, Vol. I, Department of Civil Engineering and Engineering Mechanics, Columbia University, 1986.
16. Harada, T. & Shinozuka, M., The scale of correlation for stochastic fields. Technical Report, Department of Civil Engineering and Engineering Mechanics, Columbia University, 1986.
17. Tatarski, V. I., *Wave Propagation in a Turbulent Medium*, McGraw-Hill, New York, 1961.
18. Monin, A. S. & Yaglom, A. M., *Statistical Fluid Mechanics: Mechanics of Turbulence*, ed. J. L. Lumley. MIT Press, Cambridge, Mass., 1971.
19. Lumley, J. L., *Stochastic Tools in Turbulence*, Academic Press, New York, 1970.
20. Vanmarcke, E., *Random Fields—Analysis and Synthesis*, MIT Press, Cambridge, Mass., 1983.
21. Zienkiewicz, O. C., *The Finite Element Method*, 3rd edn, McGraw-Hill, London, 1977.
22. Bathe, K.-J., *Finite Element Procedures in Engineering Analysis*, Prentice-Hall, Englewood Cliffs, N.J., 1982.
23. Inference Corporation, *SMP—A Symbolic Manipulation Program Reference Manual*, Version 1, 1983.

24. Yamazaki, F., Shinozuka, M. & Dasgupta, G., Neumann expansion for stochastic finite element analysis. *Stochastic Mechanics*, Vol. I, Department of Civil Engineering and Engineering Mechanics, Columbia University, 1986.
25. Shinozuka, M. & Wen, Y-K., Monte Carlo simulation of nonlinear vibrations. *AIAA Journal*, **10**(1) (1972) 37–40.
26. Vaicaitis, R., Jan, C-M. & Shinozuka, M., Nonlinear panel response and noise transmission from a turbulent boundary layer by a Monte Carlo approach. *AIAA Journal*, **10**(7) (1972) 895–9.
27. Vanmarcke, E. & Grigoriu, M., Stochastic finite element analysis of simple beams. *Journal of the Engineering Mechanics Division, ASCE*, **109**(5) (1983) 1203–14.
28. Yamazaki, F. & Shinozuka, M., Digital generation of non-Gaussian stochastic fields. *Stochastic Mechanics*, Vol. I, Department of Civil Engineering and Engineering Mechanics, Columbia University, 1986.
29. Shinozuka, M., Stochastic fields and their digital simulation. Lecture Notes for the CISM Course on Stochastic Methods in Structural Mechanics, Udine, Italy, 1985.

15

Noise Transmission Through Nonlinear Sandwich Panels

R. Vaicaitis

Department of Civil Engineering, Columbia University, New York, USA

&

H.-K. Hong

Department of Civil Engineering, National University of Taiwan, Taipei, Taiwan

ABSTRACT

An analytical study is presented to predict noise transmission through nonlinear double wall sandwich panel systems subject to random loading. Viscoelastic and nonlinear spring dashpot models are chosen to characterize the behavior of the soft core. The noise transmission through this panel system is determined into an acoustic enclosure of which the interiors are covered with porous absorption materials. The absorbent boundary conditions of the enclosure are accounted for by a two-step transformation of the boundary effect into a wave equation which governs the acoustic pressure field. The nonlinear panel response and interior acoustic pressure are obtained by utilizing modal analyses and Monte Carlo simulation techniques. Numerical results include the nonlinear response time histories and noise reduction. It is found that by proper selection of the dynamic parameters and damping characteristics, the noise transmission can be significantly reduced by the double wall sandwich construction.

NOTATION

a, b, d	Dimensions of rectangular enclosure
A^B_{mn}	Generalized coordinates of bottom plate

B	Bulk reaction coefficient of absorbing layer
c	Acoustic damping coefficient
c_0	Speed of sound
c_1, c_2, c_3	Damping coefficients (per unit area) of nonlinear core
$G(z)$	Function chosen to modify acoustic boundary conditions
h_B, h_S, h_T	Thickness of bottom plate, core, and top plate, respectively
i, j, k, m, n, r	Structural and acoustic modal indices
$i = \sqrt{-1}$	Imaginary unit
k_1, k_2, k_3	Stiffness coefficients (per unit area) of nonlinear core
k_A, k_F, k_S	Stiffness coefficients (per unit area) of absorbing layer, flexible wall, and soft core, respectively
K_S	Constitutive law operator of viscoelastic core
L_x, L_y	Dimensions of plate
NR	Noise reduction
$NR_{1/3}$	1/3-octave noise reduction
p	Acoustic pressure field inside enclosure
p_0	Reference pressure $= 2 \cdot 9 \times 10^{-9}$ psi $= 20 \, \mu N/m^2$
p_E	External noise pressure acting on the double wall system
$s(t)$	Volume flow rate of a point source in enclosure
S_p	Spectral density of acoustic pressure inside enclosure
S_p^E	Prescribed spectral density of external pressure
SPL	Sound pressure level measured in decibels relative to the reference pressure p_0
t	Time
w	Displacement at fluid absorbing layer interface
w_A, w_S	Deformations (or relative displacements) of absorbing layer and soft core, respectively
w_B, w_F, w_T	Deflections of bottom plate, flexible wall, and top plate, respectively
$\mathbf{x} = x, y, z$	Cartesian coordinate system for enclosure; see Fig. 1(a)

$\mathbf{x}_0 = x_0, y_0, z_0$ Location of point source of volume flow

X_{mn} Modes of top or bottom plate

Y_{ijk} Acoustic modes

Z Impedance of absorbing layer

Z_F Impedance (operator) of flexible wall

∇^2 (3-D) Laplacian operator

∇_s^2 (2-D) Laplacian operator to be taken on boundary surface

∇^4 Biharmonic operator

ω Circular frequency

ω_l, ω_u Lower and upper bound frequencies, respectively

$^-$ Fourier transform

1 INTRODUCTION

The information available in the literature and from ongoing research on noise transmission into various transportation vehicles exposed to high intensity inputs indicate a need to reduce these noise levels to acceptable limits. This is especially important for the new proposed fuel efficient advanced (ATP) prop-fan aircraft where the exterior noise levels are very high in the low frequency region (below 1000 Hz). It is well known that acoustic absorption materials used for noise treatments are not very effective at these low frequencies.[1-5]

Preliminary noise transmission studies tend to indicate that an aircraft sidewall constructed from multilayered treatments might give the required noise attenuation in the low frequency region.[6-9] However, double walls, multilayered and sandwich type constructions show inherent resonances in the low frequency region resulting in large vibration amplitudes and low noise attenuation. By minimizing the structural dynamic coupling between the face layers, using soft and highly dissipative core materials, the response at resonance might be reduced. Past studies have demonstrated that sandwich panels can achieve a significant amount of response and noise reduction.[10,11] However, these studies were limited to linear structures and idealized acoustic models. Under high intensity inputs such as prop-fan noise, soft core materials and/or thin face plate constructions, linear theory becomes invalid. A preliminary analytical study of noise transmission through nonlinear sandwich panels is undertaken in this paper.

The noise transmission through the double wall sandwich construction shown in Fig. 1(a) is obtained by solving the linearized wave equation for the interior sound pressure field. Time domain and frequency domain formulations are considered. In the time domain approach, the time dependent boundary conditions are transformed into the governing equation and then the solution of the resulting

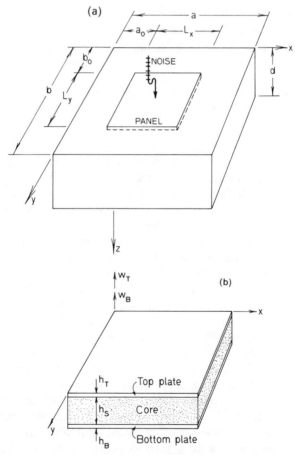

Fig. 1. Geometries of (a) an acoustic enclosure and (b) a double wall sandwich panel system.

nonhomogeneous differential equation with homogeneous boundary conditions is obtained. The effect of wall absorption is accounted for by utilizing point impedance and bulk reacting models.[12,13] The interior acoustic pressure is ultimately expressed in terms of spectral density functions and a quantity relating the external pressure spectral density to the pressure spectral density at points inside the enclosure is defined and called noise reduction.

The objective of the present work is to develop an analytical model capable of predicting noise transmission through a nonlinear double wall sandwich construction and to demonstrate that such a model could be used to minimize interior noise. To achieve this, the noise field in the enclosure is coupled to the vibrations of the interior (bottom) plate which, in turn, is coupled via the soft core to the vibration of the exterior (top) plate. Through proper selection of the dynamic parameters and damping characteristics, the response of the bottom plate can be reduced even though vibration levels of the top plate are large. The ultimate result is lower noise levels in the enclosure in comparison to the noise levels of the linear case under similar conditions.

2 NOISE TRANSMISSION THROUGH DOUBLE WALL SANDWICH PANELS

Consider a rectangular acoustic space occupying a volume $V = abd$ as shown in Fig. 1(a). The interior surface of the enclosure is assumed to be covered with absorption materials for which the impedance characteristics are known. A portion of the surface $z = 0$ is taken to be flexible while the remaining walls are assumed to be acoustically rigid. The flexible surface is the double wall sandwich construction shown in Fig. 1(b). Noise enters the acoustic enclosure through the vibration of the double wall panel system. The perturbation pressure p inside the enclosure satisfies the linear acoustic wave equation

$$\nabla^2 p - c\dot{p} - \ddot{p}/c_0^2 + \rho\dot{s}\delta(\mathbf{x} - \mathbf{x}_0) = 0 \qquad (1)$$

where ∇^2 is the Laplacian operator $\nabla^2 = \partial^2/\partial x^2 + \partial^2/\partial y^2 + \partial^2/\partial z^2$ and δ is the Dirac delta function which locates the point source of volume flow.

The types of boundary conditions to be satisfied by eqn (1) depend on the interior surface conditions of the walls. These could range from those of acoustically hard walls to those of highly absorbent walls which are treated with acoustic insulation materials. On the rigid boundaries of an acoustic region

$$\partial p / \partial n = 0 \qquad (2)$$

where n is the outward normal to the boundary. At the interface, the fluid velocity and the wall motion are equal. The momentum equation gives the following boundary condition

$$\partial p / \partial n = -\rho \ddot{w} \qquad (3)$$

where $\ddot{w}$ is the acceleration normal to the boundary.

However, the boundary conditions given in eqns (2) and (3) do not allow for the acoustic energy dissipation at the boundary. For interior surfaces treated with acoustic insulation materials, the effect of absorption at the boundary needs to be included. One of the most commonly used is the point reaction (or impedance) model

$$Z(\omega)\dot{w}_A = p \qquad (4)$$

where $\dot{w}_A$ is the normal velocity at the absorbing boundary and $Z(\omega)$ is the specific acoustic point impedance and because of the time lag involved, it is composed of real and imaginary parts. Combining eqns (3) and (4) yields the boundary conditions of the point reacting wall

$$\partial p / \partial n = -\rho \dot{p} / Z(\omega) \qquad (5)$$

The point reaction model assumes that the fluid–solid interaction is only locally dependent and ignores the interaction from other surface regions. Such an idealization might not properly simulate the fluid–solid interaction of many porous materials.[12,13] A local bulk reaction model has been proposed to account for the aforementioned phenomenon.[12] In this formulation, a complex bulk reaction coefficient $B(\omega)$ is introduced such that

$$Z(\omega)\dot{w}_A = p + B(\omega) \nabla_s^2 p \qquad (6)$$

where ∇_s^2 is the Laplacian at the surface of the enclosure. Then, from eqns (3) and (6),

$$\partial p / \partial n = -\rho(\dot{p} + B(\omega) \nabla_s^2 \dot{p}) / Z(\omega) \qquad (7)$$

At the boundaries where a sound absorbing material is attached to a flexible vibrating surface as shown in Fig. 2(a), the interface displace-

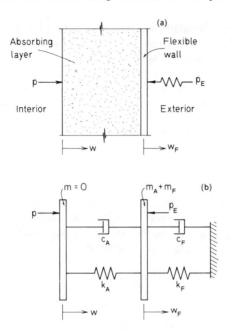

Fig. 2. (a) Absorbing acoustic material mounted on a flexible wall and (b) lumped-mass idealization of an acoustically treated wall.

ment can be written as

$$w = w_F + w_A \tag{8}$$

where w_F is the flexible wall motion and w_A is the deformation of the absorption material. From eqns (3), (4) and (8), the boundary condition for a point reacting material mounted on a flexible wall is

$$\partial p / \partial n = -\rho \ddot{w}_F - \rho \dot{p} / Z(\omega) \tag{9}$$

For a bulk reaction model, the condition is determined from eqns (3), (6) and (8)

$$\partial p / \partial n = -\rho \ddot{w}_F - \rho (\dot{p} + B(\omega) \, \nabla_s^2 \dot{p}) / Z(\omega) \tag{10}$$

Either eqn (9) or (10) represents the boundary condition needed to obtain solutions to the interior acoustic pressure for flexible vibrating walls and absorbing interiors. The boundary condition given in eqn (10) can be interpreted as a general representation of a flexible boundary with a porous material–fluid interface. The solutions for the interior pressure p will be obtained utilizing this boundary condition.

With regard to some special cases, it is advantageous to express the acoustic pressure p in terms of the flexible wall accelerations $\ddot{w}_F$.[14] To do this, replace the system shown in Fig. 2(a) by a simple lumped-mass model as illustrated in Fig. 2(b) where p is the interior enclosure pressure and p_E is the external random pressure acting on the exterior surface of the flexible wall. From this model, two equations of motion are formulated

$$p - c_A \dot{w}_A - k_A w_A = 0 \tag{11}$$

$$p_E + (m_A + m_F)\ddot{w}_F + c_F \dot{w}_F + k_F w_F - c_A \dot{w}_A - k_A w_A = 0 \tag{12}$$

where m_A, c_A and k_A are the mass, damping and stiffness per unit surface area of the absorbing layer and the subscript F refers to the flexible vibrating wall. By definition the impedances of the absorbing layer and the flexible wall can be written as

$$Z \equiv i\omega m_A + c_A + k_A/i\omega \tag{13}$$

$$Z_F \equiv i\omega m_F + c_F + k_F/i\omega \tag{14}$$

respectively. Then, from eqns (11)–(14),

$$p = p_E + (i\omega m_A + Z_F)\dot{w}_F \tag{15}$$

Substituting eqn (15) into eqns (9) and (10) results in boundary conditions at the fluid–absorption material interface

$$\partial p/\partial n = -\rho \dot{p}_E/Z - \rho(i\omega m_A + Z + Z_F)\ddot{w}_F/Z \tag{16}$$

$$\partial p/\partial n = -\rho(\dot{p}_E + B\,\nabla_s^2 \dot{p}_E)/Z - \rho(i\omega m_A + Z + Z_F)\ddot{w}_F/Z$$
$$- \rho B(i\omega m_A + Z_F)\,\nabla_s^2 \ddot{w}_F/Z \tag{17}$$

If the flexible wall is modeled as a linear elastic thin plate, its impedance becomes an operator

$$Z_F = i\omega m_F + c_F + D_F\,\nabla^4/i\omega \tag{18}$$

where D_F is the plate stiffness and ∇^4 is the biharmonic operator.

Equations (9) and (10) demonstrate that the boundary conditions for the acoustic pressure p are inhomogeneous and time dependent. Due to the nonlinear response of the sandwich panel, a time domain analysis is employed where a homogeneous differential equation with nonhomogeneous boundary conditions is transformed into a non-homogeneous differential equation with homogeneous boundary conditions.[15,16] The details of the linear response solution can be found in Refs 10, 11 and 17.

The perturbation pressure inside the enclosure is determined by solving eqn (1) together with the boundary conditions

$$-\partial p/\partial z = -\rho[\dot{p} + B(\omega)(\partial^2/\partial x^2 + \partial^2/\partial y^2)\dot{p}]/Z(\omega) - \rho\ddot{w}_B \quad \text{at} \quad z = 0 \tag{19}$$

$$\partial p/\partial n = -\rho[\dot{p} + B(\omega)\nabla_s^2\dot{p}]/Z(\omega) \quad \text{otherwise} \tag{20}$$

Equations (19) and (20) demonstrate that the boundary conditions for the pressure p on $z = 0$ is nonhomogeneous. The solution can be achieved by first transforming the time dependent boundary acceleration input $\rho\ddot{w}_B$ into the governing equation, eqn (1), by introducing a function $G(z)$ such that

$$p(x, y, z, t) = q(x, y, z, t) + \rho\ddot{w}_B(x, y, t) \cdot G(z) \tag{21}$$

where $G(z)$ is a function chosen to modify the given boundary conditions. Thus, a boundary value problem of an inhomogeneous differential equation in q is established which can be solved using the resulting homogeneous boundary conditions. Furthermore, by utilizing Green's theorem, the effect of the absorption materials as given in eqns (19) and (20) can also be transferred into a governing equation in q.[17]

Let the function $G(z)$ have the following form

$$G(z) = z - 2z^2/d + z^3/d^2 \tag{22}$$

Substituting eqn (21) into eqns (19) and (20) yields

$$\nabla^2 q - c\dot{q} - \ddot{q}/c_0^2 - \rho\ddot{f} + \rho\dot{s}\delta(\mathbf{x} - \mathbf{x}_0) = 0 \tag{23}$$

in the enclosure and

$$\partial q/\partial n = -\rho[\dot{q} + B(\omega) \nabla_s^2\dot{q}]/Z(\omega) \tag{24}$$

at all the boundaries where a forcing function has been defined as

$$f(x, y, z, t) = (-\partial^2 w_B/\partial x^2 - \partial^2 w_B/\partial y^2 + \ddot{w}_B/c_0^2 + c\dot{w}_B) \cdot G - w_B \, \mathrm{d}^2G/\mathrm{d}z^2 \tag{25}$$

In obtaining the boundary conditions given by eqn (24) at the walls $x = 0$, a and $y = 0$, b, it was assumed that the flexible wall motions w_B do not extend to these boundaries. Thus, the flexible plate can be located anywhere in the region $\varepsilon_0 \leqslant x \leqslant a - \varepsilon_0$, $\varepsilon_0 \leqslant y \leqslant b - \varepsilon_0$ where ε_0 could be a small positive number but $\varepsilon_0 \neq 0$.

The hard wall acoustic modes Y_{ijk} satisfy

$$\nabla^2 Y_{ijk} + (\omega_{ijk}/c_0)^2 Y_{ijk} = 0 \tag{26}$$

in the enclosure and

$$\partial Y_{ijk}/\partial n = 0 \tag{27}$$

at all the boundaries where ω_{ijk} are the corresponding modal frequencies. From Green's theorem

$$\int_V (q \, \nabla^2 Y_{ijk} - Y_{ijk} \, \nabla^2 q) \, dV = \int_S \left(q \frac{\partial Y_{ijk}}{\partial n} - Y_{ijk} \frac{\partial q}{\partial n} \right) dS \tag{28}$$

where V and S indicate volume and surface integrals, respectively. Then utilizing eqns (23), (24) and (26)–(28), it can be shown that

$$\ddot{Q}_{ijk} + 2\zeta_{ijk}\omega_{ijk}\dot{Q}_{ijk} + \omega_{ijk}^2 Q_{ijk} + \rho c_0^2 \ddot{F}_{ijk} - \rho c_0^2 \dot{S}_{ijk}$$
$$+ \frac{\rho c_0^2}{Z(\omega)} \frac{\partial}{\partial t} \int_A [q + B(\omega) \, \nabla_s^2 q] Y_{ijk} \, dA = 0 \tag{29}$$

where

$$Q_{ijk}(t) = \int_V q Y_{ijk} \, dV \tag{30}$$

$$S_{ijk}(t) = \int_V s\delta(\mathbf{x} - \mathbf{x}_0) Y_{ijk} \, dV = s(t) Y_{ijk}(x_0, y_0, z_0) \tag{31}$$

$$F_{ijk}(t) = \int_V f Y_{ijk} \, dV \tag{32}$$

$$\zeta_{ijk} = cc_0^2/(2\omega_{ijk}) \tag{33}$$

and A indicates that the integration is taken over the absorbing interior surface.[17,18] Upon expanding the function q in terms of the acoustic modes, utilizing the orthogonality principle and using eqn (30)

$$q = \frac{8}{abd} \sum_{i=0}^{\infty} \sum_{j=0}^{\infty} \sum_{k=0}^{\infty} Q_{ijk} Y_{ijk}/(e_i e_j e_k) \tag{34}$$

where

$$e_i = \begin{cases} 2 & i = 0 \\ 1 & i \neq 0 \end{cases} \tag{35}$$

Substitution of eqn (34) into eqn (29) results in

$$\ddot{Q}_{ijk} + 2\zeta_{ijk}\omega_{ijk}\dot{Q}_{ijk} + \omega_{ijk}^2 Q_{ijk} + (2\rho c_0^2/Z)\Bigg\{(b_{0jk}/a)\sum_{r=0}^{\infty}[1+(-1)^{r+i}]\dot{Q}_{rjk}/e_r$$

$$+ (b_{i0k}/b)\sum_{s=0}^{\infty}[1+(-1)^{s+j}]\dot{Q}_{isk}/e_s + (b_{ij0}/d)\sum_{u=0}^{\infty}[1-(-1)^{u+k}]\dot{Q}_{iju}/e_u\Bigg\}$$

$$+ \rho c_0^2\ddot{F}_{ijk} - \rho c_0^2\dot{S}_{ijk} = 0 \tag{36}$$

where

$$b_{ijk} = 1 - B(\omega)(\omega_{ijk}/c_0)^2 \tag{37}$$

The modal forcing functions F_{ijk} can be obtained from eqns (25), (32) and the modal solutions of the flexible panel motion w_{B}.

$$w_{\mathrm{B}}(x, y, t) = \sum_{m=1}^{\infty}\sum_{n=1}^{\infty} A_{mn}^{\mathrm{B}}(t)X_{mn}(x, y) \tag{38}$$

where A_{mn}^{B} is the generalized coordinate of the bottom plate, and $X_{mn} = \sin(m\pi x/L_x)\sin(n\pi y/L_y)$.

Hence

$$F_{ijk}(t) = \sum_{m=1}^{\infty}\sum_{n=1}^{\infty} L_{ijmn}\{C_k A_{mn}^{\mathrm{B}}(t) + D_k[((m\pi/L_x)^2$$

$$+ (n\pi/L_y)^2)A_{mn}^{\mathrm{B}}(t) + \ddot{A}_{mn}^{\mathrm{B}}(t)/c_0^2 + 2\zeta_{ijk}\omega_{ijk}\dot{A}_{mn}^{\mathrm{B}}/c_0^2]\} \tag{39}$$

where

$$C_k = \begin{cases} 6[1-(-1)^k]/(k\pi)^2 & k \neq 0 \\ 1 & k = 0 \end{cases} \tag{40}$$

$$D_k = \begin{cases} d^2(C_k - 1)/(k\pi)^2 & k \neq 0 \\ d^2/12 & k = 0 \end{cases} \tag{41}$$

The solution for the generalized coordinates A_{mn}^{B} corresponding to the bottom plate motion of the double wall system is given in Refs 19 and 20.

The information that is usually available on the point impedance Z and the bulk reaction coefficient B is given in a frequency domain. Thus it is advantageous to express eqn (36) in a frequency domain.

Taking the Fourier transformation of eqns (36) and (39)

$$\bar{Q}_{ijk}(\omega_{ijk}^2 - \omega^2 + 2i\omega\zeta_{ijk}\omega_{ijk}) + (2\rho c_0^2 i\omega/Z)$$

$$\left\{(b_{0jk}/a) \sum_{r=0}^{\infty} [1 + (-1)^{r+i}]\bar{Q}_{rjk}/e_r\right.$$

$$+ (b_{iok}/b) \sum_{s=0}^{\infty} [1 + (-1)^{s+j}]\bar{Q}_{isk}/e_s$$

$$\left. + (b_{ij0}/d) \sum_{u=0}^{\infty} [1 + (-1)^{u+k}]\bar{Q}_{iju}/e_u\right\}$$

$$= \rho c_0^2\{i\omega\bar{S}_{ijk} + \omega^2\bar{F}_{ijk}\} \qquad (42)$$

where

$$\bar{F}_{ijk}(\omega) = \sum_{m=1}^{\infty} \sum_{n=1}^{\infty} L_{ijmn}\bar{A}_{mn}^{B}\{C_k + D_k[(m\pi/L_x)^2 + (n\pi/L_y)^2$$

$$- (\omega/c_0)^2 + 2i\omega\zeta_{ijk}\omega_{ijk}/c_0^2]\} \qquad (43)$$

and a bar indicates a transformed quantity. When the response of the double wall system is linear, the amplitudes $\bar{A}_{mn}^{B}$ can be obtained in a closed form. However, for nonlinear vibrations, the amplitudes are determined first in the time domain by numerical procedures and then an FFT technique is applied to calculate $\bar{A}_{mn}^{B}$.[19]

Finally, from eqns (21) and (34), the solution for the interior pressure is

$$\bar{p}(x, y, z, \omega) = \frac{8}{abd} \sum_{i=0}^{\infty} \sum_{j=0}^{\infty} \sum_{k=0}^{\infty} Y_{ijk}(x, y, z)\bar{Q}_{ijk}(\omega)/e_ie_je_k)$$

$$- \omega^2\rho G(z)\bar{w}_B(x, y, \omega) \qquad (44)$$

where the solutions to $\bar{Q}_{ijk}$ are determined from eqn (42). The spectral density S_p of the acoustic pressure given in eqn (44) can be obtained by procedures similar to those presented in Refs 10 and 21. The sound pressure given in eqn (44) is a function of the nonlinear vibrations of the bottom plate, w_B. A detailed analysis on nonlinear response of double wall sandwich panels is presented in Refs 19 and 20.

The sound pressure levels inside the enclosure measured in decibels relative to a reference pressure p_0 are determined by

$$\text{SPL}(x, y, z, \omega) = 10 \log\{S_p(x, y, z, \omega) \, \Delta\omega/p_0^2\} \qquad (45)$$

where $\Delta\omega$ is a selected bandwidth at which the spectral density S_p is estimated. A quantity relating the spectral density of the enclosure pressure S_p to the spectral density of the external pressure $S_p^E(\omega)$ is noise reduction NR which is defined as

$$NR(x, y, z, \omega) = 10 \log\{S_p^E(\omega)/S_p(x, y, z, \omega)\} \qquad (46)$$

It is convenient to define the sound pressure level and noise reduction on a 1/3-octave scale. Thus

$$NR_{1/3}(x, y, z, \omega) = 10 \log\left\{ \frac{\displaystyle\int_{\omega_l}^{\omega_u} S_p^E(\omega)\, d\omega}{\displaystyle\int_{\omega_l}^{\omega_u} S_p(x, y, z, \omega)\, d\omega} \right\} \qquad (47)$$

where ω_l and ω_u are the lower and upper bounds, respectively, of each 1/3-octave band.

3 RANDOM PRESSURE INPUT

In the formulations of the present study, the external noise acting on a structural panel is described, in general, by random pressure process $p_E(x, y, t)$. To provide the random response analysis of structures with nonlinearities and/or nonstationary inputs, a time domain Monte Carlo type analysis is needed. Simulation techniques for multidimensional and multivariate Gaussian random processes with prescribed power spectral densities or evolutionary spectra can be utilized to generate sample inputs.[22,24] These simulations are greatly expedited by incorporating the Fast Fourier Transform (FFT) technique.[23,25] In the present study, the random input pressures are assumed to be Gaussian and are simulated according to the procedure given in Ref. 22.

4 NUMERICAL RESULTS

The numerical results presented herein correspond to the acoustic enclosure and double wall sandwich panel system shown in Figs 1(a) and 1(b). It is assumed that noise enters only through the double wall panel; noise leakage or noise generated directly inside the enclosure or at the boundaries is not included. The following set of parameters are selected for the study.

The dimensions of the double wall panel and the enclosure are $L_x = 20$ in, $L_y = 10$ in, $a = 72$ in, $b = 52$ in, $d = 50$ in, (Fig. 1(a)). The panel is located on the sidewall at $z = 0$, $a_0 = 26$ in, $b_0 = 21$ in. The structural response is computed at the center of the plate, i.e. $x = 10$ in, and $y = 5$ in. The transmitted noise is calculated at $x = 36$ in, $y = 26$ in, $z = 10$ in. The top and bottom face plates are assumed to be made of aluminum with material densities $\rho_T = \rho_B = 0.000\,251$ lbf-sec^2/in^4, moduli of elasticity $E_T = E_B = 10^7$ psi and Poisson's ratio $\nu_T = \nu_B = 0.3$. The structural modal damping ratios of each plate are calculated from

$$\zeta_{mn} = \zeta_{11}(\omega_{11}/\omega_{mn}) \tag{48}$$

where the ω_{mn} are the uncoupled natural frequencies.[19,20] The fundamental modal damping ratio is taken to be $\zeta_{11} = 0.02$.

The air density and speed of sound in the enclosure are $\rho = 1.147 \times 10^{-7}$ lbf/sec^{-2}/in^4 and $c_0 = 13\,540$ in/sec, respectively. The acoustic modal damping ratios are taken as

$$\zeta_{ijk} = \zeta_0(\omega_0/\omega_{ijk}) \tag{49}$$

where ω_0 is the lowest acoustic modal frequency chosen from ω_{001}, ω_{010} and ω_{100}. The damping ratio corresponding to the fundamental acoustic mode, $\zeta_0 = 0.005$, takes into account the contributions of all the damping effects in the interior of the acoustic space. The effect of the absorption materials mounted on the six walls is allowed for separately and will be discussed later.

The input random pressure p_E acting on the top plate of the double wall panel system is taken to be that of truncated Gaussian white noise for which the spectral densities are

$$S_P^E = \begin{cases} 8.41 \times 10^{-7} (\text{psi})^2/\text{Hz}, & 0 \leq f < 1024\,\text{Hz} \\ 0 & \text{otherwise} \end{cases} \tag{50}$$

The spatial distribution of the random pressure is assumed to be uniform over the exterior surface of the top plate. The spectral densities given in eqn (50) correspond to a 110 dB sound level. The overall input sound pressure level is about 130 dB.

The core material of the sandwich construction shown in Fig. 1(b) is modeled as a combination of nonlinear springs and nonlinear dampers. The stiffness and damping terms which couple the two plate motions are taken as[19,20]

$$K_S[w_T - w_B] = k_1(w_T - w_B) + k_2(w_T - w_B)^2 + k_3(w_T - w_B)^3$$
$$+ c_1(\dot{w}_T - \dot{w}_B) + c_2(\dot{w}_T - \dot{w}_B)^2 + c_3(\dot{w}_T - \dot{w}_B)^3 \qquad (51)$$

where k_1, k_2, k_3 and c_1, c_2, c_3 are the stiffness and damping coefficients, respectively. The hard spring behavior is simulated by a positive cubic coefficient k_3. In this case, when the relative core displacement $(w_T - w_B)$ exceeds $(k_1/k_3)^{1/2}$, a state of strong nonlinearity is present. When the coefficient k_3 is negative, the effect of nonlinearity is that of a soft spring. By excluding the quadratic terms in eqn (51), a system with asymmetric nonlinearity can be studied. It should be noted that a symmetric nonlinearity at the equilibrium position might be forced into asymmetric behavior if the system is excited about some mean position.

To demonstrate the accuracy of the time domain solution, a direct comparison of results was made between the power spectral density (PSD) method and the Monte Carlo approach.[19,20] The agreement between these two methods was found to be very good when the response is linear. In obtaining these results, the following data were used: equal top and bottom plate thicknesses $h_T = h_B = 0.032$ in, core material—air spring with density $\rho_S = 1.147 \times 10^{-7}$ lbf-sec^2/in^4, thickness $h_S = 1$ in, spring constant $k_1 = 19.48$ lbf/in^3, and $k_2 = k_3 = c_1 = c_2 = c_3 = 0$.

The results corresponding to the nonlinear case are obtained by simulating the input random pressure in a time domain and then solving the governing differential equations for the panel responses and noise transmission. Shown in Fig. 3 are segments of the simulated time history of the input pressure and the response time histories of the top and bottom plates. From these results it was found that the nonlinear response of the top plate tends to exhibit nonGaussian charactertistics. The response of the bottom plate is linear and its amplitude distribution is very close to a Gaussian one.

For the calculations of noise transmission into the enclosure, it is assumed that the absorption material is uniformly applied on all the interior surfaces of the six walls and it can be represented by the point impedance model

$$Z(\omega) = Z_R(\omega) + iZ_I(\omega) \qquad (52)$$

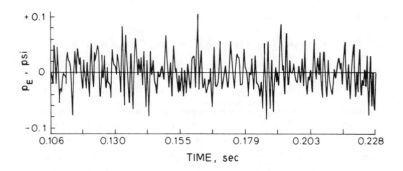

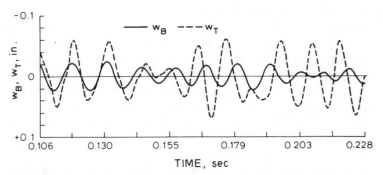

Fig. 3. Time histories of simulated input pressure and plate response.

where for porous materials, the resistance Z_R and the reactance Z_I are given by[2]

$$Z_R(\omega) = \rho c_0\{1 + 0.0571[2\pi R_1/(\rho\omega)]^{0.754}\} \qquad (53)$$

$$Z_I(\omega) = -\rho c_0\{0.087[2\pi R_1/(\rho\omega)^{0.732}\} \qquad (54)$$

in which ρc_0 is the characteristic impedance of the air and R_1 is the flow resistivity of the absorbing layer. The results in Fig. 4 show the effect of the flow resistivity R_1 on noise reduction. The noise reduction NR is calculated from eqn (46). A high flow resistivity value corresponds to an acoustically hard wall resulting in low damping values for the acoustic disturbance.

A parametric study to determine the effects of variations in the

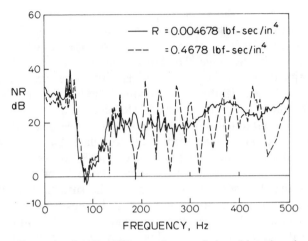

Fig. 4. Noise reduction for different characteristics of interior absorption.

system parameters on the double wall panel response and noise transmission is performed. The effect of variations in the top plate thickness h_T is illustrated in Fig. 5 where a Monte Carlo type analysis was performed to obtain these results. For the case of $h_T = 0.064$ in, the response of the top plate is linear and distinct resonant peaks are observed in the noise reduction. However, when the response reaches

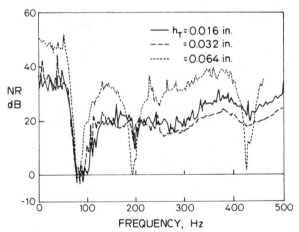

Fig. 5. Noise reduction for different thicknesses of top plate.

the nonlinear range, the resonant peaks about the fundamental mode are suppressed due to decoupling of the top and bottom plate modes. Furthermore, the width of the resonant peaks increases when response reaches the nonlinear range. The flexural modes are most affected and the dilatational modes, except the fundamental one, are excited to a lesser extent. Thus, significant reduction of the bottom plate response and noise transmission is achieved. However, the response at the fundamental dilatational mode does not seem to be affected much by the nonlinear response variations of the top plate. Since the response is dominated by the fundamental mode, the RMS response for the three cases considered is about the same. It should be noted that by decreasing the thickness of the top plate, the surface density of the double wall sandwich construction is reduced. Thus, favorable gains in noise reduction can be expected for a smaller amount of added weight through a design consisting of a thin top plate and a thicker bottom plate.

The effects of decreasing the stiffness of the core appear in Fig. 6. The response of the bottom plate is reduced over almost the entire range of frequencies and the resonant peaks are shifted to lower frequencies. A softer core material, therefore, can accommodate more deformation. As a result, a significant decrease in the bottom plate response and reduction of noise transmission may be obtained by reducing the thickness of the top plate and decreasing the core

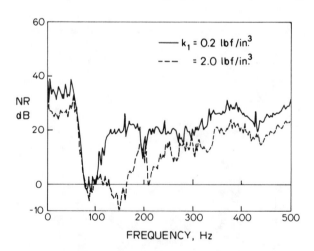

Fig. 6. Effect of core stiffness on noise reduction.

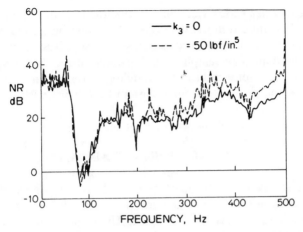

Fig. 7. Noise reduction for different values of core stiffness.

stiffness. This could be attributed to large deflections of the top plate and the core, which contribute to most of the deformation of the coupled system and thus alleviate the motions of the bottom plate.

The effects of nonlinear stiffness and nonlinear damping of the core are shown in Figs 7 and 8. The results shown in Fig. 7 indicate that a positive nonlinearity might have a positive effect on noise reduction

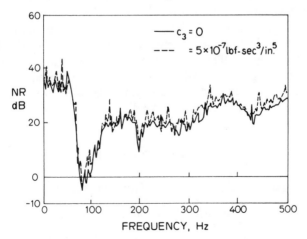

Fig. 8. Effect of nonlinear damping on noise reduction.

312   *R. Vaicaitis & H.-K. Hong*

above frequencies of 250 Hz, while in the low frequency range (below 200 Hz) no definite trend is detected. This could be attributed to the hardening effect of the core stiffness on suppressing dilatational modes. The nonlinear damping, as illustrated in Fig. 8, has very little influence on the noise reduction throughout the frequency range. The damping in the present system seems to be dominated by the linear term.

5 CONCLUDING REMARKS

The noise transmission through a coupled (sandwich) panel system into an enclosure of which the interiors are covered with porous absorption materials has been obtained. The results indicate that noise transmission is strongly dependent on the geometric and material properties of the double wall sandwich construction. The deflection response of the bottom plate and, in turn, the noise transmission into the acoustic enclosure can be controlled by proper selection of the core stiffness and top plate thickness. For a soft core and nonlinear response of the top plate, no distinct resonance peaks were observed at frequencies above the fundamental resonance frequency of the coupled system. By increasing damping in the core, the response peak at the fundamental frequency can also be suppressed. This suggests that the viscoelastic damping factor of the core should be large in the vicinity of the fundamental natural frequency of the coupled system.

The effect of nonlinear stiffness and nonlinear damping of the soft core on noise reduction are in general favorable, but the contributions are not very large. The absorption materials can be used very effectively to suppress acoustic resonances. Since the added damping and mass of these materials to the flexible panels are very small, their effect on the structural response is negligible.

REFERENCES

1. Cockburn, J. A. & Jolly, A. C., Structural acoustic response, noise transmission losses, and interior noise levels of an aircraft fuselage excited by random pressure fields. AFFDC-TR-68-2, Air Force Flight Dynamics Laboratory, Technical Report, August 1968.
2. Beranek, Leo L. (Ed.), *Noise and Vibration Control.* McGraw-Hill, New York, 1971.

3. Wilby, J. F. & Scharton, T. O., Acoustic transmission through a fuselage sidewall. NASA CR-132602, July 1974.
4. Dowell, E. H., Master plan for prediction of vehicle interior noise. *AIAA Journal*, **18**(4) (1980) 353–66.
5. Vaicaitis, R., Recent research on noise transmission into aircraft. *The Shock and Vibration Digest*, **14**(8) (1982) 13–18.
6. Vaicaitis, R. & Slazak, M., Cabin noise control for twin engine general aviation aircraft. NASA CR-165833, February 1982.
7. Vaicaitis, R., Grosveld, F. W. & Mixson, J. S., Noise transmission through aircraft panels. *Journal of Aircraft, AIAA,* **22**(4) (1985) 303–10.
8. Revell, J. D., Balena, F. J. & Koval, L. R., Analytical study of interior noise control by fuselage design techniques on high-speed, propeller driven aircraft. NASA CR-159222, July 1978.
9. Rennison, D. C., Wilby, J. F., March, A. M. & Wilby, E. G., Interior noise control prediction study for high-speed, propeller-driven aircraft. NASA Contract Report 159200, September 1979. (Also AIAA-80-0998, presented at the AIAA 6th Aeroacoustics Conference, Hartford, CT., June 4–6, 1980.)
10. Vaicaitis, R., Noise transmission by viscoelastic sandwich panels. NASA TND-8516, August 1977.
11. Slazak, M. & Vaicaitis, R., Response of stiffened sandwich panels. AIAA/ASME/ASCE/AHS, 22nd Structures, Structural Dynamics and Materials Conference, Atlanta, GA., Paper No. 81-0557, 1981.
12. Bliss, D. B., Study of bulk reacting porous sound absorbers and a new boundary condition for thin porous layers. *Journal of the Acoustical Society of America*, **71**(3) (March 1982) 533–45.
13. Bliss, D. B. & Burke, S. E., Experimental investigation of the bulk reaction boundary condition. *Journal of the Acoustical Society of America*, **71**(3) (March 1982) 546–51.
14. Dowell, E. H., Chao, C.-F. & Bliss, B., Absorption material mounted on a moving wall—fluid/wall boundary condition. *Journal of the Acoustical Society of America*, **70**(1) (July 1981) 244–5.
15. Courant, R. & Hilbert, D., *Methods of Mathematical Physics, Vol. 1.* Interscience, New York, 1953.
16. McDonald, W. B., Vaicaitis, R. & Myers, M. K., Noise transmission through plates into an enclosure. NASA Technical Paper 1173, May 1978.
17. Dowell, E. H., Gorman, G. F., III & Smith, D. A., Acoustoelasticity: general theory, acoustic natural modes and forced response to sinusoidal excitation, including comparisons with experiment. *Journal of Sound and Vibration*, **52**(4) (1977) 519–42.
18. Chao, C.-F., Modal analysis of interior noise fields. MAE Technical Report 1499-T, Princeton, University, December 1980.
19. Hong, H.-K. & Vaicaitis, R., Nonlinear response of double wall sandwich panels. *Journal of Structural Mechanics*, **12**(4) (1985) 483–503.
20. Vaicaitis, R. & Hong, H.-K., Nonlinear random response of double wall sandwich panels, *AIAA/ASME/ASCE/AHS SDM Conference,* Paper no. 83-1037, Lake Tahoe, Nevada, 2–4 May 1983, American Institute of Aeronautics and Astronautics, New York, USA.

21. Lin, Y. K., *Probabilistic Theory of Structural Dynamics*. McGraw-Hill, New York, 1967.
22. Shinozuka, M. & Jan, D.-M., Digital simulation of random processes and its applications. *Journal of Sound and Vibration*, **25**(1) (1972) 111–28.
23. Shinozuka, M., Monte Carlo solution of structural dynamics. *International Journal of Computers and Structures*, **2** (1972) 855–74.
24. Vaicaitis, R., Jan, C.-M. & Shinozuka, M., Nonlinear panel response from a turbulent boundary layer. *AIAA Journal*, **10**(7) (July 1972) 895–9.
25. Yang, J.-N., Simulation of random envelope processes. *Journal of Sound and Vibration*, **21**(1) (1972) 73–85.

16

Lyapunov Exponents of Stochastic Systems and Related Bifurcation Problems

WALTER V. WEDIG

Institute for Technical Mechanics, University of Karlsruhe, FRG

ABSTRACT

The parametric oscillator under white noise leads to a stochastic bifurcation problem in the sense that in the presence of a progressive nonlinearity there exist a trivial solution and a stationary branching process. The bifurcation points are calculated by means of linear stochastic transformations. The square means of the branching processes follow from the multiplicative expectation theorem. They are physically existent if and only if the linear system is almost sure unstable.

1 INTRODUCTION

Bifurcation problems in mechanical vibration theory can be introduced by a nonlinear oscillator subjected to parametric white noise. Because of its homogeneity the oscillator equation possesses the trivial solution of the state space. For an increasing intensity of the parametric excitation this trivial solution becomes unstable. It branches to nontrivial processes which are stationary in presence of a progressive restoring. Consequently, the investigation of such problems involves linear stochastic stability theory and nonlinear diffusion process theory. From the mathematical point of view, those bifurcations are related to stochastic center manifolds and stochastic eigenvalues[1,2] giving a qualitative theory of an increasing development and importance. However, in the engineering application area, there are still

315

open questions[3] concerning physical contradictions between moment and almost sure stability of linear stochastic systems. How to compute the stationary characteristics of nonlinear branching processes of vibrating systems?

According to Ref. 4, this paper presents some first steps in this direction under the following key words:

(i)   Stochastic   transformations   for   Lyapunov   exponents calculations.

(ii) Multiplicative expectation theorem for stochastic bifurcations.

Exact stability boundaries of linear stochastic systems are derived by means of stochastic transformations of the state norm processes. They are determined by a deterministic second order eigenvalue problem defined on the stationary phase process of the system. The eigenvalue problem is solved by orthogonal Fourier expansions leading to an infinite determinant whose convergence properties are numerically investigated. If the noise intensity goes beyond the bifurcation point the branching process is stabilized by a progressive restoring of cubic form. To calculate associated square means we need the multiplicative expectation theorem which enables a statistical separation of the trivial solution and the branching process. Thus, this theorem is a key for stochastic bifurcation investigations. It is proved for the scalar nonlinear system.

2 MULTIPLICATIVE EXPECTATION THEOREM

To introduce practical calculations of stochastic stability and bifurcation problems let us consider the scalar system described by the following Ito differential equation:

$$dX_t = (\alpha + \tfrac{1}{2}\sigma^2 - \gamma X_t^2)X_t \, dt + \sigma X_t \, dW_t,$$
$$X_{t0} = X_0 \qquad (\alpha \geq 0, \, \gamma \geq 0) \tag{1}$$

Here, X_t denotes a scalar stochastic process dependent on time t and the initial value X_0. W_t is the normed Wiener process with intensity σ. The parameters α and γ determine the linear term of the system and the cubic restoring, respectively. In the linear case ($\gamma \equiv 0$), we get the well-known solutions of (1)

$$X_t = \exp(\alpha t + \sigma W_t)X_0, \qquad E(X_t^p) = \exp[p(\alpha + \tfrac{1}{2}\sigma^2 p)t]X_0^p \tag{2}$$

and therewith a different growth behaviour for the samples X_t and their p-order moments $E(X_t^p)$. More precisely, since W_t grows like $[t \log \log t]^{1/2}$ with probability one, the sample stability property is decided only by the deterministic parameter α. The noise intensity σ, however, has no influence on the bifurcation point $\alpha = 0$. On the other hand, the moment behaviour in (2) indicates an earlier bifurcation dependent on σ and the expectation order p.

Those different bifurcation points can be eliminated by means of nonlinearly completed systems and corresponding stationary moment calculations. For this purpose, we investigate the increment of the squared state process Q_t for $\gamma > 0$.

$$dQ_t = 2(\alpha + \sigma^2 - \gamma Q_t)Q_t \, dt + 2\sigma Q_t \, dW_t, \qquad Q_t = X_t^2 \qquad (3)$$

$$d(\log Q_t) = 2(\alpha - \gamma Q_t) \, dt + 2\sigma \, dW_t, \qquad Q_t > 0 \quad \text{(w.p.1)} \qquad (4)$$

Provided that $Q_t > 0$ with probability one, we are allowed to take the natural logarithm of Q_t defined by the Ito equation (4). If there exists a stationary process $Q_t > 0$, the expectation of the log-process is finite and time-invariant such that its increment has a vanishing mean value.

$$E[d(\log Q_t)] = 0, \qquad \text{for } Q_t > 0 \text{ with probability one} \qquad (5)$$

Relation (5) is called the multiplicative expectation theorem. It is the key to stochastic bifurcation problems. It can be extended to the multidimensional case[4] by introducing a suitable norm of the associated system vector.

In the scalar case (1), the multiplicative expectation theorem can easily be proved. For this purpose, we set up the associated Fokker–Planck equation.

$$\frac{\partial p}{\partial t} + \frac{\partial}{\partial q}[2q(\alpha + \sigma^2 - \gamma q)p(q)] - \frac{1}{2}\frac{\partial^2}{\partial q^2}[4\sigma^2 q^2 p(q)] = 0 \quad (q \geq 0) \qquad (6)$$

Obviously, there exists a stationary solution $(\partial p/\partial t = 0)$ of (6) calculated to be

$$p(q) = \frac{(\gamma/\sigma^2)^{\alpha/\sigma^2}}{\Gamma(\alpha/\sigma^2)} q^{-1+\alpha/\sigma^2} \exp\left(-\frac{\gamma}{\sigma^2}q\right) \qquad (\alpha, \gamma > 0) \qquad (7)$$

$$E(\log Q_t) = \psi\left(\frac{\alpha}{\sigma^2}\right) - \log\left(\frac{\gamma}{\sigma^2}\right), \qquad \psi(x) = \frac{d}{dx}\log\Gamma(x) \qquad (8)$$

Integrating[5] the density distribution (7) we arrive at (8), where ψ

denotes Euler's psi function. For almost sure instability $\alpha > 0$ of the linear system and for positive definite cubic parameter $\gamma > 0$, there exist an invariant measure μ derivable from (7) and a time-invariant expectation value (8) of the stationary log-process. Consequently, its time increment has a vanishing mean, as formulated in (5). In particular, it follows from (3) and (4)

$$E(Q_t) = \frac{\alpha}{\gamma}, \qquad E(Q_t^2) = \frac{\alpha}{\gamma^2}(\alpha + \sigma^2), \qquad Q_t = X_t^2 \qquad (\alpha, \gamma > 0) \quad (9)$$

simply by taking the expectation in both equations under the stationary conditions (5) and $E(dW_t) = E(dQ_t) = 0$. Naturally, the results (9) can be computed from (7) as well. Coincident with the sample instability, all stationary moments (9) are physically existent for $\alpha > 0$. They have to be replaced by $E(Q_t) = E(Q_t^2) = 0$ in the stable case $\alpha < 0$. This eliminates the physical contradiction between mean square calculations and sample stability properties of linear systems.

To give more physical insights into the multiplicative expectation theorem (5) we perform two different simulations shown in Fig. 1. They are obtained from the corresponding one-step recursion formulas as follows.

$$L_{n+1} = L_n + 2(\alpha - \gamma Q_n)\,\Delta t + 2\sigma\,\Delta W_n, \qquad L_n = \log(Q_n) \quad (10)$$

$$Q_{n+1} = Q_n \exp[2(\alpha - \gamma Q_n)\,\Delta t + 2\sigma\,\Delta W_n], \qquad n = 0, 1, 2, \ldots \quad (11)$$

Given an initial value Q_n we get its logarithm L_n and the next one

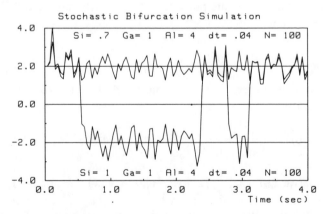

Fig. 1. Time-discrete simulations of the unstable system.

L_{n+1} by applying a time-increment Δt and the difference of two independent random numbers in (10). The recurrence formula (11) is just the inversion of (10). Both result in the same simulation for $Q_n > 0$. Although we need only the squared processes for the verification of the multiplicative expectation theorem (5), it is of interest to simulate the linear process $X_n = v\sqrt{Q_n}$, itself. Herein, the sign v of the square root can be computed from the expanded recursion formula

$$v = \begin{cases} +1 & \text{if} \quad [1 + (\alpha + \tfrac{1}{2}\sigma^2 - \gamma Q_n)\,\Delta t + 2\sigma\,\Delta W_n]X_n > 0 \\ -1 & \text{if} \quad [1 + (\alpha + \tfrac{1}{2}\sigma^2 - \gamma Q_n)\,\Delta t + 2\sigma\,\Delta W_n]X_n < 0 \end{cases} \quad (12)$$

which is valid in the zero range $Q_n \ll 1$ corresponding to (1).

In Fig. 1, we applied (12), (11) or (10) for the noted values of α, γ, Δt, and σ. For small intensities σ, the time-discrete process X_n exists mainly in one half side of the state space. If σ increases, jumps over the zero position are possible so that the simulation (11) can be stopped by an underflow. In those extreme cases, the recursion (11) has to be restarted with new initial conditions and with smaller $\Delta t > 0$. In the limiting case $\Delta t \to 0$ of a time-continuous simulation, zero-crossings are excluded with probability one. This is just the inversion of the almost sure stability property, i.e. if X_t reaches small values near the zero position, we observe an exponential growth in the unstable case $\alpha > 0$ which is stopped and then inverted by an increasing nonlinear restoring in the higher state range.

3 LYAPUNOV EXPONENTS OF THE LINEAR OSCILLATOR

The stability investigations of the scalar system are extended to the linear oscillator subjected to parametric excitation by white noise.

$$\ddot{X}_t + 2D\omega_1\dot{X}_t + \omega_1^2(1 + \sigma\dot{W}_t/\sqrt{\omega_1})X_t = 0 \qquad (D, \omega_1 > 0) \quad (13)$$

Here, ω_1 denotes the natural frequency of the oscillator, D and σ are dimensionless parameters for the system damping and for the excitation intensity. For $\sigma \equiv 0$ (13) represents a deterministic system. Its equilibrium position $X_t, \dot{X}_t \equiv 0$ is asymptotically stable for $D, \omega_1 > 0$. For increasing noise intensity we reach a bifurcation point to be calculated in the following.

For this purpose, we introduce the state processes $X_{1,t} = X_t$ and $X_{2,t} = \dot{X}_t/\omega_1$ in arriving at the Ito differential equations

$$dX_{1,t} = \omega_1 X_{2,t}\, dt \qquad E[(dW_t)^2] = dt \qquad (14)$$

$$dX_{2,t} = -2D\omega_1 X_{2,t}\, dt - \omega_1 X_{1,t}\, dt - \sigma\sqrt{\omega_1}\, X_{1,t}\, dW_t \qquad (15)$$

Subsequently, we go over to polar coordinates

$$X_{1,t} = A_t \cos \Psi_t, \qquad X_{2,t} = A_t \sin \Psi_t \qquad (16)$$

$$P_t = A_t^p = (X_{1,t}^2 + X_{2,t}^2)^{p/2} \qquad (0 \leqslant p < \infty) \qquad (17)$$

and calculate the increments of the phase process Ψ_t and of the norm process P_t which is the amplitude process A_t raised to the power p.

$$d\Psi_t = -\omega_1 \varphi(\Psi_t)\, dt - \sigma\sqrt{\omega_1} \cos^2 \Psi_t\, dW_t$$
$$\varphi(\Psi_t) = 1 + 2D \sin \Psi_t \cos \Psi_t + \sigma^2 \sin \Psi_t \cos^3 \Psi_t \qquad (18)$$

$$dP_t = pP_t\omega_1 f(\Psi_t)\, dt - pP_t\sigma\sqrt{\omega_1} \sin \Psi_t \cos \Psi_t\, dW_t$$
$$f(\Psi_t) = \tfrac{1}{2}\sigma^2[\cos^4 \Psi_t + (p-1)\cos^2 \Psi_t \sin^2 \Psi_t] - 2D \sin^2 \Psi_t \qquad (19)$$

For $p = 2$, P_t is the square mean of the state processes $X_{1,t}$ and $X_{2,t}$. For $p \to 0$, we obtain the zeroth mean which describes the sample properties of the system processes determining the almost sure stability conditions of interest. This relation has been shown in Ref. 6 and recently proved for the general situation of stationary ergodic parameter excitations in linear stochastic systems.[7]

Following Ref 8, we perform the linear stochastic transformation

$$S_t = T(\Psi_t)P_t, \qquad P_t = T^{-1}(\Psi_t)S_t \qquad (20)$$

introducing the new norm process S_t by means of the scalar function $T(\Psi_t)$ which is defined on the stationary phase process Ψ_t in the range $-\pi/2 \leqslant \psi \leqslant \pi/2$.

$$dS_t = \{\omega_1[\tfrac{1}{2}\sigma^2 \cos^4 \Psi_t T_{\psi\psi}(\Psi_t) + (p\sigma^2 \sin \Psi_t \cos^3 \Psi_t - \varphi(\Psi_t))T_\psi(\Psi_t)$$
$$+ pf(\Psi_t)T(\Psi_t)]\, dt - \sigma\sqrt{\omega_1} \cos \Psi_t[\cos \Psi_t T_\psi(\Psi_t)$$
$$+ p \sin \Psi_t T(\Psi_t)\, dW_t\}P_t \qquad (21)$$

If the transformation function $T(\psi)$ is bounded and nonsingular, both processes P_t and S_t possess the same stability behavior. Therefore, it is reasonable to look for such a transformation $T(\psi)$ that the Ito differential equation (21) degenerates to a time-invariant drift term which does not depend on the phase process Ψ_t.

$$dS_t = -\omega_1 \lambda S_t\, dt + \sigma\sqrt{\omega_1}\, S_t g(\Psi_t)\, dW_t \Rightarrow \dot{E}(S_t) = -\omega_1 \lambda E(S_t) \quad (22)$$

In this form we are able to take the expectation of S_t whose stability behavior is simply determined by the dimensionless Lyapunov exponent λ. The time-invariant postulation in (21) or (22) results in the following differential equation:

$$\tfrac{1}{2}\sigma^2 \cos^4 \psi\, T_{\psi\psi}(\psi) - [1 + 2D \sin\psi \cos\psi + \sigma^2(1-p)\sin\psi \cos^3\psi]T_\psi(\psi)$$
$$+ \tfrac{1}{2}p[(p-1)\sigma^2 \sin^2\psi \cos^2\psi + \sigma^2 \cos^4\psi - 4D\sin^2\psi]T(\psi)$$
$$= -\lambda T(\psi) \quad (23)$$

It represents a deterministic second order eigenvalue problem for the determination of the unknown transformation function $T(\psi)$ and associated eigenvalue λ or Lyapunov exponent of the pth mean.

The solution of (23) can be calculated by an orthogonal expansion applying the complex-valued Fourier series (24).

$$T(\psi) = \sum_{n=-\infty}^{\infty} \exp(i2n\psi)z_n, \qquad n = 0, \mp 1, \mp 2, \ldots \quad (24)$$

$$c_{n-2}^n z_{n-2} + b_{n-1}^n z_{n-1} + a_n z_n + b_{n+1}^p z_{n+1} + c_{n+2}^p z_{n+2} = 0 \quad (25)$$

Inserting (24) into (23) and comparing all coefficients, we obtain the algebraic equations (25) with

$$a_n = (p^2 + 2p - 12n^2)\sigma_2/4 - pD + \lambda - i2n, \qquad \sigma_2 = \sigma^2/4$$
$$b_n^{p(n)} = \tfrac{1}{2}\{(p(\pm)2n)D + \sigma_2[p - 4n^2(\pm)2(1-p)n]\} \quad (26)$$
$$c_n^{p(n)} = [p(2-p) - 4n^2(\pm)4n(1-p)]\sigma_2/8, \qquad n = 0, \mp 1, \mp 2, \ldots$$

Equation (25) results in an infinite determinant for the calculation of the Lyapunov exponent λ and associated nontrivial solutions z_n.

$$\Delta(\lambda, p) = \begin{vmatrix} a_{-3} & b_{-2}^p & c_{-1}^p & 0 & 0 & 0 & 0 \\ b_{-3}^n & a_{-2} & b_{-1}^p & c_0^p & 0 & 0 & 0 \\ c_{-3}^n & b_{-2}^n & a_{-1} & b_0^p & c_1^p & 0 & 0 \\ 0 & c_{-2}^n & b_{-1}^n & \boxed{a_0} & b_1^p & c_2^p & 0 \\ 0 & 0 & c_{-1}^n & b_0^n & a_1 & b_2^p & c_3^p \\ 0 & 0 & 0 & c_0^n & b_1^n & a_2 & b_3^p \\ 0 & 0 & 0 & 0 & c_1^n & b_2^n & a_3 \end{vmatrix} \quad (27)$$

It is evaluated by cutting off at $m = 2n + 1$ elements as indicated in (27). For $n = 0$, for example, we obtain

$$\Delta^{(0)} = a_0 = \frac{\sigma_2}{4} p(p + 2) - pD + \lambda = 0, \quad \Rightarrow \frac{\lambda}{p} = -\frac{\sigma_2}{4}(p + 2) + D \quad (28)$$

and therewith the well known result of different zeroth approximations performed in Ref. 9 or in Ref. 10 for small values of D, $\sigma_2 \ll 1$.

In the limiting case of stability boundaries, the top Lyapunov exponents are real-valued and vanishing for all p, noted in (17). Therefore, higher order evaluations of (27) can be restricted to $\lambda \equiv 0$. The corresponding results are shown in Fig. 2 for $n = 0, 1, 2, 3$ and 4, marked by different line types. The applied norm powers are $p = 0, 1, 2$ and 4. In the discussed parameter range we observe a fast convergence rate so that higher approximations $(n > 4)$ coincide graphically with the noninterrupted line results of the approximation order $n = 4$. In particular, there are only small deviations from the well-known stability limits of second and fourth order moment equations given by Ref. 11

$$\sigma^2_{(2MO)} = 4D, \quad \sigma^2_{(4MO)} = D(8 + 24D^2)/(3 + 18D^2) \quad (29)$$

Naturally, these deviations are explained by the application of

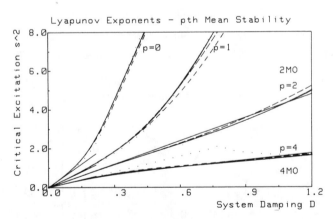

Fig. 2. Sample, moment and pth mean stability of the parametric oscillator under white noise.

different norms. The second order moments of the system (15) are described by a finite set of three differential equations, meanwhile the square mean of the norm process in (19) is determined by an infinite set of moment equations. Finally, it is worthwhile to check the almost sure stability condition $(p \to 0)$ of the first approximation of (27). Equating the corresponding 3×3 determinant we obtain the following cubic polynomial in σ_2.

$$\lim_{p \to 0} \frac{1}{p} \Delta^{(1)}(\sigma, p) = \tfrac{1}{2}[3\sigma_2^3 - 18D\sigma_2^2 + 2(2 + 3D^2)\sigma_2 - 8D] \qquad (\sigma_2 = \sigma^2/4)$$

(30)

Its numerical evaluation leads to the dotted stability line in Fig. 2. We need this result for comparison with corresponding bifurcation investigations in the next section.

4 BIFURCATION OF THE PARAMETRIC OSCILLATOR

The multiplicative expectation theorem is now applied to the parametric oscillator (13) which is nonlinearly completed by the cubic restoring $\gamma\omega_1^2 X_t^3$.

$$\ddot{X}_t + 2D\omega_1\dot{X}_t + \omega_1^2(1 + \gamma X_t^2 + \sigma\dot{W}_t/\sqrt{\omega_1})X_t = 0 \qquad (\omega_1, D, \gamma > 0) \quad (31)$$

For this purpose, we again introduce the amplitude A_t and the phase Ψ_t in arriving at the following Ito differential equations:

$$d\Psi_t = -\omega_1\,dt - D\omega_1\sin 2\Psi_t\,dt - \frac{\gamma}{8}\omega_1(3 + 4\cos 2\Psi_t + \cos 4\Psi_t)A_t^2\,dt$$

$$- \frac{\sigma^2}{8}\omega_1(2\sin 2\Psi_t + \sin 4\Psi_t)\,dt - \frac{\sigma}{2}\sqrt{\omega_1}(1 + \cos 2\Psi_t)\,dW_t \qquad (32)$$

$$dA_t = -D\omega_1(1 - \cos 2\Psi_t)A_t\,dt - \frac{\gamma}{8}\omega_1(2\sin 2\Psi_t + \sin 4\Psi_t)A_t^3\,dt$$

$$+ \frac{\sigma^2}{16}\omega_1(3 + 4\cos 2\Psi_t + \cos 4\Psi_t)A_t\,dt - \frac{\sigma}{2}\sqrt{\omega_1}A_t\sin 2\Psi_t\,dW_t$$

(33)

From (32) we derive the increments of sine and cosine of the phase Ψ_t

$$d(\sin 2\Psi_t) = -2\omega_1 \cos 2\Psi_t \, dt - D\omega_1 \sin 4\Psi_t \, dt$$

$$-\frac{\sigma}{2}\sqrt{\omega_1}(1 + 2\cos 2\Psi_t + \cos 4\Psi_t) \, dW_t$$

$$-\frac{\gamma}{8}\omega_1(4 + 7\cos 2\Psi_t + 4\cos 4\Psi_t + \cos 6\Psi_t)A_t^2 \, dt$$

$$-\frac{\sigma^2}{4}\omega_1(3\sin 2\Psi_t + 3\sin 4\Psi_t + \sin 6\Psi_t) \, dt$$

$$d(\cos 2\Psi_t) = 2\omega_1 \sin 2\Psi_t \, dt + D\omega_1(1 - \cos 4\Psi_t) \, dt$$

$$+\frac{\sigma}{2}\sqrt{\omega_1}(2\sin 2\Psi_t + \sin 4\Psi_t) \, dW_t$$

$$+\frac{\gamma}{8}\omega_1(5\sin 2\Psi_t + 4\sin 4\Psi_t + \sin 6\Psi_t)A_t^2 \, dt$$

$$-\frac{\sigma^2}{4}\omega_1(1 + 3\cos 2\Psi_t + 3\cos 4\Psi_t + \cos 6\Psi_t) \, dt$$

and taking the expectation we get the corresponding moment equations. Here, we are allowed to neglect higher order terms in the sense of an orthogonal Fourier and Laguerre expansion.

$$\dot{E}(\sin 2\Psi_t) + \omega_1 \frac{3\sigma^2}{4} E(\sin 2\Psi_t) + 2\omega_1 E(\cos 2\Psi_t) + \frac{\gamma}{2}\omega_1 E(A_t^2) = 0$$

$$\tag{34}$$

$$\dot{E}(\cos 2\Psi_t) - 2\omega_1 E(\sin 2\Psi_t) + \omega_1 \frac{3\sigma^2}{4} E(\cos 2\Psi_t) + \omega_1\left(\frac{\sigma^2}{4} - D\right) = 0$$

$$\tag{35}$$

Both equations (34) and (35) are inhomogeneous and possess stable solutions given by $\dot{E}(.) = 0$ in the stationary case.

Subsequently, we derive a third equation by means of the multiplicative expectation theorem. From (33) we obtain

$$d[\log(A_t^2)] = -2D\omega_1(1 - \cos 2\Psi_t) \, dt - \frac{\gamma}{4}\omega_1(2\sin 2\Psi_t + \sin 4\Psi_t)A_t^2 \, dt$$

$$+\frac{\sigma^2}{4}\omega_1(1 + 2\cos 2\Psi_t + \cos 4\Psi_t) \, dt - \sigma\sqrt{\omega_1}\sin 2\Psi_t \, dW_t$$

As before, we take the expectation in neglecting higher order terms.

$$E[d(\log A_t^2)] = 0, \qquad \Rightarrow \left(\frac{\sigma^2}{2} + 2D\right)E(\cos 2\Psi_t) + \frac{\sigma^2}{4} - 2D = 0 \quad (36)$$

Together with (34) and (35), eqn (36) is sufficient to calculate a first approximation of the branching oscillator processes. According to Ref. 4, the squared amplitude expectation has a stationary value, as follows.

$$E(A_t^2) = \frac{1}{2\gamma(\sigma_2 + D)} [3\sigma_2^3 - 18D\sigma_2^2 + 2(2 + 3D^2)\sigma_2 - 8D] \qquad \left(\sigma_2 = \frac{\sigma^2}{4}\right)$$

(37)

Similar to the scalar case (9), the almost sure stability condition (30) of the linear oscillator appears in the numerator of the stationary result (37) calculated from the nonlinear system (31). From this it follows that the square mean $E(A_t^2)$ is positive definite and physically existent if and only if the linear parametric oscillator is almost sure unstable. Inversely, the calculated value (37) has to be replaced by a zero-valued mean square if the almost sure stability condition (30) is satisfied in case of sufficiently low parametric excitation intensities σ. In Fig. 3 we

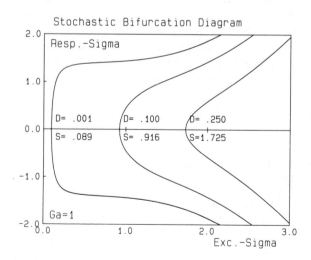

Fig. 3. Stationary square means of the nonlinear oscillator—bifurcations at the sample instability point.

show some numerical evaluations of (37) by plotting the stationary response variance over the excitation variance σ for different damping values D. If both the damping D and the excitation σ are vanishing simultaneously then $E(A_t^2)$ goes to $2/\gamma$ or $-4/\gamma$ including the zero position of the deterministic case. For higher parametric excitations $E(A_t^2)$ increases with $3\sigma_2^2/2\gamma$. Naturally, the starting solution (37) can be extended to higher approximations applying orthogonal expansions of Fourier series for Ψ_t and Laguerre polynomials for A_t^2 and calculating higher order Laguerre and Fourier moments.

5 CONCLUDING REMARKS

Stochastic bifurcation problems are of increasing interest and importance. So far, these problems have been studied by looking at qualitative changes of Fokker–Planck solutions in the additional presence of additive noise or multiplicative harmonic perturbations. In our paper, however, we investigated structural instabilities at the bifurcation points of the systems. The linearized oscillator, for example, is asymptotically stable if there is no multiplicative noise. The associated density distribution is concentrated around the equilibrium position. For increasing noise intensity we reach a bifurcation point with a change from stable to unstable system determined by a vanishing Lyapunov exponent. Beyond this point there exists a nonconcentrated density distribution which is stationary in the presence of a progressive cubic restoring. The paper presents first results of the oscillator mean square calculated via the multiplicative expectation theorem.

REFERENCES

1. Arnold, L., Eigenvalues, bifurcations and center manifolds in the presence of noise. To appear in *Proceedings of the 1987 Equadiff Conference, Lecture Notes in Pure and Applied Mathematics*, eds C. Dafermos, G. Ladas & G. Papanicolaou. Marcel Dekker, New York.
2. Kliemann, W., Qualitative Theorie nichtlinearer stochastischer Systeme. Report No. 34, Forschungsschwerpunkt Dynamische Systeme, Universität Bremen, 1980.
3. Lin, Y. K., Kozin, F., Wen, Y. K., Casciati, F., Schuëller, G. I., Der Kiureghian, A., Ditlevsen, O. & Vanmarcke, E. H., Methods of stochastic structural dynamics. *Structural Safety*, **3** (1986) 167–94.

4. Wedig, W., Stability of nonlinear stochastic systems. To appear in *Proceedings of the 1987 Equadiff Conference, Lecture Notes in Pure and Applied Mathematics,* ed C. Dafermos, G. Ladas & G. Papanicolaou, Marcel Dekker, New York.

5. Gradshteyn, I. S. & Ryzhik, I. M., *Table of Integrals, Series and Products,* Academic Press, New York, 1980, p. 576, No. 4.352 and p. 943, No. 8.360.

6. Kozin, F. & Sugimoto, S., Relations between sample and moment stability for linear stochastic differential equations. *Proc. Conf. Stochastic Diff. Equations,* ed. D. Mason, Academic Press, 1977, pp. 145–62.

7. Arnold, L., A formula connecting sample and moment stability of linear stochastic systems. Report No. 92, Forschungsschwerpunkt Dynamische Systeme, Bremen, 1983.

8. Wedig, W., Berechnung des *p*-ten Lyapunov Exponenten. GAMM-Tagung 1987. *ZAMM,* **68**(4) (1988) T135–8.

9. Pardoux, E. & Wihstutz, V., Lyapunov exponent and rotation number of two dimensional linear stochastic systems with small diffusion. Report No. 151, Forschungsschwerpunkt Dynamische Systeme, Bremen, 1986.

10. Wedig, W., Stabilität stochastischer Systeme. *ZAMM,* **55** (1975) 185–7.

11. Wedig, W., Moments and probability densities of parametrically excited systems and continuous systems. In *VII. Internationale Konferenz über Nichtlineare Schwingungen, Vol.* II, 2, Akademie-Verlag, Berlin, 1978, pp. 469–92.

17

Random Vibration Application to Soil-Structural Problems

Y. K. WEN & A. H-S. ANG

Department of Civil Engineering, University of Illinois, Urbana, Illinois, USA

ABSTRACT

The stress–strain relationship of soil deposit under severe cyclic excitation is known to be highly nonlinear and hysteretic. The shear stress may deteriorate at a fast rate in a saturated sand deposit resulting in liquefaction. Also, when the soil serves as support for structures, it needs to be treated as a compliant rather than a rigid medium. The effect of foundation movement may be significant and the resulting restoring force is again highly nonlinear and hysteretic. This paper summarizes the application of a recently developed random vibration method to these classes of problems when the excitation is a random earthquake motion. The methodology is outlined; examples on applications to actual soil and structural systems are given and comparisons are made with field data when available.

1 INTRODUCTION

Soil as a construction material exhibits complicated behavior under loading, which can be highly nonlinear and hysteretic. Under earthquake excitation, the potential hazards caused by soil failure include liquefaction and foundation failure of a structure. To evaluate soil and structural performance under earthquake loadings, such nonlinear behavior of soil needs to be taken into consideration. Also, earthquake excitation is random in nature and needs to be treated as

329

such. The purpose of this paper is to apply a recently developed random vibration method to the problem of liquefaction and soil–structure interaction, whereby the statistics of the soil and structural responses, including displacement and energy dissipation can be obtained, and the system performance can be described in more realistic and quantitative terms.

2 MODELING AND ANALYSIS OF SOIL AS AN INELASTIC SYSTEM

The nonlinear behavior of a soil can be described by a typical cyclic shearing stress–strain loop of the soil, the associated skeleton curve (dashed line), and the equivalent viscous damping ratio as shown in Fig. 1. In general, the secant shear modulus G decreases with the strain amplitude γ, whereas the equivalent viscous damping ratio increases with γ (Fig. 2). To model this behavior, a smooth hysteresis based on the differential equation model of Wen[16] and Baber & Wen[2] was developed. It closely matches the cyclic response behavior of soil as indicated in Fig. 1. It also allows deterioration as oscillation progresses. Details of this model are available in Pires et al.[8] The comparison of the model secant shear modulus and equivalent viscous damping ratio with experimental data (Tatsuoka et al.[14]) as a function of the shear strain amplitude are given in Fig. 2. Under random excitation such as earthquake ground motion, the solution of the response of the inelastic system as described in the foregoing can be obtained based on an equivalent linearization procedure (Wen[16]). The response statistics obtained include the covariance matrix of all the response variables, the first two moments of the hysteretic energy dissipation, which is a good measure of cumulative damage suffered by an inelastic system. In the case of soil deposits, the hysteretic shear-strain energy is related to the excess pore pressure, which has direct bearing on the occurrence of liquefaction.

The ground motion is modeled by a Gaussian random process, characterized by a power spectral density function of the Kanai–Tajimi type.

Details of the solution procedure of an inelastic system under random excitation can be found in Wen[16] and Baber & Wen.[2] The results have been verified by extensive Monte Carlo simulations.

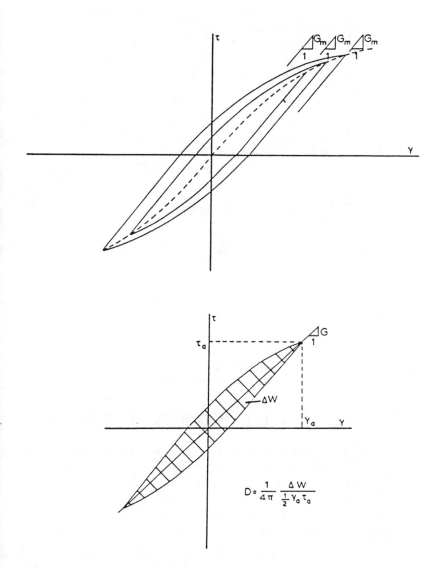

Fig. 1. Soil hysteresis and equivalent viscous damping ratio D, G = secant shear modulus, G_m = max G.

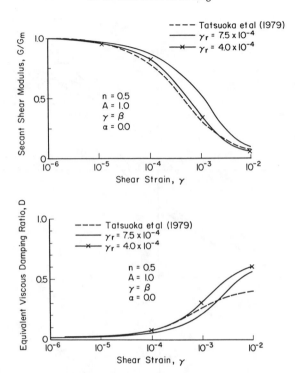

Fig. 2. Dependence of secant shear modulus and equivalent viscous ratio on shear strain amplitude.

3 LIQUEFACTION UNDER EARTHQUAKE

Under random seismic loadings, the excess pore pressure may rise in saturated sand deposits; it is assumed that liquefaction occurs when the excess pore pressure becomes equal to the initial effective vertical stress, i.e. when the sand stiffness has deteriorated to zero.

Under uniform cyclic loading, the work performed in rearranging the particles of sand in saturated and undrained conditions is given by

$$\Delta W = \frac{\bar{v}\sigma_{c0}'}{\eta_w} \frac{e_0}{g(e_0 - e_m)} \int_0^{r_u} \frac{\mathrm{d}r_u'}{f(1 + r_u')} \tag{1}$$

in which $r_u = \bar{u}/\sigma_{v0}'$ is the excess pore pressure ratio, σ_{c0}' is the initial effective confining pressure, e_0 is the initial void ratio of the sand, e_m is the minimum void ratio for the sand, η_w is the bulk modulus of the water, and $\bar{v}$ is a parameter independent of the void ratio of the sand; $f(1 + r_u)$ and $g(e_0 - e_m)$ are nondecreasing functions (Nemat-Nasser & Shokooh[6]). Liquefaction occurs when $r_u = 1$, and the work performed, $\Delta W(r_u = 1)$ is a constant for a given initial state of the sand. The work done when the excess pore pressure rises from 0 to r_u, $\Delta W(r_u)$, can be normalized with respect to $\Delta W(r_u = 1)$; thus, the need to measure $\bar{v}$, σ'_{c0}, η_W, e_0 and $g(e_0 - e_m)$ is eliminated. The normalized measure of work is then defined as $r_W = \Delta W(r_u)/\Delta W(1)$.

Let the energy dissipated by hysteresis in one cycle of amplitude $\bar{\tau} = \tau/\sigma'_{c0}$ be denoted as $E_c(\bar{\tau})$. The value of ΔW after N cycles of constant amplitude loading may be considered proportional to the number of cycles of loading for large $\bar{\tau}$ (Nemat-Nasser & Shokooh[6]); thus

$$\Delta W = h(\bar{\tau})N E_c(\bar{\tau}) \tag{2}$$

where $h(\bar{\tau})$ is a weight function that depends only on $\bar{\tau}$. Liquefaction occurs when N reaches a critical value N_l, that is,

$$N_l h(\bar{\tau}) E_c(\bar{\tau}) = \Delta W(r_u = 1) \tag{3}$$

The energy ratio r_W, according to eqns (2) and (3) is given alternatively by $r_W = N/N_l$. The function $h(\bar{\tau})$ as well as $\Delta W(1)$ are obtained from the test results of undrained resistance to liquefaction under uniform cyclic stress loading.

It has been suggested (Seed *et al.*[12]) that N/N_l and r_u may be related by

$$\frac{N}{N_l} = \sin^{2\theta}\left(\frac{\pi r_u}{2}\right) \tag{4}$$

where θ is an empirical constant. Substituting N/N_l in eqn (4) by r_W and differentiating, the differential equations relating the excess pore pressure rise to the energy ratio r_W, can be written as

$$\frac{dr_u}{dr_W} = \frac{1}{\theta\pi \cos\left(\dfrac{\pi r_u}{2}\right) \sin^{2\theta-1}\left(\dfrac{\pi r_u}{2}\right)} \tag{5}$$

Under random seismic loading, the work ΔW at any time t after the start of the excitation is given by

$$\Delta W = \int_0^t X(t')\dot{E}_T(t')\,dt' \tag{6}$$

where $\dot{E}_T(t')$ is the rate of shear-strain energy dissipated at time t', and $X(t')$ is an equivalent weighting function considering the randomness of the stress amplitude based on the result of random vibration analysis (Pires *et al.*[8]). Using the chain rule of differentiation together with eqns (5) and (6), the differential equation describing the excess pore pressure rise can be written as

$$\frac{dr_u}{dt} = \frac{X(t)\dot{E}_T(t)}{\Delta W(1) \cos\left(\dfrac{\pi r_u}{2}\right) \sin^{2\theta-1}\left(\dfrac{\pi r_u}{2}\right)} \tag{7}$$

Therefore, eqn (7) governs the excess pore pressure rise as a function of hysteretic energy dissipation rate. Under random excitation, $\dot{E}_T(t)$ is a random function of time and can be obtained from the random vibration analysis outlined earlier. Under the excitation of an earthquake with given intensity and duration, liquefaction occurs when the excess pore pressure ratio r_u reaches unity, or alternatively when the cumulative energy dissipation ΔW reaches the critical value $\Delta W(1)$. Using the latter criterion, the probability of liquefaction is, therefore,

$$P[\Delta W(1) - \Delta W < 0] \tag{8}$$

in which $\Delta W(1)$ and ΔW are random variables. The statistics of ΔW are obtained from the random vibration analysis, including the uncertainties in ground motion parameters and soil properties through sensitivity analysis, and first-order-second-moment approximation (Pires *et al.*[8]). The statistics of $\Delta W(1)$ are obtained from the uncertainty analysis of the undrained resistance against liquefaction under uniform cyclic stress loading.

4 CASE STUDIES

The probabilities of liquefaction predicted with the proposed methodology are compared with the field performance of sand deposits during past earthquakes. The historical data for the in situ resistance of the sand are conventionally plotted against the intensity of the earthquake load, and a boundary separating the cases of liquefaction and no-liquefaction is determined (Seed & Idriss[10]). A convenient parameter to represent the intensity of an earthquake is the ratio of the average shear stress τ_{ave} developed on horizontal surfaces of the sand to the initial effective vertical stress σ'_{v0}. Values of these shear-stress ratios known to be associated with some evidence of liquefaction or no-liquefaction in the field are plotted as a function of the standard penetration resistance (SPT), N_1, corrected to a value of σ'_{v0} equal to 1 ton/sq-ft (Seed & Idriss[10]).

A graphical representation of some of the available historical data is shown in Fig. 3, where the boundary separating the cases of liquefaction and no-liquefaction (Seed *et al.*[11]) is shown by a solid line; information about these data points can be found in Table 1.

The proposed methodology is used to calculate the probability of liquefaction for some of the data points in Fig. 3, namely: (i) at three locations in the city of Hachinohe (Japan) during the Tokachioki earthquake of May 16, 1968 (points 5, 6 and 9 in Fig. 3); (ii) at three locations in the city of Niigata (Japan) during the Niigata earthquake of 1964 (points 1, 3 and 4 in Fig. 3); and (iii) at one location in the city of Niigata (Japan) for two historical earthquakes of magnitude 6·1 and 6·6 (points 10 and 11 in Fig. 3).

The probabilities of liquefaction were calculated for each case considering two different values of the c.o.v.'s of the shear stress ratio that causes liquefaction in a given number of uniform loading cycles;

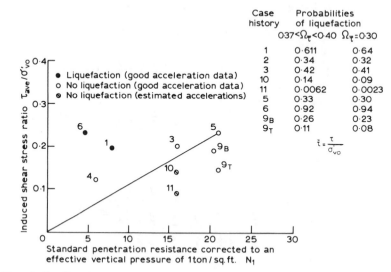

Fig. 3. Predicted liquefaction probabilities and comparison with historical data.

Table 1
Historical data on liquefaction (partial data)

Case history	Date	Site	M	Distance (miles)	$\bar{T}_E$ (s)	COV T_E	a_{max} (g)	Depth of water table (ft)	D (ft)	σ'_{v0} (psf)	N-SPT	N_1	D_r	$\frac{\tau_{ave}}{\sigma'_{v0}}$	Field behavior	Reference
1	1964 Niigata		7·5	32	15	0·8	0·17	3	20	1200	6	8	53	0·195	Liq.	Seed & Idriss[10]
2	1964 Niigata		7·5	32	15	0·8	0·17	3	25	1500	15	18	64	0·195	Liq.	Kishida[18]
3	1964 Niigata		7·5	32	15	0·8	0·17	3	20	1200	12	16	64	0·195	No-liq.	Seed & Idriss[10]
4	1964 Niigata		7·5	32	15	0·8	0·17	12	25	2000	6	6	53	0·12	No-liq.	Seed and Idriss[10]
5	1968 Hachinohe		7·8	45–100	15	0·8	0·21	3	12	800	14	21	78	0·23	No-liq.	Ohsaki[19]
6	1968 Hachinohe		7·8	45–100	15	0·8	0·21	3	12	800	<4	<6	~45	0·23	Liq.	Ohsaki[19]
7	1968 Hachinohe		7·8	45–100	15	0·8	0·21	5	10	800	15	23	80	0·185	No-liq.	Ohsaki[19]
8	1968 Hakodate		7·8	100	15	0·8	0·21	3	15	1000	6	9	55	0·205	Liq.	Kishida[20]
9_B	1968 Hachinohe		7·8	45–100	15	0·8	0·21	7	47	2900	25	21	75	0·19	No-liq.	Ohsaki[19]
9_T	1968 Hachinohe		7·8	45–100	15	0·8	0·21	7	9	850	15	22	80	0·16	No-liq.	Ohsaki[19]
10	1802 Niigata		6·6	24	10	0·8	0·12	3	20	1200	12	16	64	0·135	No-liq.	Seed & Idriss[10]
11	1887 Niigata		6·1	29	8	0·8	0·08	3	20	1200	12	16	64	0·09	No-liq.	Seed & Idriss[10]

these probabilities are summarized in Fig. 3. The model by Fardis & Veneziano[4] is used to characterize the undrained resistance to liquefaction under uniform cyclic stress loading. The statistics of the strong-motion duration shown in Table 1 were obtained with the correlations between magnitude and epicentral distance proposed by Lai[5] and Shinozuka *et al.*[13] Details concerning the modeling of the soil profiles for each case as well as the assumptions underlying the analysis can be found in Pires *et al.*[8]

Comparison with the observed historical data shows that the proposed methodology appears to be a viable procedure for predicting the seismic reliability of sand deposits against liquefaction, and for assessing the relative reliability of design alternatives.

5 NONLINEAR SOIL–STRUCTURE INTERACTION

When the soil that supports the structural system is considered as a compliant medium rather than a rigid one, the soil–structure interaction needs to be included in the structural response and damage analysis. Such consideration is generally advisable for rigid and massive structures such as power plant on soft ground. The restoring force due to foundation movement is generally nonlinear and hysteretic because of (1) the generally significant nonlinear stress–strain relationship of the soil under large-amplitude oscillations such as during earthquakes, and (2) possible separation (uplifting) of the foundation from the soil (Roesset & Tassoulas[9]).

To include the effects of soil–structure interaction, additional degrees of freedom (translation and rotational motions) of the foundation of the structure were introduced and a substructure (for the soil medium) technique was used (Chu *et al.*[3]). It is well known that the impedance function of the substructure is frequency-dependent. Such characteristics of the restoring force can be accurately modeled by representing the rocking motion as a third order system and the translation as a Voigt-type system shown in Fig. 4.

The interacting foundation shear force $H(t)$ and rocking moment $M(t)$ are given by

$$H(t) = c_x \dot{X} + \alpha_x k_{xn} x + (1 - \alpha_x) k_{xn} z_x$$
$$M(t) = m_{rn} \ddot{\theta} + c_{r1n} \dot{\theta} + \alpha_\theta k_{r1} \theta + (1 - \alpha_\theta) k_{r1} z_\theta + k_{r2n}(\theta - \theta_1) \quad (9)$$

in which x is foundation lateral displacement, θ is foundation rotation

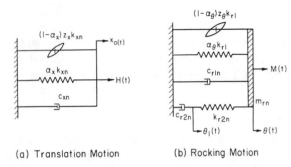

(a) Translation Motion   (b) Rocking Motion

Fig. 4. Substructure model for nonlinear soil half-space.

(rocking motion), k_{xn} is a lateral stiffness, k_{r1} is rotational stiffness, c's are viscous damping coefficients, and m_{rn} is the added mass in rotation. These coefficients are determined from a half-space analysis, depending on the foundation size and soil properties. z_x and z_θ provide the hysteretic parts of the restoring force and moment, based on the foregoing smooth hysteresis model. k_{r2n} and θ_1 are introduced to model the frequency-dependence of the impedance

$$\theta_1 = \frac{k_{r2n}}{c_{r2n}}(\theta - \theta_1) \tag{10}$$

When separation of the foundation from the supporting soil occurs, the primary effect is a decrease in the restoring force, i.e. lateral force as well as restoring moment. Figure 5 shows such softening of the 'stiffness' with respect to the foundation rocking motion due to foundation uplifting. The result is based on analysis developed by Wolf[17] using a Winkler-type foundation. Again, the hysteresis of foundation restoring force, largely a softening system with little damping, can be accurately described by the foregoing hysteretic restoring force model. A comparison of the proposed analytical model with a small scale experimental result (Akiyoshi et al.[1]) is shown in Fig. 6.

Consider a soil–structure system shown in Fig. 7. The structure is idealized as a shear beam system, i.e. with rigid girder. However, the columns may go into the inelastic range with the hysteresis given by the differential equation model described earlier. It is assumed that the foundation is of the surface type with little embedment. Therefore, the kinematic interaction, i.e. modification of the earthquake ground

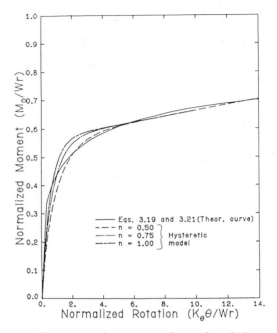

Fig. 5. Nonlinear restoring moment due to foundation uplifting.

motion due to the presence of the foundation, is negligible and only the inertial interaction due to the oscillation of the structure–foundation system is considered. The equations of motion of such a system with an *n*-story superstructure building may be written as follows:

$$m_0(\ddot{x}_0 + \ddot{b}) + \{1\}[M]\{\ddot{y}\} + H(t) = 0 \qquad (11)$$

$$J_0\ddot{\theta} + \{h\}^{\mathrm{T}}[M]\{\ddot{y}\} + M(t) = 0 \qquad (12)$$

$$m_i\ddot{y}_i + q_i - q_{i+1}(1 - \delta_{in}) = 0 \qquad (i = 1, n) \qquad (13)$$

where

m_0 = the mass of the foundation;
b = the free-field ground displacement in the absence of the building;
$\{1\}$ = the unit vector of order $(n \times l)$;
$[M]$ = the diagonal mass matrix of order n of the superstructure, m_i is the mass of the *i*th floor;

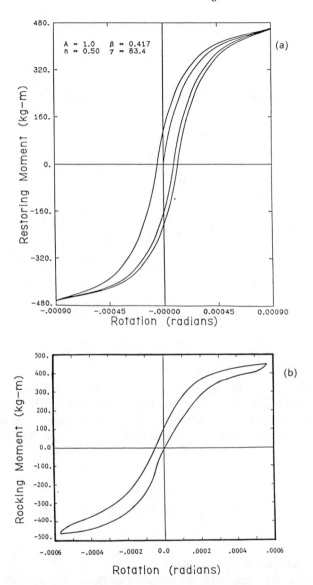

Fig. 6. Comparison of analytical model (a) for foundation restoring moment (with uplifting) with experimental result (b).

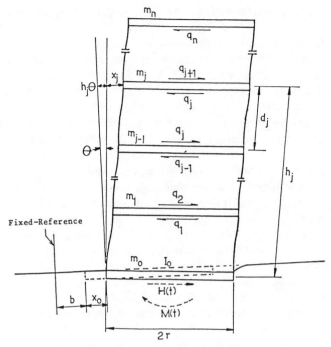

Fig. 7. Idealized building–foundation system.

$J_0 =$ the sum of the mass moment of inertia of all floors including the foundation about their respective centroidal axes;

$\{h\} =$ vector of the heights of the floor masses above the foundation;

$\delta_{in} =$ Kronecker delta;

$q_i = c_i\dot{u}_i + a_i k_i u_i + (1 - a_i)k_i z_i$, the ith interstory inelastic restoring force including viscous damping; where z_i is the hysteretic restoring force, and u_i is the interstory displacement;

$\{y\} =$ the absolute displacement vector of order $(n \times l)$, given by

$$\{y\} = \{x\} + \{l\}x_0 + \{h\}\theta + \{l\}b \qquad (14)$$

in which $\{x\}$ is the deformation vector of order $(n \times l)$ relative to the foundation; x_i is the ith floor deformation.

In applying the equivalent linearization solution procedure, because of the mismatch of the structural stiffness with that of the foundation (which is much larger), the algorithm proposed by Parlett & Reinsch[7] is used. Details of the analysis and solution procedure can be found in Chu *et al.*[3]

Extensive parametric studies were carried out to examine the effect of soil–structure interaction and the importance of various nonlinear effects under random excitations. It is found that the soil–structure interaction may increase or decrease the root mean square (r.m.s.) response depending on the structural height-to-base aspect ratio. Also, the nonlinearity in the superstructure contributes a dominant role in determining the structural response. Nonlinearities in the soil material and the effect of uplifting are relatively unimportant as far as displacement response is concerned. For example, the r.m.s. displacements of reinforced concrete buildings modeled as single-degree-of-freedom systems under an earthquake excitation with maximum acceleration of $0.6g$ are shown in Fig. 8; r indicates the effective foundation radius (size), $\lambda_2 = h/r$ is the building height to width (aspect) ratio. $\lambda_1 = Kh/(\pi r^2 G)$ is the stiffness ratio of the building to that of the soil, K is building stiffness, and G is soil shear modulus. The r.m.s. displacement is normalized by that of a fixed-base structure which is otherwise identical. The deviation of this r.m.s. displacement ratio from unity is, therefore, a measure of the significance of the soil structure interaction. It is seen that as the stiffness ratio increases, the interaction becomes more important. Depending on the aspect ratio

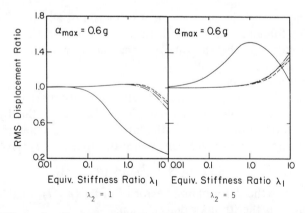

Fig. 8. Effect of soil-structural interaction and system nonlinearities. ———, linear system; - - - - -, nonlinear structure with linear foundation; —·—, nonlinear structure with foundation uplifting; —··—, nonlinear structure, soil with foundation uplifting.

and assumption of structural and foundation restoring forces being linear or nonlinear, the response may increase or decrease because of the interaction effect. The lateral movement of the foundation, in general, cushions the earthquake impact. On the other hand, the rocking motion of the foundation amplified the response, especially for tall buildings (large λ_2). A salient feature of the effect of various nonlinearities in the system is the dominance of the nonlinearity in the structure, i.e. when the structure is allowed to become inelastic, the interaction effect diminishes dramatically. Soil nonlinearity as well as nonlinearity from foundation uplifting are of minor significance.

The response of a ten-storey steel frame building designed according to the Uniform Building Code Specifications (Fig. 9) is also examined. The soil is assumed to have a shear modulus of $1.94\,k/in^2$. The equivalent radii of the foundation are 537 in. in translation and 510 in. in rocking motion. The r.m.s. story drifts under two different levels of earthquake excitation with and without interaction effect are shown in Fig. 10. It is seen that the interaction effect is generally small. It is pointed out that the above findings are based on systems with surface foundation. Embedments, specially deep ones, may change the response behavior significantly.

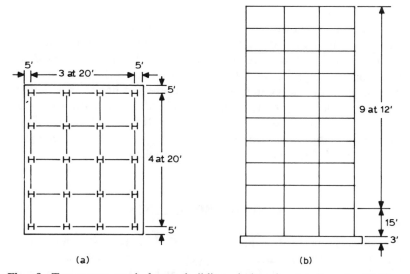

Fig. 9. Ten-storey steel frame building designed according to Uniform Building Code in the USA. (a) Floor framing plan and (b) typical frame section.

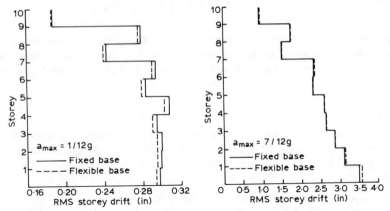

Fig. 10. Effect of soil–structure interaction on r.m.s. storey drift.

Uncertainty and reliability studies have also been carried out for the building–foundation systems. Details can be found in Chu et al.[3]

6 SUMMARY AND CONCLUSION

Soil and soil–structure systems subject to random excitations have been important and challenging problems because of the related safety consideration and the extremely complicated behavior of soil under cyclic loads. Empirical and time history methods have been relied upon to evaluate the system response. Presented herein is a random vibration method to predict the response and quantify the variability which is an integral part of the problem. Results generally indicate that it is a viable approach. Further efforts, however, are needed in relaxing some of the idealizations of the problem.

ACKNOWLEDGEMENTS

The study is part of a program on the safety and reliability of structures to seismic hazard, supported by the National Science Foundation under grants CEE 82-13729 and ECE 85-11972. Contributions of former graduate students, in particular, Dr J. A. Pires and Dr C. T. Chu are gratefully acknowledged.

REFERENCES

1. Akiyoshi, Y., Akio, H., Yoshiharu, K. & Naito, S., Laboratory test of foundation rocking characteristics. *Research for Vibration Test of Structure-Foundation Model*, No. 3, *Trans. A. I. J.*, No. 2049, 1981, p. 156.
2. Baber, T. T. & Wen, Y. K., Random vibration of hysteretic degrading systems. ASCE, Vol. 107, No. EM6, Proc. Paper 16712, December 1981, pp. 1069–87.
3. Chu, C. T., Wen, Y. K. & Ang, A. H-S., Random vibration of nonlinear building–foundation systems. Civil Engineering Studies, SRS No. 517, University of Illinois, Urbana, IL, 1985.
4. Fardis, M. N. & Veneziano, D., Statistical analysis of sand liquefaction. ASCE, Vol. 107, No. GT10, Proc. Paper 16604, October 1981, pp. 1361–77.
5. Lai, S-S. P., Overall safety assessment of multistorey steel buildings subjected to earthquake loads. Department of Civil Engineering, Publication No. R80-26, MIT, 1980.
6. Nemat-Nasser, S. & Shokooh, A., A unified approach to densification and liquefaction of cohesionless sand under cyclic shearing. *Canadian Geotechnical Journal*, **16**(4) (1979) 659–78.
7. Parlett, B. N. & Reinsch, C., Balancing a matrix for calculation of eigenvalues and eigenvectors. *Numerical Mathematics*, **13** (1969) 293–504.
8. Pires, J. E. A., Wen, Y. K. & Ang, A. H-S., Stochastic analysis of liquefaction under earthquake loading. Civil Engineering Studies, SRS No. 450, University of Illinois, Urbana, IL, 1983.
9. Roesset, J. M. & Tassoulas, J. L., Nonlinear soil structure interaction: an overview. *Earthquake Ground Motion and Its Effect on Structures*, AMD, ASME, **53** (1982) 57–76.
10. Seed, H. B. & Idriss, I. M., Evaluation of liquefaction potential of sand deposits based on observations of performance in previous earthquakes. Preprint 81–544, ASCE, October 1981.
11. Seed, H. B., Arango, I. & Chan, C. K., Evaluation of soil liquefaction potential during earthquakes. Report EERC 75–28, University of California, Berkeley, CA, October 1975.
12. Seed, H. B., Martin, P. P. & Lysmer, J., Porewater pressure changes during soil liquefaction. ASCE, Vol. 102, No. GT4, Proc. Paper 12074, April 1976, pp. 323–46.
13. Shinozuka, M., Kameda, H. & Koike, T., Ground strain estimation for seismic risk analysis. ASCE, *Journal of Engineering Mechanics*, Vol. 109, No. EM1, February 1983, pp. 175–91.
14. Tatsuoka, F. *et al.*, Shear modulus and damping by drained tests on clean sand specimens reconstructed by various methods. *Soils and Foundation*, **19**(1) (March 1979) 39–54.
15. Vanmarcke, E. H. & Lai, S-S. P. Strong-motion duration and RMS amplitude of earthquake records. *Bulletin of the Seismological Society of America*, **70**(4) (August 1980) 1293–307.

16. Wen, Y. K., Equivalent linearization for hysteretic systems under random excitation. *Journal of Applied Mechanics*, Transactions of the ASME, **47** (March 1980) 150–4.
17. Wolf, J. P., Soil–structure interaction with separation of base mat from soil. *Nuclear Engineering Design*, **38** (August 1976) 357–84.
18. Kishida, H., Damage to reinforced concrete buildings in Niigata City with special reference to formulation engineering, *Soils and Foundations, Tokyo, Japan*, **VII** (1) (1966) 71–86.
19. Ohsaki, Y., Niigata earthquake, 1964: building damage and soil condition, *Soils and Foundations*, **6**(2) (1966) 14–37.
20. Kishida, H., Characteristics of liquefaction of level sandy ground during the Tokachioki earthquakes, *Soils and Foundations, Tokyo, Japan*, **X** (2) (June 1970).

18

Modal Analysis of Nonclassically Damped Structural Systems Using Canonical Transformation

J. N. YANG, S. SARKANI & F. X. LONG

Department of Civil, Mechanical and Environmental Engineering, The George Washington University, Washington, DC, USA

ABSTRACT

An alternate modal decomposition method for dynamic analysis of nonclassically damped structural systems is developed. The resulting decoupled equations contain only real parameters. Hence, the solution can be obtained in the real field. Several procedures are outlined to solve these equations for both deterministic and nonstationary random ground excitations. Prior work has shown that the effect of nonclassical damping may be significant for the response of light equipment attached to a structure. Therefore, the proposed solution technique is applied to find the response of a light equipment that is attached to a multidegree-of-freedom structure.

Numerical results obtained from deterministic and nonstationary random vibration analyses indicate that the effect of nonclassical damping on the response of tuned equipment is significant only when the mass ratio and damping ratio of the equipment are small. Under this circumstance, the approximate classically damped solution, i.e. the solution obtained using the undamped modal matrix and disregarding the off-diagonal terms of the resulting damping matrix, is usually unconservative.

For detuned equipment, neglecting the effect of nonclassical damping generally results in an equipment response that is close to the exact solutions.

347

1 INTRODUCTION

In seismic analysis of linear multidegree-of-freedom viscously damped structures, it is quite common to assume that the damping matrix is of the classical (proportional) form (i.e. the form specified by Caughey & O'Kelly[1]). This assumption enables one to decouple the equations of motion by using the undamped eigenvectors of the system. After the equations are decoupled, the response quantities of interest can be obtained by either the response spectrum approach or numerical integration of the decoupled equations.[2] Solution of each decoupled equation represents the contribution of a particular mode of vibration of the structure to the total response. Furthermore, the response in most situations can be approximated by contributions from only a few dominant modes.

In general, however, real structures are not classically damped. Therefore, the damping matrix can not be diagonalized by the eigenvectors of the undamped system. Under this circumstance, one may still decouple the equations of motion of a nonclassically damped structural system using the eigenvectors of the damped system. However, for such structures, the damped eigenvectors are complex valued. This procedure, first proposed by Foss[3] and Traill-Nash,[8] decouples the equations of motion of an n degree-of-freedom nonclassically damped system into a set of $2n$ first order equations. These decoupled equations contain complex parameters. Recently, Singh & Ghafory-Ashtiany[7] developed a step by step numerical integration algorithm to solve these decoupled complex valued equations. Igusa et al.,[4] used a random vibration approach to obtain statistical moments of the response of a nonclassically damped system subjected to stationary white noise excitations. Velectos & Ventura[9] investigated the dynamic properties of nonclassically damped systems, and compared the exact solutions of such systems with those computed using the undamped eigenvectors and disregarding the off-diagonal terms of the resulting damping matrix under deterministic excitations.

In this paper an alternate modal decomposition approach employing canonical transformation is presented. The resulting decoupled equations involve only real parameters, thus avoiding computations to be carried out in the complex field. Several procedures are outlined to solve these equations for both deterministic and nonstationary random ground excitations. Prior work[4] has shown that the effect of nonclassical damping may be significant for the response of light equipment attached to a structure. The proposed canonical modal analysis technique is employed to carry out a parametric study for the effect of nonclassical damping on the response of a secondary system that is attached to a primary structure. Numerical results indicate that for tuned light equipment, neglecting nonclassical damping can result in a significantly unconservative equipment response.

2 BACKGROUND

The response of a linear n degree-of-freedom structure to a ground acceleration, $\ddot{x}_g$, can be obtained by solving the following matrix equation of motion

$$\mathbf{M}\ddot{\mathbf{X}} + \mathbf{C}\dot{\mathbf{X}} + \mathbf{K}\mathbf{X} = -\mathbf{M}\mathbf{r}\ddot{x}_g \tag{1}$$

in which $\mathbf{M}$, $\mathbf{C}$ and $\mathbf{K}$ denote the $(n \times n)$ mass, damping and stiffness matrices of the structure, respectively, and $\mathbf{X}$ is an n displacement vector relative to the moving base. The vector $\mathbf{r}$ is the influence coefficient vector obtained by statically displacing the base of the structure by unity in the direction of the ground motion. A super dot $(\cdot)$ represents differentiation with respect to time and a bold symbol denotes a vector or a matrix.

Caughey & O'Kelly[1] showed that if the damping matrix satisfies the identity $\mathbf{C}\mathbf{M}^{-1}\mathbf{K} = \mathbf{K}\mathbf{M}^{-1}\mathbf{C}$, the eigenvectors of the undamped system can be used to transform the equations of motion, eqn (1), into a set of n decoupled equations. The system with a damping matrix satisfying this condition is said to be classically damped. However, the eigenvectors of the undamped system will no longer diagonalize the damping matrix that is not of the classical form.

For the nonclassically damped system, the approach proposed by Foss[3] and Traill-Nash[8] can be used. In this approach, the n second order equations of motion are converted into $2n$ first order equations. The eigenvalue problem is solved leading to n pairs of complex conjugate eigenvalues and eigenvectors. Finally, using the complex eigenvectors, the equations of motion are decoupled into $2n$ first order equations. The decoupled equations contain complex parameters, although the solution for the structural response is real. In order to avoid the numerical solution for the complex differential equations, we propose the following alternate approach.[10,13] In this approach, the only complex calculations are the determination of the eigenvalues and eigenvectors.

3 CANONICAL MODAL ANALYSIS

The n second order equations of motion, eqn (1), are converted into $2n$ first order equations as follows:

$$\dot{\mathbf{Y}} = \mathbf{A}\mathbf{Y} + \mathbf{W}\ddot{x}_g \tag{2}$$

where $\mathbf{Y}$ is a $2n$ vector, referred to as the state vector,

$$\mathbf{Y} = \left\{ \begin{matrix} \dot{\mathbf{X}} \\ \cdots \\ \mathbf{X} \end{matrix} \right\}$$

and

$$\mathbf{A} = \left[\begin{array}{c|c} -\mathbf{M}^{-1}\mathbf{C} & -\mathbf{M}^{-1}\mathbf{K} \\ \hline \mathbf{I} & \mathbf{0} \end{array} \right]; \quad \mathbf{W} = \left\{ \begin{matrix} -\mathbf{r} \\ \cdots \\ \mathbf{0} \end{matrix} \right\} \tag{3}$$

The eigenvalues and eigenvectors of the matrix $\mathbf{A}$ consist of n pairs of complex conjugates. The jth pair of eigenvalues, denoted by λ_{2j-1} and λ_{2j}, and the jth pair of eigenvectors, denoted by $\boldsymbol{\phi}_{2j-1}$ and $\boldsymbol{\phi}_{2j}$, can be expressed as[6]

$$\lambda_{2j-1} = -\xi_j\omega_j + i\omega_{Dj}, \qquad \lambda_{2j} = -\xi_j\omega_j - i\omega_{Dj} \tag{4}$$

$$\boldsymbol{\phi}_{2j-1} = \mathbf{a}_j + i\mathbf{b}_j \tag{5a}$$

$$\boldsymbol{\phi}_{2j} = \mathbf{a}_j - i\mathbf{b}_j, \qquad j = 1, 2, \ldots, n \tag{5b}$$

in which ξ_j, ω_j and $\omega_{Dj} = \omega_j(1 - \xi_j)^{1/2}$ are real values, and $\mathbf{a}_j$ and $\mathbf{b}_j$ are $2n$ real vectors.

The $(2n \times 2n)$ real matrix $\mathbf{T}$ constructed in the following

$$\mathbf{T} = [\mathbf{a}_1, \mathbf{b}_1, \mathbf{a}_2, \mathbf{b}_2, \ldots, \mathbf{a}_j, \mathbf{b}_j, \ldots, \mathbf{a}_n, \mathbf{b}_n] \tag{6}$$

will transform the matrix $\mathbf{A}$ into a canonical form $\boldsymbol{\Lambda}$ (Ref. 5), i.e.

$$\boldsymbol{\Lambda} = \mathbf{T}^{-1}\mathbf{A}\mathbf{T} \tag{7}$$

in which $\mathbf{T}^{-1}$ is the inverse of the $\mathbf{T}$ matrix and

$$\boldsymbol{\Lambda} = \begin{bmatrix} \boldsymbol{\Lambda}_1 & & & \mathbf{0} \\ & \boldsymbol{\Lambda}_2 & & \\ & & \ddots & \\ \mathbf{0} & & & \boldsymbol{\Lambda}_n \end{bmatrix} \tag{8}$$

where

$$\boldsymbol{\Lambda}_j = \begin{bmatrix} -\xi_j\omega_j & \omega_{Dj} \\ -\omega_{Dj} & -\xi_j\omega_j \end{bmatrix}, \qquad j = 1, 2, \ldots, n \tag{9}$$

Let the transformation of the state vector be

$$\mathbf{Y} = \mathbf{Tv} \tag{10}$$

Substituting eqn (10) into eqn (2) and premultiplying it by the inverse of the $\mathbf{T}$ matrix, $\mathbf{T}^{-1}$, one obtains

$$\dot{\mathbf{v}} = \mathbf{\Lambda v} + \mathbf{F}\ddot{x}_g \tag{11}$$

in which $\mathbf{\Lambda}$ is given by eqn (8) and

$$\mathbf{F} = \mathbf{T}^{-1} \left\{ \begin{matrix} -\mathbf{r} \\ \cdots \\ \mathbf{0} \end{matrix} \right\} \tag{12}$$

Equation (11) consists of n pairs of decoupled equations. Each pair of equations represents one vibrational mode, and it is uncoupled with other pairs. However, the two equations in each pair are coupled. The transformation given in eqn (10) is referred to as the canonical transformation.[5,10,11] All the parameters in eqn (11) are real.

The jth pair of coupled equations in eqn (11), corresponding to the jth vibrational mode, is given as follows:

$$\dot{v}_{2j-1} = -\xi_j \omega_j v_{2j-1} + \omega_{Dj} v_{2j} + F_{2j-1} \ddot{x}_g \tag{13a}$$

$$\dot{v}_{2j} = -\omega_{Dj} v_{2j-1} - \xi_j \omega_j v_{2j} + F_{2j} \ddot{x}_g \tag{13b}$$

in which F_{2j-1} and F_{2j} are the $2j$-1th and the $2j$th elements of the $\mathbf{F}$ vector, respectively. Solutions of eqns (13) together with the transformation of eqn (10) yield the response state vector $\mathbf{Y}(t)$ of the structural system.

The advantage of the formulation given above is that the computations for the solutions are all in the real field. The modal decomposition approach described above is referred to as the canonical modal decomposition. When the structure is classically damped, it can be shown analytically that eqn (13a) reduces to the solution for the jth mode as expected.

Equation (13) can be solved easily using either the impulse response function approach or frequency response function approach to be described later. Likewise, it can further be decoupled, if one wishes to obtain one equation in terms of each unknown, although this procedure is unnecessary.

In the following sections different approaches that can be used to solve eqns (13a,b) for both deterministic and nonstationary stochastic ground excitation are described.

4 DETERMINISTIC RESPONSE TO SPECIFIC EXCITATION

4.1 Direct Numerical Integration

Given the record of ground excitation $\ddot{x}_g$, eqns (13a,b) can be numerically integrated and the response state vector $Y(t)$ can be obtained by using the transformation of eqn (10). Again, all the parameters in these equations are real.

4.2 Impulse Response Function

One can obtain the impulse response vector $h_v(t)$ for mode j, $v_j(t) = [v_{2j-1}, v_{2j}]'$, due to the ground acceleration $\ddot{x}_g(t) = \delta(t)$ where $\delta(t)$ is the Dirac delta function and a prime denotes the transpose of a vector or matrix. The jth modal impulse response vector is given by

$$h_{v_j}(t) = \left\{ \begin{array}{c} (e^{-\xi_j\omega_j t}\cos\omega_{Dj}t)F_{2j-1} + (e^{-\xi_j\omega_j t}\sin\omega_{Dj}t)F_{2j} \\ -(e^{-\xi_j\omega_j t}\sin\omega_{Dj}t)F_{2j-1} + (e^{-\xi_j\omega_j t}\cos\omega_{Dj}t)F_{2j} \end{array} \right\} \quad (14)$$

and the response vector, $v_j(t)$, corresponding to mode j at time t is

$$v_j(t) = \int_0^t h_{v_j}(t - \tau)\ddot{x}_g(\tau)\,d\tau \quad (15)$$

Again eqn (15) can be integrated numerically and the response state vector $Y(t)$ can be obtained through the transformation of eqn (10).

4.3 Frequency Response Function

Let $H_{v_j}(\omega)$ be the complex frequency response vector for mode j due to ground acceleration, $\ddot{x}_g = e^{i\omega t}$. Then, $H_{v_j}(\omega)$ can be obtained easily from eqns (13a,b) as follows

$$h_{v_j}(\omega) = \left\{ \begin{array}{c} \dfrac{\xi_j\omega_j F_{2j-1} + \omega_{Dj}F_{2j} + i\omega F_{2j-1}}{-\omega^2 + 2i\xi_j\omega_j\omega + \omega_j^2} \\ \dfrac{\xi_j\omega_j F_{2j} - \omega_{Dj}F_{2j-1} + i\omega F_{2j}}{-\omega^2 + 2i\xi_j\omega_j\omega + \omega_j^2} \end{array} \right\} \quad (16)$$

Note that the vectors $H_{v_j}(\omega)$ and $h_{v_j}(t)$ are related through the Fourier transform pair, i.e.

$$h_{v_j}(t) = \frac{1}{2\pi}\int_{-\infty}^{\infty} H_{v_j}(\omega)e^{i\omega t}\,d\omega; \qquad H_{v_j}(\omega) = \int_{-\infty}^{\infty} h_{v_j}(t)e^{-i\omega t}\,dt \quad (17)$$

Finally, the jth modal response vector in the time domain can be obtained as follows

$$\mathbf{v}_j(t) = \frac{1}{2\pi} \int_{-\infty}^{\infty} \mathbf{H}_{v_j}(\omega)\ddot{x}_g(\omega)e^{i\omega t}\, d\omega \tag{18}$$

in which $\ddot{x}_g(\omega)$ is the Fourier transform of the ground acceleration, $\ddot{x}_g(t)$. The above calculation can be carried out efficiently using the Fast Fourier Transform (FFT) algorithm.

5 STOCHASTIC RESPONSE TO RANDOM EXCITATION

The canonical modal decomposition method presented above can be used conveniently to obtain the response of a nonclassically damped structural system to a nonstationary random ground acceleration.[10-12] Often, the earthquake ground excitation, $\ddot{x}_g(t)$, can be modeled as a uniformly modulated nonstationary random process with zero mean $\ddot{x}_g(t) = \alpha(t)\ddot{x}_0(t)$, in which $\alpha(t)$ is a deterministic nonnegative modulating or envelope function and $\ddot{x}_0(t)$ is a stationary random process with zero mean and a power spectral density, $\phi_{\ddot{x}\ddot{x}}(\omega)$.

Since $\ddot{x}_g(t)$ has a zero mean, the mean value of the state response vector, $\mathbf{Y}(t)$, is zero. After carrying out the random vibration analysis,[13] it can be shown that the variance vector $\sigma_Y^2(t)$ of $\mathbf{Y}(t)$, which is equal to the mean square response vector of $\mathbf{Y}(t)$, is given as follows

$$\sigma_Y^2(t) = \int_{-\infty}^{\infty} |\mathbf{M}_Y(t, \omega)|^2\, \phi_{\ddot{x}\ddot{x}}(\omega)\, d\omega \tag{19}$$

in which $|\mathbf{M}_Y(t, \omega)|^2 \phi_{\ddot{x}\ddot{x}}(\omega)$ is the evolutionary power spectral density of the state vector, $\mathbf{Y}(t)$, and $|\mathbf{M}_Y(t, \omega)|^2$ is a vector in which its elements are the square of the absolute value of the corresponding element of $\mathbf{M}_Y(t, \omega)$ given by

$$\mathbf{M}_Y(t, \omega) = \int_0^t \mathbf{T}\, \mathbf{h}_v(\tau)\alpha(t - \tau)e^{-i\omega\tau}\, d\tau \tag{20}$$

in which $\mathbf{h}_v(\tau) = [\mathbf{h}'_{v_1}(\tau), \mathbf{h}'_{v_2}(\tau), \ldots, \mathbf{h}'_{v_n}(\tau)]'$ is the impulse response vector.

The nonstationary mean square response given by eqn (20) can be computed easily using the FFT technique.[13]

6 NUMERICAL EXAMPLES

The canonical modal analysis presented in this paper will be used to study the response behavior of nonclassically damped structural systems subjected to earthquake-type excitations. Parametric studies will be conducted to determine the response behavior of the primary-secondary structural system. In particular, these studies will ascertain the conditions under which nonclassical damping becomes significant. Frequently, it may be assumed that the structure and equipment are classically damped individually. However, because of different damping characteristics of the equipment and the structure, the combined equipment–structure system generally is not classically damped.

Igusa et al.[4] considered a single degree-of-freedom equipment attached to a single degree-of-freedom structure and subjected to a stationary white noise ground excitation. They showed that at tuning (when frequencies of the equipment and structure coincide) the effect of nonclassical damping on the response of light equipment becomes significant. Here we investigate the effect of nonclassical damping on somewhat complex equipment–structure systems excited by an earthquake ground acceleration. Yong and Lin[14] conducted an extensive parametric study for the frequency response function of the primary-secondary structural system. Here we examine the response of the equipment–structure system to both deterministic and nonstationary stochastic ground excitations. Of particular interest is the equipment–structure system in which the primary structure is nonclassically damped.

The example problem studied is subjected to either a simulated sample ground motion or the ground motion that is modeled as a nonstationary random process as described in the previous section. The envelope function, $\alpha(t)$, and the spectral density, $\phi_{\ddot{x}\ddot{x}}(\omega)$, of the earthquake model are given in the following:

$\alpha(t) = t^2/9$ for $0 \le t \le 3$ s, $\alpha(t) = 1$ for $3 \le t \le 13$ s, and $\alpha(t)$

$= \exp[-0.26(t - 13)]$ for $t > 13$ s;

$$\phi_{\ddot{x}\ddot{x}}(\omega) = \frac{1 + 4\xi_g^2(\omega/\omega_g)}{[1 - (\omega/\omega_g)^2]^2 + 4\xi_g^2(\omega/\omega_g)^2} S^2$$

in which $\omega_g = 3.0$ Hz., $\xi_g = 0.65$ and $S^2 = 74.7 \times 10^{-4}$ m^2/s^3/rad. A simulated sample function of the earthquake ground acceleration is shown in Fig. 1.

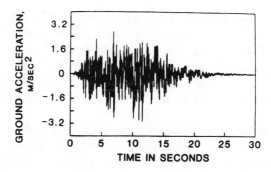

Fig. 1. Sample time history of simulated ground acceleration.

Considered is a two degree-of-freedom shear beam type structure with a single degree-of-freedom light equipment attached to it as shown in Fig. 2. This primary structure is classically damped if $c_1/k_1 = c_2/k_2$ and the combined equipment–structure system is classically damped if $c_1/k_1 = c_2/k_2 = c_e/k_e$ in which the subscript e refers to the equipment. The mass and stiffness of each storey unit of the primary structure are: $m_1 = m_2 = m = 30$ t; and $k_1 = k_2 = k = 19\cdot379$ kN/m. The natural frequencies of the primary structure are $2\cdot5$ and $6\cdot5$ Hz, respectively. Parametric studies will be carried out by varying the distribution of the dampings of the primary structure and the equipment.

Let the values of c_1 and c_2 be equal to $123\cdot4$ kN/m/s, so that the primary structure is classically damped with a first modal damping ratio of approximately 5%. Given a mass ratio, the damping ratio of the equipment is varied and the response of the equipment, i.e. displacement relative to the attachment point, is evaluated by the

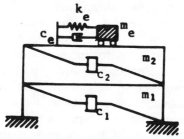

Fig. 2. Primary structure and single degree-of-freedom equipment.

exact methods presented in this paper. The results are compared with those obtained using the approximate classically damped approach. The approximate classically damped approach is to decouple the second order equations of motion using eigenvectors of the undamped system and disregard the off-diagonal terms of the $\boldsymbol{\Phi}C\boldsymbol{\Phi}$ matrix, where $\boldsymbol{\Phi}$ is the $(n \times n)$ modal matrix of the undamped primary-secondary system.

Let the equipment frequency, ω_e, be tuned to the fundamental frequency of the primary system, i.e. $\omega_e = 2.5\,\text{Hz}$. The response of the equipment to the simulated deterministic ground acceleration shown in Fig. 1 is presented in Fig. 3 for different values of equipment damping ratio, ξ_e, and mass ratio, γ. The mass ratio, γ, of the equipment is the ratio of the equipment mass to the first modal mass of the primary structure that is equal to 30 t. In Fig. 3, the ordinate is the maximum response of the equipment relative to the attachment point in 30 s of earthquake episode, and the abscissa, η, is the ratio of the equipment damping ratio, ξ_e, to the unique damping ratio of the equipment, ξ_{ec}, that would make the combined equipment–structure system classically damped. For the classically damped primary system in which $c_1/k_1 = c_2/k_2$, the damping ratio of the equipment, ξ_{ec}, that results in a classically damped primary-secondary system can be obtained using the Caughey–O'Kelly identity, $\xi_{ec} = (\omega_e/2)(c_1/k_1)$. Likewise, when the primary system is classically damped, ξ_{ec} is equal to the jth modal damping ratio of the primary structure if the equipment is tuned to the jth mode of the primary structure. Hence in the present example, ξ_{ec} is equal to the first modal damping ratio of the primary structure, which is 5%.

Figure 3 presents the results for three different values of the equipment mass ratio, γ. As expected, the equipment response increases as its damping ratio ξ_e reduces. The results based on the approximate classically damped approach start to deviate from the exact solution only when $\xi_e < \xi_{ec}$. The deviation increases as the equipment damping ratio or the mass ratio reduces. Further, the solutions obtained using the approximate classically damped approach in the region $\xi_e < \xi_{ec}$ are nonconservative. In other words, the effect of nonclassical damping becomes significant only when the equipment damping ratio, ξ_e, is smaller than ξ_{ec} that results in a classical damping for the combined equipment–structure system. It is also evident that the smaller the mass ratio or the equipment damping, the more pronounced is the effect of neglecting the off-diagonal terms of the $\boldsymbol{\Phi}'C\boldsymbol{\Phi}$ matrix.

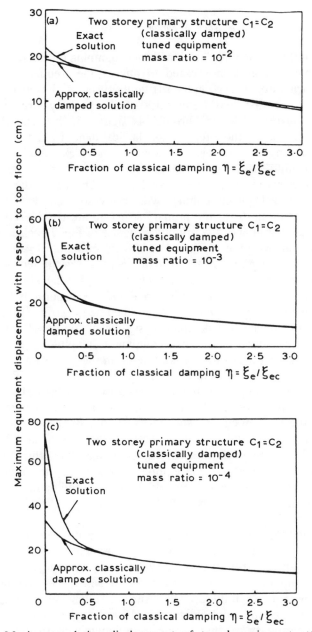

Fig. 3. Maximum relative displacement of tuned equipment attached to classically damped primary structure as function of equipment damping, $\eta = \dfrac{\xi_e}{\xi_{ec}}$. (a) Mass ratio $\gamma = 10^{-2}$; (b) mass ratio $\gamma = 10^{-3}$; (c) mass ratio $\gamma = 10^{-4}$.

Figure 4 shows the effect of nonclassical damping on the response of the equipment that is not tuned to any of the frequencies of the primary structure. The equipment frequency ω_e is chosen to be the average of the first two frequencies of the primary structure, i.e. $\omega_e = (\omega_1 + \omega_2)/2 = 4 \cdot 5$ Hz. From this figure, it is clear that for detuned equipment, the effect of nonclassical damping can be ignored without causing any problem. Likewise, the maximum equipment response is not sensitive to the mass ratio.

Next, the response of the secondary system when the primary structure is nonclassically damped will be investigated. Two different damping distributions for the primary structure are considered. In the first case, all damping of the primary structure of Fig. 2 is placed in the first story unit; with the results $c_1 = 246 \cdot 8$ kN/m/s and $c_2 = 0 \cdot 0$. In the second case, all damping of the structure is placed in the second storey unit, leading to the results $c_1 = 0 \cdot 0$ and $c_2 = 246 \cdot 8$ kN/m/s. The damping ratio of the equipment is varied and the response behavior of the equipment will be examined. It has been demonstrated above that the effect of nonclassical damping is significant for tuned equipment with small mass ratio γ and small equipment damping ξ_e. Equipment with such characteristics will be considered in the following.

Figures 5(a) and 5(b) show the maximum response in 30 s of the time history for equipment that is attached to a nonclassically damped primary structure described above. The equipment frequency, ω_e, is tuned to the undamped fundamental frequency of the primary structure, and the simulated sample earthquake shown in Fig. 1 is considered for the excitation. In these figures, the ordinate is the maximum displacement of the equipment relative to the attachment point and the abscissa, $\eta' = \xi_e/\xi'_{ec}$, is the ratio of the damping ratio of the equipment to the approximate ξ_{ec}, denoted by ξ'_{ec} as described in the following.

For tuned equipment attached to a classically damped primary structure, the damping ratio ξ_{ec} of the equipment which results in a classically damped primary-secondary system is the same as the modal damping ratio of the primary structure in which the equipment is tuned. However, when the primary structure is nonclassically damped, one can obtain an approximate value for ξ_{ec}, denoted by ξ'_{ec}, by treating the primary system as being approximately classically damped, i.e. neglecting the off-diagonal terms of the $\tilde{\Phi}'\mathbf{C}\tilde{\Phi}$ matrix of the primary system. In other words, ξ'_{ec} is obtained by dividing the diagonal terms of the $\tilde{\Phi}'\mathbf{C}\tilde{\Phi}$ matrix of the primary structure by twice

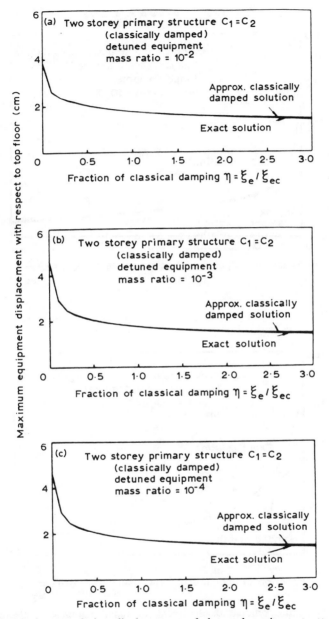

Fig. 4. Maximum relative displacement of detuned equipment attached to classically damped primary structure as function of equipment damping, $\eta = \dfrac{\xi_e}{\xi_{ec}}$. (a) Mass ratio $\gamma = 10^{-2}$; (b) mass ratio $\gamma = 10^{-3}$; (c) mass ratio $\gamma = 10^{-4}$.

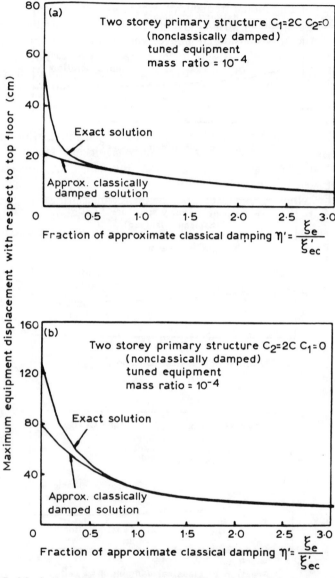

Fig. 5. Maximum relative displacement of tuned equipment attached to nonclassically damped primary structure as function of equipment damping,
$$\eta = \frac{\xi_e}{\xi_{ec}}. \text{ (a) } c_1 = 2c, \ c_2 = 0; \text{ (b) } c_2 = 2c, \ c_1 = 0.$$

the corresponding modal frequencies, where $\tilde{\Phi}$ is the modal matrix of the undamped primary structure. For the two nonclassically damped primary structures considered above, the approximate first modal damping ratios are $\xi'_{ec} = 7\%$ ($c_1 = 2c$, $c_2 = 0$) and $\xi'_{ec} = 2\cdot 8\%$ ($c_1 = 0$, $c_2 = 2c$), respectively. Recall that for this primary structure, distributing the damping equally between the two storey units results in a classically damped structure with the first modal damping ratio of approximately 5%, i.e. $\xi_{ec} = 5\%$.

From the results shown in Fig. 5, it is clear that the distribution of damping in the primary structure has a significant effect on the response of the equipment. Likewise, the effect of nonclassical damping is significant only when the equipment damping is small. To examine whether ξ'_{ec} is a useful parameter for measuring the effect of nonclassical damping, results in Fig. 5 are replotted in Fig. 6 in a different form. In this figure the ordinate is the ratio of the maximum

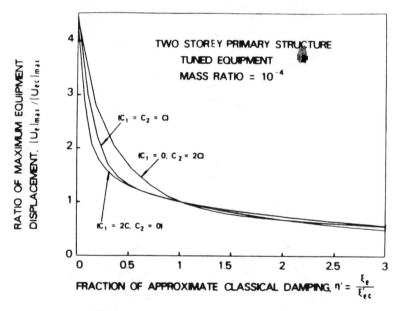

Fig. 6. Ratio of maximum relative displacement of tuned equipment attached to nonclassically damped primary structure as function of equipment damping,

$$\eta' = \frac{\xi_e}{\xi_{ec}}.$$

exact equipment response, denoted by $(U_e)_{max}$, to that exact equivalent response obtained by setting the equipment damping equal to ξ'_{ec}, denoted by $(U_{ec})_{max}$. When the damping of the equipment is equal to ξ'_{ec}, i.e. $\eta' = \xi_e/\xi'_{ec} = 1$, the exact equipment response and the approximate classically damped results are almost identical as observed from Fig. 5 (see the solutions at $\eta' = \xi_e/\xi'_{ec} = 1\cdot0$ in Fig. 5). Also plotted in Fig. 6 are the corresponding results when the primary structure is classically damped ($c_1 = c_2 = c$). From this figure it is clear that the approximate modal damping ratios obtained from the diagonal terms of the $\mathbf{\Phi}'\mathbf{C}\mathbf{\Phi}$ matrix of the primary structure can be used as a useful measure in determining the effect of nonclassical damping on the response of tuned equipment even if this matrix is not diagonal.

The first example is reconsidered when the input process is a nonstationary stochastic ground acceleration with a power-spectral density, $\phi_{\ddot{x}\ddot{x}}(\omega)$, and an envelope function, $\alpha(t)$, described previously. Since the mean value of earthquake ground acceleration is zero, the mean value of the response quantities are all zero. Therefore, the root mean square (r.m.s.) of the response is equal to the standard

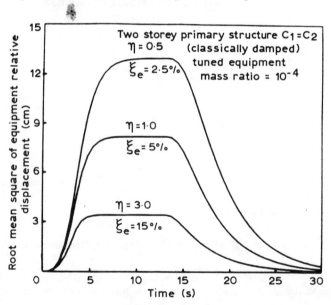

Fig. 7. Standard deviation of tuned equipment response attached to classically damped primary structure as a function of time, t.

deviation. The time dependent r.m.s. response of the equipment tuned to the first mode of the classically damped primary structure is shown in Fig. 7 for several values of equipment damping ratio. The corresponding results for detuned equipment with $\omega_e = (\omega_1 + \omega_2)/2$ are displayed in Fig. 8. As expected, the r.m.s. response increases as the equipment damping reduces and the response for tuned equipment is at least one order of magnitude larger than that of detuned equipment. However, a comparison between Figs 7 and 8 indicates that the r.m.s. response of tuned equipment is quite sensitive to the equipment damping ξ_e, whereas this is not the case for detuned equipment.

Extensive numerical results show that the effect of nonclassical damping on the r.m.s. response of detuned equipment with $\omega_e = (\omega_1 + \omega_2)/2$ is insignificant. The effect of nonclassical damping on the response of tuned equipment with $\omega_e = \omega_1$ in terms of the maximum r.m.s. value is shown in Fig. 9 for three different values of mass ratio, γ. In Fig. 9 the ordinate is the maximum r.m.s. of the equipment

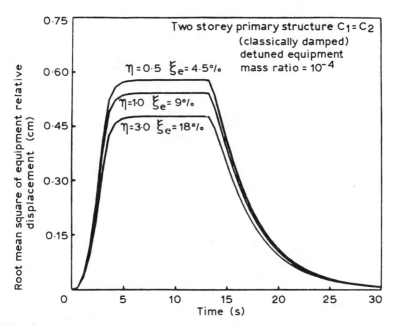

Fig. 8. Standard deviation of detuned equipment response attached to classically damped primary structure as a function of time, t.

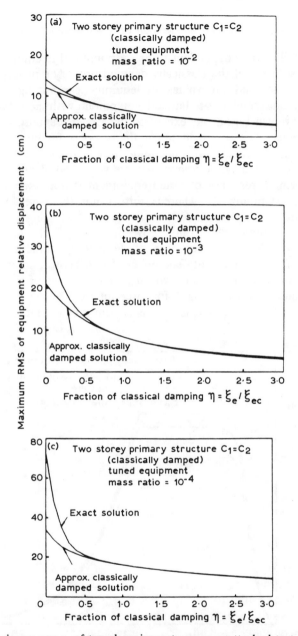

Fig. 9. Maximum r.m.s. of tuned equipment response attached to two storey classically damped primary structure as function of equipment damping, $\eta = \dfrac{\xi_e}{\xi_{ec}}$. (a) Mass ratio $\gamma = 10^{-2}$; (b) mass ratio $\gamma = 10^{-3}$; (c) mass ratio $\gamma = 10^{-4}$.

response in 30 s of earthquake episode, and the abscissa is $\eta = \xi_e/\xi_{ec} = \xi_e/0 \cdot 05$ as described in the first example. The results based on exact solution and approximate classically damped approach are presented in the figure. Similar to the response of equipment subjected to a simulated deterministic excitation presented previously, the effect of nonclassical damping for tuned equipment is pronounced only when the equipment is light and its damping ratio ξ_e is smaller than ξ_{ec}, i.e. $\eta = \xi_e/\xi_{ec} < 1$. Further, the approximate classically damped solutions are always unconservative. Further sensitivity studies for the effect of nonclassical damping on the behavior of primary-secondary structural systems have been conducted, and various interesting results have been obtained in Ref. 13.

7 CONCLUSIONS

A modal analysis approach, referred to as the canonical modal decomposition procedure, for dynamic analysis of nonclassically damped structural systems is presented. The main advantage of this procedure is that the resulting decoupled equations of motion contain only real parameters. Procedures are outlined to solve the decoupled equations for deterministic ground excitations. Also presented is a procedure to solve these decoupled equations when the ground excitation is a nonstationary random process.

The canonical modal decomposition procedure is used to perform parametric studies for the effect of nonclassical damping on the response of secondary systems. Both deterministic and nonstationary stochastic ground accelerations were considered. A single degree-of-freedom equipment attached to a classically damped multidegree-of-freedom primary structure was considered. Using the canonical modal decomposition procedures the response of the equipment for various dampings of the equipment, mass ratios, and tuning and detuning have been calculated.

Based on both deterministic and stochastic earthquake ground motion inputs, the following conclusions are obtained from our sensitivity studies for the response of primary-secondary systems, where the primary structure is classically damped. (1) The effect of nonclassical damping on the equipment response is significant when the following conditions are satisfied simultaneously: (i) the frequency of the equipment is tuned to that of any mode of the primary structure; (ii) the mass ratio is small; and (iii) the damping ratio ξ_e of the equipment is smaller than the damping ratio ξ_{ec} that results in a classically damped primary-secondary system. Under this circumstance, the approximate classically damped solutions, i.e. the solutions obtained using the undamped modal matrix and disregarding the off-diagonal terms of the resulting damping matrix, are usually unconservative. (2) When the equipment is detuned, the effect of nonclassical damping on the equipment response is negligible. Hence, the approximate classically damped approach can be used.

Also studied were equipment–structure systems in which the primary structure is nonclassically damped. Limited results indicate that for such primary-secondary systems, the effect of nonclassical damping on the equipment response can be estimated by approximating the primary structure as being classically damped. This is accomplished using the undamped modal matrix of the primary structure and disregarding the off-diagonal terms of the resulting damping matrix. Then the conclusions described above hold. Corresponding to ξ_{ec} for a classically damped primary structure, a meaningful measure of equipment damping for nonclassically damped primary systems, denoted by ξ'_{ec}, is determined using the approximate classically damped primary system.

ACKNOWLEDGEMENT

This work is supported by the National Center for Earthquake Engineering Research, State University of New York at Buffalo under grant NCEER-86-3022.

REFERENCES

1. Caughey, T. H. & O'Kelly, E. J., Classical normal modes in damped linear dynamic systems, *J. Applied Mechanics, ASME*, **32** (1965) 583–8.
2. Clough, R. W. & Penzien, J., *Dynamics of Structures*. McGraw-Hill, New York, 1975.
3. Foss, K. A., Coordinates which uncouple the equations of motion of damped linear dynamic systems. *Journal of Applied Mechanics, ASME*, **25** (1958) 361–4.
4. Igusa, T., Der Kiureghian, A. & Sackman, J. L., Modal decomposition

method for stationary response of non-classically damped systems. *Earthquake Engineering and Structural Dynamics,* **12** (1984) 121–36.

5. Perlis, S., *Theory of Matrices.* Addison Wesley, Reading, MA, 1958.
6. Singh, M. P., Seismic response by SRSS for proportional damping. *Journal of the Engineering Mechanics Division, ASCE,* **106** (1980) 1405–19.
7. Singh, M. P. & Ghafory-Ashtiany, M., Modal time history analysis of non-classically damped structures for seismic motions. *Earthquake Engineering and Structural Dynamics,* **14** (1986) 133–46.
8. Traill-Nash, R. W., Modal methods in the dynamics of systems with non-classical damping. *Earthquake Engineering and Structural Dynamics,* **9** (1983) 153–69.
9. Veletos, A. S. & Ventura, C. E., Modal analysis of non-classically damped linear systems. *Earthquake Engineering and Structural Dynamics,* **14** (1986) 217–43.
10. Yang, J. N. & Lin, M. J., Optimal critical-mode control of tall building under seismic load. *Journal of the Engineering Mechanics Division, ASCE,* **108** (EM6) (1982) 1167–85.
11. Yang, J. N. & Lin, M. J., Building critical-mode control: nonstationary earthquake. *Journal of the Engineering Mechanics, ASCE,* Paper 18429, **109** (EM6) (1983) 1375–89.
12. Yang, J. N., Lin, Y. K. & Sae Ung, S., Tall building response to earthquake excitations. *Journal of the Engineering Mechanics Division, ASCE,* **106** (EM4) (1980) 801–17.
13. Yang, J. N., Sarkani, S. & Long, F. X., Modal analysis of nonclassically damped structural systems using canonical transformation. Technical Report No. NCEER-87-00019, National Center for Earthquake Engineering Research, Sept. 1987, SUNY, Buffalo, NY, USA.
14. Yong, Y. & Lin, Y. K., Parametric studies of frequency response of secondary systems under ground acceleration excitations. National Center for Earthquake Engineering Research, Technical Report No. NCEER-87-0012, June 1987, SUNY, Buffalo, NY, USA.

Index